SUSTAINABLE WATER TREATMENT AND ECOSYSTEM PROTECTION STRATEGIES

SUSTAINABLE WATER TREATMENT AND ECOSYSTEM PROTECTION STRATEGIES

Edited by

Hossein Hariri Asli

Tamara Tatrishvili, PhD

Ann Rose Abraham, PhD

A. K. Haghi, PhD

AAP | APPLE ACADEMIC PRESS

First edition published 2025

Apple Academic Press Inc.
1265 Goldenrod Circle, NE,
Palm Bay, FL 32905 USA

760 Laurentian Drive, Unit 19,
Burlington, ON L7N 0A4, CANADA

CRC Press
2385 NW Executive Center Drive,
Suite 320, Boca Raton FL 33431

4 Park Square, Milton Park,
Abingdon, Oxon, OX14 4RN UK

Library and Archives Canada Cataloguing in Publication

Title: Sustainable water treatment and ecosystem protection strategies / edited by Hossein Hariri Asli, Tamara Tatrishvili, PhD, Ann Rose Abraham, PhD, A.K. Haghi, PhD.
Names: Asli, Hossein Hariri, editor. | Tatrishvili, Tamara, editor. | Abraham, Ann Rose, editor.
Description: First edition. | Includes bibliographical references and index.
Identifiers: Canadiana (print) 20240312058 | Canadiana (ebook) 20240312082 | ISBN 9781774915189 (hardcover) | ISBN 9781774915196 (softcover) | ISBN 9781003470014 (ebook)
Subjects: LCSH: Water—Purification—Technological innovations. | LCSH: Bioremediation—Technological innovations. | LCSH: Water—Pollution—Prevention—Case studies.
Classification: LCC TD430 .S87 2024 | DDC 628.1/62—dc23

Library of Congress Cataloging-in-Publication Data

CIP data on file with US Library of Congress

ISBN: 978-1-77491-518-9 (hbk)
ISBN: 978-1-77491-519-6 (pbk)
ISBN: 978-1-00347-001-4 (ebk)

About the Editors

Hossein Hariri Asli
Research Scholar, College of Engineering and PhD Candidate, Lamar University, Texas, USA

Hossein Hariri Asli is currently a Research Scholar and PhD candidate at Lamar University, Beaumont, Texas, USA. He has over 50 publications in international journals and has authored several books. His research activities includes flood monitoring and mapping using the Internet of Things, networked sensors, and GIS in Southeast Texas. He is collaborating and sharing the elevation and water stage data with federal agencies, including the Texas Department of Transportation, Houston TranStar, and the National Weather Service.

Tamara Tatrishvili, PhD
Main Specialist, Office of the Academic Process Management, Ivane Javakhishvili Tbilisi State University (TSU); Director of the Institute of Macromolecular Chemistry and Polymeric Materials at TSU, Tbilisi, Georgia

Tamara Tatrishvili, PhD, is the Main Specialist in the Office of Academic Process Management (Faculty of Exact and Natural Sciences) at Ivane Javakhishvili Tbilisi State University, as well as Director of the Institute of Macromolecular Chemistry and Polymeric Materials at TSU, Tbilisi, Georgia. She is also a DAAD alumni and a member of Georgian Chemical Society. Her research interests include polymer chemistry, polymeric materials, and chemistry of silicon–organic compounds. She is the executive editor of the *Journal of the Georgian Chemical Society*. Dr. Tatrishvili is the author of more than 190 scientific publications, 12 books, and several monographs.

Ann Rose Abraham, PhD

Assistant Professor, Department of Physics Sacred Heart College (Autonomous), Thevara, Kochi, Kerala, India

Ann Rose Abraham, PhD, is currently an Assistant Professor at the Department of Physics, Sacred Heart College (Autonomous), Thevara, Kochi, Kerala, India. Her PhD thesis was titled, "Development of Hybrid Mutliferroic Materials for Tailored Applications". She has expertise in the field of condensed matter physics, nanomagnetism, multiferroics, and polymeric nanocomposites, etc. She has research experience at various reputed national institutes including Bose Institute, Kolkata, India; SAHA Institute of Nuclear Physics, Kolkata, India; and UGC-DAE CSR Centre, Kolkata, India and has collaborated with various international laboratories. She is the recipient of a Young Researcher Award in the area of physics and Best Paper Awards–2020, 2021, a prestigious forum for showcasing intellectual capability. She served as assistant professor and examiner at the Department of Basic Sciences, Amal Jyothi College of Engineering, under APJ Abdul Kalam Technological University, Kerala, India. Dr. Abraham is a frequent speaker at national and international conferences. She has a good number of publications to her credit in many peer-reviewed high impact journals of international repute. She has authored many book chapters and edited more than 10 books with Taylor and Francis, Elsevier, etc. Dr. Abraham received her MSc, MPhil, and PhD degrees in Physics from the School of Pure and Applied Physics, Mahatma Gandhi University, Kerala, India.

A. K. Haghi, PhD

Research Associate, Department of Chemistry, University of Coimbra, Portugal

A. K. Haghi, PhD, is a retired professor and has written, co-written, edited or co-edited more than 1000 publications, including books, book chapters, and papers in refereed journals with over 3800 citations and h-index of 32, according to the Google Scholar database. He is currently a research associate at the University of Coimbra, Portugal. Professor Haghi has received several grants, consulted for several major corporations, and is a frequent speaker to national and international audiences. He is Founder and former Editor-in-Chief of the *International Journal of Chemoinformatics and*

Chemical Engineering and *Polymers Research Journal*. Professor Haghi has acted as an editorial board member of many international journals. He has served as a member of the Canadian Research and Development Center of Sciences & Cultures. He has supervised several PhD and MSc theses at the University of Guilan (UG) and co-supervised international doctoral projects. Professor Haghi holds a BSc in urban and environmental engineering from the University of North Carolina (USA) and holds two MSc degrees, one in mechanical engineering from North Carolina State University (USA) and another one in applied mechanics, acoustics, and materials from the Université de Technologie de Compiègne (France). He was awarded a PhD in engineering sciences at Université de *Franche-Comté* (France). He is a regular reviewer of leading international journals.

Contents

Contributors

Hossein Hariri Asli
Lamar University, Texas State University System, Beaumont, TX, USA

Subbulakshmi Ganesan
Department of Chemistry, Jain University, Bangalore, Karnataka, India

Przemysław Gnatowski
Department of Polymer Technology, Faculty of Chemistry, Gdansk University of Technology, Gdansk, Poland

G. K. Grigoryan
Scientific and Technological Center for Organic and Pharmaceutical Chemistry of the National Academy of Sciences of Armenia, Yerevan, Armenia

N. G. Grigoryan
Scientific and Technological Center for Organic and Pharmaceutical Chemistry of the National Academy of Sciences of Armenia, Yerevan, Armenia

A. A. Hovhannisyan
Scientific and Technological Center for Organic and Pharmaceutical Chemistry of the National Academy of Sciences of Armenia, Yerevan, Armenia

G. Makharadze
Ivane Javakhishvili Tbilisi State University, Tbilisi, Georgia

T. Makharadze
Ivane Javakhishvili Tbilisi State University R. Agladze Institute of Inorganic Chemistry and Electrochemistry, Tbilisi, Georgia

Ritika Mathur
Department of Chemistry, Jain University, Bangalore, Karnataka, India

A. G. Nadaryan
Scientific and Technological Center for Organic and Pharmaceutical Chemistry of the National Academy of Sciences of Armenia, Yerevan, Armenia

Izegaegbe Daniel Omoikhoje
Department of Chemistry, Jain University, Bangalore, Karnataka, India

Mrudula P.
Department of Chemistry, Jain University, Bangalore, Karnataka, India

G. Padmapriya
Department of Chemistry, Jain University, Bangalore, Karnataka, India

G. Padma Priya
Department of Chemistry, Jain University, Bangalore, Karnataka, India

Marta Razimowicz
Department of Polymer Technology, Faculty of Chemistry, Gdansk University of Technology, Gdansk, Poland

S. Rogava
Ivane Javakhishvili Tbilisi State University, Tbilisi, Georgia

Priyanshi Roshan
Department of Chemistry, Jain University, Bangalore, Karnataka, India

Sneha S.
Department of Chemistry, Jain University, Bangalore, Karnataka, India

Agampodi Sanduni
Department of Chemistry, Jain University, Bangalore, Karnataka, India

M. Soselia
Ivane Javakhishvili Tbilisi State University R. Agladze Institute of Inorganic Chemistry and Electrochemistry, Tbilisi, Georgia

J. H. Tharudini
Department of Chemistry, Jain University, Bangalore, Karnataka, India

Anupama De Zoysa
Department of Chemistry, Jain University, Bangalore, Karnataka, India

Abbreviations

ABS	ATP-binding cassette
AS	activated sludge
BCF	bioconcentration factor
BOD	biological oxygen demand
CTR	copper transporter
DMA	district metering area
EP	emulsion polymerization
FA	fulvic acids
FD	finite difference
GIS	geographic information system
IoT	Internet of Things
LIMS	Laboratory Information Management System
MBR	membrane bioreactor
MLSS	mixed liquor suspended solids
MNF	minimum night flow
MOC	method of characteristics
MT	metallothionein
NRW	nonrevenue water
PC	phytochelatin
PVC	polyvinyl chloride
RBC	rotating biological contactor
RO	reverse osmosis
ROS	reactive oxygen species
RS	remote sensing
SOD	superoxide dismutase
SVS	sludge volume index
SS	suspended solids
TS	total solids
UFM	ultrasonic flow measuring
UFW	unaccounted for water
UV	ultraviolet
WHO	World Health Organization

Preface

This reference book reports on the most relevant research studies in the field of environmental engineering. It reviews new advances in pollution control and presents the need for greener water treatments in order to prevent water pollution from affecting ecosystems.

The chapters explain recent practical techniques to be used for wastewater and water treatment. It also provides new insights in to green technologies for a clean environment.

More specifically, this volume describes in detail the consequences and outcomes of this alarming circumstance and analyzes the precautionary measures to defeat global water pollution.

Meanwhile, by reporting new technological interventions for water and wastewater treatment, the chapters identify preventive measures to overcome water pollution along with perspectives on future trends and challenges to provide a better understanding regarding the development of green technologies that are greatly important for ensuring sustainability.

This title also covers case studies that are progressively important for ensuring sustainability.

This valuable reference book will prove useful for postgraduate environmental engineering students, scientists, and research scholars who are engaged in the development of sustainable technologies.

PART I
Optimal Design of Water Networks

CHAPTER 1

Applications of Networked Sensors and Internet of Things (IoT) for Wastewater Treatment

HOSSEIN HARIRI ASLI

Lamar University, Texas State University System, Beaumont, TX, USA

ABSTRACT

The current research was conducted in the direction of wastewater treatment by pure oxygen injection into wastewater aeration tank. The model of pure oxygen (O_2) injection into wastewater is useful in the wastewater collection networks of small towns and rural areas. This model, by the rapid circulation of information according to the geographic information system (GIS), was considered. The hydraulic model of the wastewater distribution network wsas defined during this work. The laboratory model was run by a 150-bar pressure liquid O_2 cylinder. The test procedure was as follows: A steady-state flow of an O_2-wastewater mixture was established in the wastewater pipe by controlling the outflow control valves and the pressure of the injected O_2 at the inlet. The flow velocity of the O_2-wastewater mixture was maintained at a high enough rate so that slug flow could be avoided by limiting the rate of O_2 injection. Transient flow was created by a rapid control valve closure at the downstream end of the wastewater pipe. The main approach of the present work was to show the changes in the behavior of the wastewater fluids flow state and its effect on pure O_2 penetration in wastewater flow. Comparison between laboratory observations on fluid streamlines in pipes for three cases including laminar flow—transient flow and turbulent flow—showed that little error in the prediction of fluid flow software results.

Sustainable Water Treatment and Ecosystem Protection Strategies. Hossein Hariri Asli, Tamara Tatrishvili, Ann Rose Abraham, & A. K. Haghi (Eds.)

1.1 INTRODUCTION

This paper presents the computational performance of numerical methods for modeling pure oxygen (O_2) diffusion in wastewater transient flow. This model was defined by the method of the Eulerian-based expressed in a method of characteristics (MOC)-based on finite difference form. Pure O_2 has been diffused in the activated sludge wastewater treatment process (pipeline and aeration tank). This method needs low detention time and low structural space for the wastewater treatment plant. In model studies and analysis of prototype problems, the similarity law for flow in pumping is generally valid for one-phase flow. Present guidelines and standards for equivalent model and prototype analysis accept that similarity of flow in model and prototype turbomachines (pump stations) exist before the critical cavitation coefficients are reached. Between the zone of incipient cavitation and critical cavitation, the similarity is considered to be satisfactory for the analyses of internal flow in pumping. However, for flow in the downstream of the pump station and pipe system, a small quantity of O_2 gas bubbles will change system wave speed characteristics and the similarity of the model. The study of hydraulic transient and hydraulic vibration problems may be difficult to satisfy. When pump stations operate in the zone of cavitation, air bubbles or O_2 will flow through the pump resulting in a two-phase flow in the downstream of the fluid system. Though aeration devices may be used to minimize the air content, some air will still remain in the fluid system. Since it is almost impossible to predict the quantity of air getting through the pump and remain downstream of the system, systematic analysis has to be carried out. It is assumed with various amounts of O_2 content in the model analysis of the transient fluid flow problem. The first study on the effects of air content on wave speed in a transient fluid system was conducted. O_2 diffusion in wastewater first was conducted by Union Carbide Corporation.[1,2]

1.1.1 NETWORKED SENSORS AND INTERNET OF THINGS (IOT) FOR WASTEWATER TREATMENT

The purpose of this work is the possibility of controlling and monitoring the treatment plan from the monitoring center. This system must be open.

It means that one or more control systems may be added in the future. This system should be open for communication with Networked Sensors and programming languages according to the Internet of Things (IoT). The controlling system is connected to the monitoring in operation. In this case, operators can monitor the process. The wastewater collection and wastewater treatment plant including an activated sludge process, aeration, Flow measurement unit, aeration tank for the process of extensive aeration, chlorination unit, sludge return, and disposal pumping station, sludge gravity thickener, the pumping equipment and pipelines of the sludge units, condensation tank were investigated in present work. In this work, the MOC approach transforms the partial differential equations into the ordinary differential equations along the characteristic lines. It is defined as the continuity equation and the momentum equation are needed to determine V and P in a one-dimensional flow system. Solving these two equations produces a theoretical result that usually corresponds quite closely to actual system measurements if the data and assumptions used to build the numerical model are valid.[3–6]

1.1.2 *WASTEWATER PIPELINES*

It consists of urban wastewater collection pipelines, preconstruction settlements, and its connection to treatment plant systems and disposal sites. The characteristics of the collection networks in the form of open channels for collecting surface water are outside of this technical specification.[7]

1.1.3 *PIPELINE*

The pipeline generally consists of pipes, pipeline connections, accessories, valves and accessories, and finally, basins and manholes, and other related structures that are used in the complete construction of the pipeline. The construction of the pipeline as a hydraulic structure, including the construction of an open trench, the foundation if necessary and bed construction, pipe installation, first stage embankment, hydraulic testing, final embankment, washing and disinfection, starting the pipeline and finally finishing works such as the construction of ponds, manholes, and installation of valves is for the complete implementation of the work.

1.1.4 TYPES OF SEWER PIPES

1.1.4.1 FIBERGLASS PIPES

Fiberglass pipes were first produced in 1948. The first use of these pipes, which is still the main consumer, is the oil industry. According to the type of resin used and how they are reinforced, fiberglass pipes are divided into several types, one of which is synonymous with each other and has glass fiber-reinforced plastics (GRP). In this technical specification and for convenience, the names fiberglass and GRP are pipes with a composite structure of resin, glass fibers, and other materials. These pipes have the same concept in GRP. Pipes are used in things like water transfer, sewage collection, surface water collection, seawater transfer, irrigation and drainage networks, industrial wastewater, power plants, oil and petrochemical industries, chemical industries, etc.

1.1.4.2 POLYETHYLENE PIPES

Polyethylene is prepared from the polymerization of ethylene and olefin along with antioxidant carbon black. This material is commercially available in three types, low specific weight, medium specific weight, and high specific weight. The higher the specific weight of polyethylene material, the greater its resistance. The design of pipelines should be done with attention to different productions at any point in time. In the production of polyethylene pipes, an extruder machine is used, and the machine automatically heats these materials into a paste and passes them through a special mold with pressure, which after cooling comes out of the machine in the form of a pipe. Polyethylene pipes are resistant to the effects of chemicals, so they do not need an external coating against the corrosive effects of the soil in which the pipe is buried, or an internal lining to resist the corrosiveness of the fluid passing through the pipe.

1.1.4.3 PVC PIPES

The group of materials named as hard polyvinyl chloride (PVC) is one of the most important materials made in the last few decades. By adding softener to PVC, the soft pipe is produced. The properties of PVC pipes come from their raw materials. PVC resin, which is the main component of PVC pipes, is a polymer made from gas or oil along with salt water and air. After manufacturing, PVC resin is mixed with thermal registers, oily materials, and other compounds and converted into PVC pipes with extruder machines. The properties of PVC pipes are desired from several aspects, which include the type of base resin used, impact resistance, tensile strength, modulus of elasticity in tension, and deformation due to heat. In addition to that, the degree of resistance against chemicals and acids is also a factor in the selection of PVC pipes.

1.1.4.4 ASBESTOS CEMENT PIPES

They are produced from a mixture of Portland cement with asbestos fibers (fireproof cotton). This kind of pipe, from the year 1931, has been used in America since and several years before that in other countries, especially Italy. Due to the suitable characteristics of asbestos cement pipes, their use in various fields was quickly welcomed. Asbestos cement pipes are widely used both in transmission and distribution lines under pressure and gravity lines. The wall thickness of pressed cement asbestos pipes is a function of internal pressure and pipe diameter. Therefore, the greater the internal pressure or diameter of the pipe, the thickness of the pipe wall increases proportionally.

1.1.4.5 REINFORCED CONCRETE PIPES UNDER PRESSURE

Used in water supply projects, they are usually used in water transmission lines. Reinforced concrete pipes are used in water distribution networks. The use of concrete pipes in water systems is not limited to distribution networks. The concrete pipes are installed with a diameter of more than 300 mm in water transmission lines. These kinds of pipes are installed also in branches within the boundaries of cities.

1.1.4.6 ORDINARY CAST IRON PIPES UNDER PRESSURE

Ordinary cast iron pipes, which are made of gray iron 1, have been one of the oldest types of pipes used in transmission lines and water distribution networks in different countries. The main advantage of this type of pipe is its high resistance against soil and water corrosion and, as a result, its lifespan is significant, due to the fragility and heavy weight of these pipes, especially for large diameters; as a result, transportation problems and high costs and transportation and installation, the use of this type of pipe has decreased in recent years and is almost obsolete.

1.1.4.7 DUCTILE OR UNBREAKABLE CAST IRON PIPES

Ductile cast iron pipes 2 due to their remarkable advantages in terms of flexibility and resistance to impact, pressure, corrosion of water and soil, as well as less weight, have quickly replaced ordinary cast iron pipes in the last few decades. Therefore, in this specification, after mentioning the standards, only ductile cast iron pipes are discussed.

According to the manufacturing technology, steel pipes include three general types as follows:

1.1.4.8 SEAMLESS STEEL PIPES

Seamless steel pipes are produced in small diameters through hot rolling and are used for high-pressure work. Therefore, such pipes are less used in water transmission lines, water distribution networks, and urban facilities.

1.1.4.9 STRAIGHT SEAM PIPES

Straight seam steel pipe is produced by changing the shape and piping of flat steel sheets and they are made in different thicknesses. The shape is changed and the resulting seam is O and then U in this method through the press, first the flat sheet of steel is welded by the contact of two edges of the sheet along the length of the pipe.

1.1.4.10 SPIRAL TUBES

Spiral steel pipes are made by twisting and rolling steel strips by a special machine and inside the relevant mold. In this method, the seam between the two edges of the two sides of the sheet spirals around.

1.1.5 TECHNICAL DESCRIPTION OF WORK

1.1.5.1 PIPING TRENCH

It is a channel that is built according to the routes determined based on the executive plans and these technical specifications to install pipes, connections, and accessories.

- **Altitude number:** It is the height of the desired points compared to the average level of the sea according to the coordinates of the organization (BM) or the height of the points resulting from the harvest according to the number of local reference points (NCC) mapping.
- **The number under the pipe:** It refers to the number of the lowest point inside the pipe, which is not at the joint.
- **The top number of the pipe:** It refers to the number of the highest point outside the pipe, the crown of the pipe, which is not at the joint.
- **Natural land amount:** It refers to the land amount of the pipeline route and related facilities before the start of executive work.
- **Amount of finished land:** It refers to the amount of the land of piping route and related facilities after the complete completion of the work.
- **Amount of the trench floor:** Refers to the amount under the soil of the trench floor.
- **Depth of the trench:** The depth of the trench means the height difference between the bottom of the trench and the average amount of natural or modified land above the trench on both sides. When there is a need to build a foundation under the pipe bed, this depth increases accordingly.
- **Width of the trench:** The width of the trench is the width of the bottom of the trench.
- **Longitudinal slope:** Longitudinal slope means the ratio of the difference in the number under the pipe between the two desired points to the distance between these two points.
- **Trench foundation:** The meaning of foundation is to modify the ground under the trench floor of the piping route with the desired and acceptable methods to improve and stabilize it in the place of passing through lands with poor quality soil and materials, mud, and so on.

- **Trench bedding:** The meaning of trench bedding is to prepare the trench floor with selected and acceptable materials to improve and stabilize it.
- **Hard pipe:** It is a pipe that does not change its shape in front of the external load and before breaking, such as asbestos cement pipes, unbreakable cast iron, and concrete pipes.
- **Flexible pipe:** It is a pipe that changes its shape in front of the external load and before breaking, such as plastic and steel pipes.
- **Diameter of pipes, accessories, and valves:** The diameter of pipes, accessories, and valves is their nominal diameter, which is defined in international standards or manufacturers' tables.
- **Connectors:** Connectors are parts and accessories that connect two pieces of pipe with its accessories, such as sleeves, glands, mechanical couplings, washers, and flanges.
- **Accessories:** The accessories are devices and accessories that either connect the pipeline to the valves or are used to change the piping route, such as elbows, tees, conversions, etc.
- **Pipeline boundary:** It is the minimum width of the boundary where the pipe can be easily threaded, the installation and movement of pipe-laying machines can be carried out easily, and finally, the implementation of related earth operations including leveling and preparation of the boundary, digging trenches, and depositing the resulting soil. Excavation, refilling, and transportation of excess soil can be done easily.
- **The scope of operational operations of the operational road:** The width of the scope of operational operations is the width of the desired part of the pipeline boundary, which is used for the transportation of machinery, threading the pipe, digging the pipeline trench during the implementation of the operation, and maintaining the line. The pipe is in operation.
- **Selected soil:** It means selected soil or high-quality soil for re-embankment under, around, and on the pipe or pond, which has the desired characteristics and is free of organic matter, frozen soil, large stones, tree branches, and roots.
- **Stone land:** It is a land that requires the use of incendiary and explosive materials to dig or extract it or the use of heavy machinery such as bulldozers with a power of more than 300 HP.

- **Collapsing land:** It refers to a land that is not easily dug according to the plan due to the fall of the soil around the trench or pit, and to dig it, special measures such as stakes or shielding must be done in it, or the trench with a suitable slope should be dug to prevent the total fall of the soil.
- **Normal land:** Land outside the definitions of rocky land and fallow land is referred to as normal land. The definition of the desired land in this technical specification is for piping works and the construction of ponds or dams.

1.1.6 WASTEWATER TREATMENT PLANT

The design of the equipment should be based on the latest method so that on the one hand, its parts are replaceable and facilitate its inspection during operation, cleaning, lubrication, and repairs, and on the contrary, it minimizes the risk of fire and prevent the entry of dust or insects and be completely safe in terms of electrical, mechanical, and chemical operation and minimize the installation costs of the equipment.

1.1.7 THE MAIN UNITS OF THE WASTEWATER TREATMENT PLANT

The units that play a key role in the operation of the wastewater treatment plant are as follows:

- Dirt stuck
  - Grain collecting pond
  - Device for measuring the incoming sewage flow (Parshall Flume)
  - Current divider
  - Aeration pond
  - Final sedimentation basin
  - Telescopic valve manhole
  - Return sludge (RS) pumping station and extra sludge
  - Sludge thickening pond
  - Sludge drying unit (sludge drying bed or sludge pool or strip filter)
  - Device for measuring the outflow of sewage
  - Chlorination pond
  - Sludge dewatering unit

1.1.8 GEOSPATIAL INFORMATION SYSTEM (GIS), NETWORKED SENSORS, AND IOT AND COMPUTATIONAL METHOD FOR WASTEWATER TREATMENT

This work introduces geospatial information system (GIS), Networked Sensors, IoT, and computational method for Wastewater Treatment. In this work, an analytical computational process combined with a spatial analysis was used to identify wastewater treatment plant. The present work also investigated the GIS and networked sensors as strong tools with the ability to overlap a lot of queries and fields for precision and rapid identification of the wastewater treatment process.[8,9] The remote sensing (RS) facility is equipped with data loggers and IoT based on GIS and networked sensors and connects a lot of detectors and sensors for identification of zones at the least time. For the best use of a wastewater treatment plant, it is necessary to have equipment for the collection of, sludge pumps, sludge treatment units, secondary settling, aeration, grain collectors, and garbage collectors. In addition to facilitating units with a proper control system and precision instrument, this system should house and optimize the operation of different units and the sewage treatment process provides the necessary protection for the optimal functioning of the devices and equipment. The control system should be designed in such a way that each of the treatment plant's different parts units can be installed in the local control panel of each unit of the collection. Each of the local control units in the central control panel will supervise the collection. Also, it controls the devices related to it. With proper control equipment and precise tools, the control system measures the required values or communicates signals with the central control room. Each of the local control units should function in such a way that all the electric faucets such as standby motor pumps located in each of the pump stations can be started. The central control unit can be controlled by the central processing unit (CPU) equipment in such a way that it has its complete status and is fully monitored and can be started or stopped remotely. The control method is shown on the spot in the central control room on the Mimic Diagram board. The Mimic Diagram board includes all the signals. Each set will be done automatically or manually status of different parts of each set with normal signal lamps (steady light). Lamps indicate the faults and the error condition and are indicated by flashing lamps.[10–14].

1.1.9 GEOSPATIAL INFORMATION SYSTEM (GIS), NETWORKED SENSORS, AND IOT

It is necessary to predict the standards and technical specifications of the control equipment. In general, precision instruments of the sewage treatment plant work in such a way; this equipment has the best ability and working capacity in climatic and geographical conditions. The control equipment has a valid international construction standard.[15,16].

1.1.10 AERATED GRAIN CATCHER UNIT

In this unit, the inlet part of the grain collection channel is an orifice and the outlet part has an overflow. The collected granular materials must have minimum organic materials and must have minimum moisture during disposal. The main equipment of the grain collector is as follows:

- Diffuser equipment
- The following two mechanisms can be used for the system of collecting seeds from the bottom of the seed collection unit:

    a. Using the air lift in a mobile way along the grain collection channel and continuous emptying of the collected seeds from the bottom of the channel under the diffusers to the cyclone for collecting and washing the seeds.
    b. Using a moving steamer along the seeding channel, collecting the seeds at the end of the channel and emptying them by air lift or a suitable pump or fixed silicone to collect and wash the seeds.[3,4,17,18]

1.1.10.1 TYPES OF AERATION SYSTEMS

Types of activated sludge process aeration systems that have high flexibility include mold flow, complete mixing, nonuniform aeration, staged aeration, modified aeration, contact stabilization, continuous aeration, rapid aeration, Krause process, pure O_2, oxidation pond, continuous working reactor, single-stage desalination deep well reactor, and duct stage desalination.

1.1.10.2 TYPES OF AERATION SYSTEMS

The following types of aeration systems in wastewater treatment using the activated sludge method.

Normal Mold Flow

Settled wastewater and returned activated sludge enter the head of the aeration tank and are mixed with the help of diffused air or mechanical aeration. The amount of incoming air is usually the same throughout the length of the tank. During aeration, surface absorption, flocculation, and oxidation of organic materials are performed. Activated sludge solids are separated in the secondary settling tank.

Thorough Mixing

One type of aeration system implemented in the activated sludge process that has been upgraded is the complete mixing process. In this process, the flow of wastewater is similar to the flow in a stirred tank reactor with continuous flow. Settled sewage and returned, activated sludge enter the aeration tank at several points. The organic load in the aeration tank and the O_2 required are the same throughout the length of the tank.

Nonuniform Aeration

Nonuniform aeration is a modified form of the conventional mold-activated sludge process. The amount of aeration of each length of the tank is variable and depends on the amount of O_2 needed. The inlet air is more at the beginning of the aeration tank, and as the mixed solution approaches the end of the outlet tank, the amount of inlet air also decreases. Nonuniform aeration is usually done by spacing different aeration diffusers along the length of the tank.

Staged Aeration

Staged aeration is one of the types of wastewater aeration systems, which is a modification of the conventional mold flow process. In this case, after sedimentation, the wastewater enters the aeration tank at several points to reduce the food to microorganism ratio (F/M) ratio due to the peak O_2 demand, and usually, three or more parallel ducts are used. High flexibility is one of the important aspects of this process.

Improved Aeration

This process is similar in all respects to the conventional mold flow process, except that a shorter aeration time and a higher F/M ratio are used. The biochemical oxygen demand (BOD) removal efficiency is lower compared to other activated sludge processes.

1.1.10.3 CONTACT STABILIZATION

In this process, two separate tanks or a two-part tank are used for wastewater treatment and activated sludge stabilization. Stabilized activated sludge is mixed with raw or settled wastewater. This process takes place in a contact tank. The mixed solution settles in a secondary sedimentation tank and the RS is aerated to stabilize organic matter in a reaeration tank. The amount of required aeration is usually 50% less than the normal mold flow process.

Continuous Aeration

Another type of aeration system is continuous aeration, which is similar to the conventional mold flow process. Except that it works during the endogenous respiration phase of the growth curve. Working at this stage requires a small loading of organic matter and long aeration. This process is widely used in prefabricated sanitary and human sewage treatment packages.

Fast Aeration

Rapid aeration is a modified form of this process that combines high mixed liquor suspended solids (MLSS) concentrations with high volume loading.

This combination enables the use of a high F/M ratio and long average residence time with a relatively short hydraulic retention time. Adequate mixing is very important in this method.

Pure O_2

Breathing pure O_2 can cause lung disorders. However, the presence of antioxidants in the lungs can prevent pure O_2 damage. If the ability of these substances is reduced for any reason, O_2 can damage the lungs.

Oxygen is one of the chemical elements in the Mendeleev table or periodic table, whose chemical symbol is O_2 and its atomic number is 8. O_2 is a very reactive nonmetal and an oxidizing agent that easily oxidizes elements and chemical compounds. In terms of mass, after hydrogen and helium, O_2 is the third most abundant element in the universe. At standard temperature and pressure, two O_2 atoms bond with each other to produce a structure known as dioxygen, or in other words, to form an O_2 molecule. The O_2 molecule in its gaseous state is colorless, odorless, and has the formula O_2. O_2 molecules make up 20.95% of the earth's atmosphere. Considering the O_2 in the chemical formula of the O_2-containing compounds in the earth's crust, O_2 is almost half of the elements that make up the earth's crust.

Molecular O_2 produces energy in the process of burning and respiration of aerobic cells. Many organic molecules in living organisms such as proteins, nucleic acids, carbohydrates, and fats, as well as a wide range of mineral compounds of animal shells, teeth, and bones have O_2 atoms. Most of the mass of living organisms consists of O_2, because the body of animals is mainly composed of water, and O_2 is the main component of water. O_2 in the earth's atmosphere is continuously supplied by the process of photosynthesis, a process in which sunlight converts water and Carbon dioxide into O_2. Chemically, O_2 is very active and as a result cannot remain free and atomic in the open air. Apart from molecular O_2, O_2 has other forms, ozone is one of them. The ozone molecule is able to absorb the ultraviolet (UV) light emitted by the sun, and as a result, the ozone layer covering the earth protects the life on earth from harmful ultraviolet rays. However, ozone that is produced near the surface of the earth is a byproduct of the O_2 and is considered as a useful method for water treatment.

O_2 is also used to kill bacteria and disinfect. Also, this gas is sometimes used to reduce the accumulation of bacteria in the body. The aim of the present work was to show how pure O_2 can decrease BOD5 to an amount of 80–90%.

1.2 MATERIALS AND METHODS

In this work, a four-stage closed reactor using high-purity O_2 was used. So, the partial pressure of O_2 in the length of the tank was from 40 to 60% in the first stage to 20% in the last stage. In high partial pressure of O_2 in hydraulic transient pressure, a higher volumetric rate of O_2 transfer was possible, so pure O_2 systems had higher MLSS concentration than conventional processes and can be used in shorter hydraulic retention time and higher organic volumetric loading. In this process, the amount of O_2 added is 2–3 times more than in conventional aeration systems. To provide pure O_2 in the process, on-site O_2 production equipment is needed.

The present work also used the MOC to solve virtually any hydraulic transient problems of wastewater flow in conventional activated sludge system. The MOC is based on a finite difference technique, where pressures are computed along the pipe for each time step.

Two cases are considered for modeling:

1. *Mathematical Model*

The combined elasticity of both the wastewater and the pipe walls is characterized by the pressure wave speed (1.1–1.18):

2. *Field Tests Model Criteria*

In this work water hammer numerical modeling and simulation processes was handled by nonlinear heterogeneous model.

1. *Mathematical model*

Approaches to Transient Flow (MOC model). This model was defined by method of characteristics or "MOC." The MOC is based on a finite difference technique where pressures are computed along the pipe for each time step. The combined elasticity of both the water and the pipe walls is characterized by the pressure wave speed.

The MOC approach transforms the water hammer partial differential equations into ordinary differential equations along the characteristic lines.

Balancing the energy across two points in the system yields the energy or Bernoulli equation for steady-state flow. The components of the energy equation can be combined to express two useful quantities, the hydraulic grade, and the energy grade:

$$\left(P_1/\gamma\right)+Z_1\left(V_1^2/2g\right)+h_P=\left(P_2/\gamma\right)+Z_2+\left(V_2^2/2g\right)+h_L \quad (1.1)$$

$$(g/a)\left(dH/dt\right)+dv/dt+\left(f\,v|v|2d\right)=0 \Rightarrow (ds/dt)=c^{+} \quad (1.2)$$

$$-(g/a)\left(dH/dt\right)+dv/dt+\left(f\,v|v|2d\right)=0 \Rightarrow (ds/dt)=c^{-} \quad (1.3)$$

The MOC is a finite difference technique where pressures were computed along the pipe for each time step. Calculation automatically subdivided the pipe into sections (intervals) and selected a time interval for computations.

$$\left(dp/dt\right)=\left(\partial p/\partial t\right)+\left(\partial p/\partial s\right)\left(ds/dt\right) \quad (1.4)$$

$$\left(dv/dt\right)=\left(\partial v/\partial t\right)+\left(\partial v/\partial s\right)\left(ds/dt\right) \quad (1.5)$$

P and *V* changes due to time are high and due to coordination are low then it can be neglected for coordination differentiation:

$$\left(\partial v/\partial t\right)+\left(1/\rho\right)\left(\partial p/\partial s\right)+g\left(dz/ds\right)+\left(f/2D\right)v|v|=0 \text{ (Euler equation)}(1.6)$$

$$C^2\left(\partial v/\partial s\right)+\left(1/P\right)\left(\partial P/\partial t\right)=0 \text{ (Continuity equation)} \quad (1.7)$$

By linear combination of Euler and continuity equations in characteristic solution Method:

$$\lambda\left[\left(\partial v/\partial t\right)+\left(1/\rho\right)\left(\partial p/\partial s\right)+g\left(dz/ds\right)+\left(f/2D\right)v|v|\right]+C^2\left(\partial v/\partial s\right) \\ +\left(1/p\right)\left(\partial p/\partial t\right)=0, \lambda=^{+}c\ \&\ \lambda=^{-}c \quad (1.8)$$

$$\left(dv/dt\right)+\left(1/cp\right)\left(dp/ds\right)+g\left(dz/ds\right)+\left(f/2D\right)v|v|=0 \quad (1.9)$$

$$\left(dv/dt\right)-\left(1/cp\right)\left(\partial p/\partial s\right)+g\left(dz/ds\right)+\left(f/2D\right)v|v|=0 \quad (1.10)$$

MOC drawing in (s-t) coordination:

$$\left(dv/dt\right)-\left(g/c\right)\left(dH/dt\right)=0 \quad (1.11)$$

$$dH=(c/g)\,dv, \text{ (Joukowski Formula)} \quad (1.12)$$

By finite difference method:

$$c+:((vp-v_{Le})(Tp-0))+((g/c)(Hp-H_{Le})/(Tp-0)) \\ +(fv_{Le}\,|\,v_{Le}\,|/2D)=0| \tag{1.13}$$

$$c-:((vp-vRi)(Tp-0))+((g/c)(Hp-HRi)/(Tp-0)) \\ +((fvRi\,|\,vRi\,|)/2D)=0| \tag{1.14}$$

$$c+:(vp-v_{Le})+(g/c)(Hp-H_{Le})+(f\Delta t)(fv_{Le|}|v_{Le}|)/2D)=0 \tag{1.15}$$

$$c-:(vp-vRi)+(g/c)(Hp-HRi)+(f\Delta t)(v_{Ri}|v_{Ri}|)/2D)=0 \tag{1.16}$$

$$V_P=1/2\begin{pmatrix}(V_{Le}+V_{ri})+(g/c)(H_{Le}-H_{ri}) \\ -(f\,\Delta t/2D)(V_{Le}|V_{Le}|+V_{ri}|V_{ri}|)\end{pmatrix} \tag{1.17}$$

$$H_P=1/2\begin{pmatrix}C/g(V_{Le}-V_{ri})+(H_{Le}+H_{ri}) \\ -C/g(f\,\Delta t/2D)(V_{Le}|V_{Le}|-V_{ri}|V_{ri}|)\end{pmatrix} \tag{1.18}$$

2. *Field Tests Model Criteria*

Transient analysis results that are not comparable with actual system measurements are generally caused by inappropriate system data (especially boundary conditions) and inappropriate assumptions.

The behavior of the wastewater fluids flow state as a combination of the diffusing process and remixing process has been studied. In this process, high-speed treatment has been achieved. Also, CO_2 is released and O_2 is utilized by microorganisms as a result of their activity. In this work, the formulation of process (1.19–1.29) in the wastewater transmission line and aeration tank is as follows[21–23]:

$$V=LO/BV \tag{1.19}$$

$$BV=MLSS\times MLVSS \tag{1.20}$$

$$R_t=H/VS \tag{1.21}$$

$$R_t=V/Q$$

In the case of low F/M, microorganisms feed by organic material in wastewater or feed by other microorganisms (Figure 1.1).

To save balancing condition, we need to high RS and high MLSS.

On the basis of European standard, for days without rain,[26–29] the ratio of RS per maximum influent become 100% and can be achieved from the following relations:

$$RS / Q_{\max} = \left[MLSS / \left(SMLSS - MLSS \right) \right] \times 100 \quad (1.22)$$

There was a problem in the treatment process when the sludge volume index (SVI) was more than 200. In this condition, sedimentation was failed. By decreasing SVI, aeration time was decreased.

Sludge volume index (SVI): The amount of surplus sludge that must be removed from the settling tank was related to biochemical O_2 demand "BOD5" of the influent entrance to the aeration tank and settling tank output.

Sludge age

$$\text{Sludge age} = (MLSS) \times (V) / (SSe \times Q + SLS) \quad (1.23)$$

Oxygenation calculation: A portion of absorbed pure O_2 spent informs of energy consumption and multiplying of bacteria. The other part of O_2 spent for oxidation of organic carbonate and organic nitrogenous material.[33]

$$Q_V = A_Y B_V + B\left(MLSS\right) + 3 - 4\left(ON\right) \quad (1.24)$$

Process design

$$O_C = \left[C_S / A\left(F\right)\left(C_S - C_X\right) \right] \times \left[.5YB_V + .1MLSS + 3.4ON \right] \quad (1.25)$$

Design criteria

Influent production per person = 200 (L/day)

Maximum influent factor = 1.71 (per 14 h)

$BV = .5\left(KG.BOD5 / M3DAY\right),$

$BV / MLVSS = F / M = .15$

$BOD5 = 60gr / p - d = .06kg / p - d$

Wastewater temperature = 20°C

1.3 RESULTS AND DISCUSSION

In this work, pure O_2 was introduced into the wastewater by a 150-bar pressure, 40 L volume pure O_2 cylinder. By submerged porous diffusers and by air nozzles, pure O_2 entered to aeration tank. Then microorganisms, by received pure O_2, grew quickly.

Research approach: Pollution decreased (Figures 4.2 and 4.3) to 20 (mg/L) for BOD5

Influent pollution calculation

- Fixed population = 700
- Office workers = 500
- Total population = 1200

$$Q_m^d = (700)(.2) + 500(.05) = 165\left(m^3 / day\right)$$

Maximum influent per hour

$Q^d_m = 165/14 = 12\ (m^3/h) = 3.3\ (L/s)$

Total amount of influent pollution

0.06 (1200) = 72 (kg/day)

Average amount of influent pollution

$72/165 = 0.45\ (kg/m^3) = 450\ (mg/L)$

Due to pollution–detention curves, decrease of influent pollution relation with detention time is as follows:

Detention time

$R_t = V/Q$ = Tank volume/Influent flow rate

$R_t + 200/12 = 1$ h

From pollution decreasing percent = 0.35

450 (1 – 0.35) = 293 (mg/L)

Average amount of influent pollution

(Per 24 h) = 0.293 (165) = 48.5 (kg BOD/day)

O_2 Required calculation (Figure 4.3):

$C_S = 9.17$ and $A = 0.9$

$Y = 0.925$ and $F = 0.85$

$MLSS = 3.3\ kg/m^3$

and

$3.4\ (ON) = 0.23$

$$O_C = \left[C_S / A(F)(C_S - C_X)\right] \times \left[.5(Y)B_V + .1(MLSS) + 3.4ON\right] \quad (1.26)$$

$$O_C = \left[9.17 / .9(.85)(9.17 - 1.5)\right] \times \left[.5(.925).5 + .1(3.3) + 3.4(ON)\right] \quad (1.27)$$

$$O_C = 1.2366 \text{ (kg O2/m}^3\text{.day)} \quad (1.28)$$

$$\sum O_{C=} [200\ (1.2366)]/24 = 10.3 \text{ (kg } O_2\text{/h)} \quad (1.29)$$

1.3.1 RESULTS

In this work pure O_2 was introduced in wastewater by a 40 L pure O_2 container. By submerged porous diffusers and air nozzles pure O_2 was entered to aeration tank. Microorganisms by received pure O_2 grow quickly. The important aim in this research was to decrease BOD5 amount equal to 80–90%. In this research, miscible wastewater interpenetration happened when they move themselves in the separated pipes toward the common joint and pipe in aeration tank. Fluid condition, for example, velocity, pressure, temperature, and the other properties in the pipes were homogeneous. The main approach of present work was the changes study on behavior of the wastewater fluids flow state and its effect on pure O_2 penetration in wastewater flow. Comparison between laboratory observations on fluid streamlines in pipes for three cases including laminar flow, transient flow, and turbulent flow showed that the little error in the (prediction of fluid flow software) results (Figures 1.1–1.9).

1.3.2 FLOW MEASUREMENT UNIT

The ultrasonic flow measuring devices [ultrasonic flow meter (UFM)] was installed on the pipe under the inlet pressure to reduce the drop. These devices were equipped with all the necessary devices for current measurement and recording such as totalizer and stability (Tables 1.1–1.4).

In this work, the UFM was installed on the inlet flow pipe. This device measured the wastewater flow rate. The data logger recorded wastewater inflow variations (Tables 1.1–1.4).

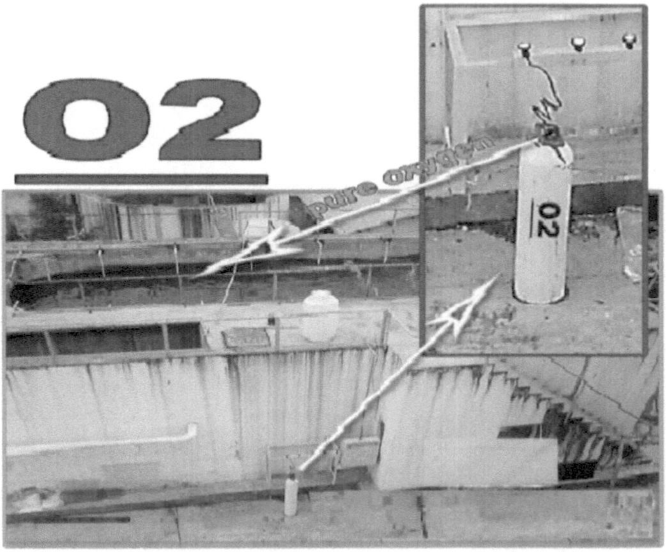

FIGURE 1.1 Pure O_2 diffusion in wastewater transient flow.

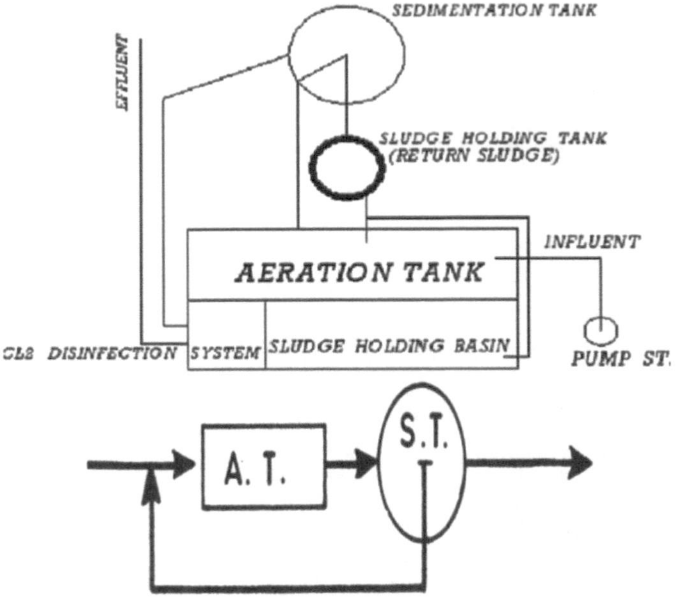

FIGURE 1.2 Schematic for wastewater treatment in conventional activated sludge system. (A.T. =Aeration tank, S.T. = Sludge tank).

FIGURE 4.3 Aeration basin in conventional activated sludge system.

FIGURE 1.4 Digester basin in conventional activated sludge system.

FIGURE 1.5 Aeration basin and digester basin in conventional activated sludge system.

FIGURE 1.6 Station of pure O_2 diffusion in wastewater transient flow.

FIGURE 1.7 Wastewater flow in aeration basin.

FIGURE 1.8 Baffles in wastewater basin for chlorination of conventional activated sludge system.

FIGURE 1.9 Station of pure O_2 diffusion in wastewater transient flow.

TABLE 1.1 Saturation Dissolved O_2 (Cs) in Water at 760 mm-hg (mg/L)

Sea water			Brackish water	Water	Water temperature (degree)
Dissolved salts in water (mg/L)					
20,000	**15,000**	**10,000**	**5000**	**0**	
11.32	12.14	12.97	13.74	14.62	0
10.01	10.7	11.39	12.09	12.8	5
8.98	9.55	10.13	10.73	11.33	10
8.14	8.63	9.14	9.65	10.15	15
7.42	7.86	8.3	8.73	9.17	20
6.74	7.15	7.56	7.96	8.38	25
6.13	6.49	6.86	7.25	7.63	30

1.3.3 CHLORINATION AND ULTRAVIOLET (UV) UNIT EQUIPMENTS

The chlorination and UV building are separated from other buildings. The chlorination equipment includes three main parts such as chlorine supply, regulation, and injection systems. The chlorine building consists of two separate parts, including the chlorine warehouse and the chlorine room. The entrance doors to the chlorine building are far from the outside and independent of each other. The chlorine storage has enough light, but the

TABLE 1.2 Various Aeration Types Technical Data.

Efficiency (%)	Air volume m^3/kgBOD	Returned sludge RS (%)	Suspended solid weight in tank MLSS (kgSS/m^3)	Pollution load sludge F/M (kgBOD5/kg MLVSS.d)	Pollution load tank volume BV (kgBOD5/ M^3.d)	Type and method of aeration
75–95	90–125	50–150	3–6	0.5–0.15	0.1–0.4	Extended aeration
85–95	45–90	15–50	1.5–3	0.2–0.4	0.3–0.6	Conventional activated sludge
85–95	45–90	15–50	1.5–3	0.2–0.4	0.3–0.6	Tapered aeration
85–95	45–90	20–75	2–3.5	0.2–0.4	0.6–1	Step aeration
80–90	45–90	25–100	2–5	0.2–0.4	1–2	Contact stabilization A.S.
75–90	25–90	100–500	4–10	0.4–1.5	1.6–6	Two stage activated sludge

TABLE 1.3 Aeration Tank Design Coefficient.

Nitrification O_2 required	Effluent pollution	Aeration tank efficiency	O_2 transfer coefficient	Detention time in without rain case	Suspended solid weight in tank	Pollution load Sludge	Pollution load tank volume	Biological treatment type
$kgO_2/m^3.d$	MgBOD5/L	%	kgO_2/kgBOD5	h	$kgSS/m^3$	kgBOD5/kg MLVSS.d	$kgBOD5/m^3.d$	
3.4ON	Le	y	OB	t	MLSS	F/M	BV	
0.05	12–15	96	2–2.5	7 <	4–5	.05–0.1	025–032	Extended aeration
0.23	15–20	92.5	2–2.5	4–7	3.3	.15	05	Complete treatment with carbon and nitrite oxidation
0.34	20–25	90	1.5–2	2–4	3.3	.3	1	Complete treatment with carbon and a part of nitrite oxidation
0.26	30	85	1–1.5	1–2	3.3	.6	2	Complete Treatment with nitrite and a part of carbon oxidation
Low	30 <	85 >	1 >	1 >	3.3	1 <	2 <	In complete treatment with a part of carbon and nitrite oxidation

TABLE 1.4 Degree of Sewage and Amount of Solids in 200 (L/24 h) of Wastewater.

Degree of sewage		Solids summations		Organic materials		Inorganic materials		Wastewater solids types
g/P-Day	g/m^3	g/(P-Day)	g/m^3	g/(P-Day)	g/m^3	g/(P-Day)	g/m^3	
8	7	**6**	**5**	4	3	2	1	
20	100	40	200	30	150	10	50	Settleable suspended solids
10	50	15	75	10	50	5	25	Nonsettleable suspended solids
30	150	125	625	50	250	75	375	Dissolved solids (DS, TDS)
60	300	180	900	90	450	90	450	Total solids (TS)

sunlight should not shine directly on the chlorine cylinders. In the storage of one set of working cylinders, one set of cylinders will be ready for operation and one set will be kept as a reserve. One-ton cylinder will be used to supply chlorine. The cylinders are arranged horizontally so that one of the guiding pipes is in the gas phase and the other is in the liquid phase.[30–33] To change the capsules, an automatic changing device (tower crane) is used. The chlorination and UV process affected by O_2 required for BOD5 decreasing due to detention time (Figure 1.10).

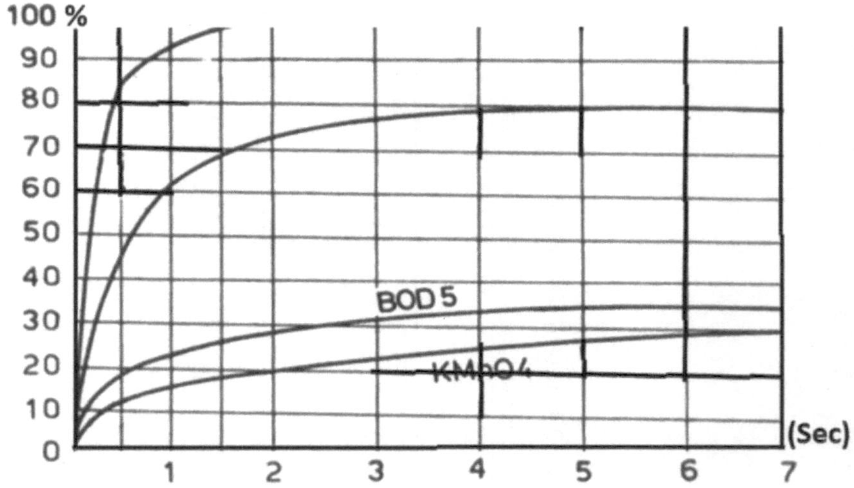

FIGURE 1.10 Pollution decreasing due to detention time.

Aeration tank volume (real value) =200 (*M3*)

Detention time (real value): $R_t = V/Q = 200/12 = 17$ h

Theoretical required volume (existence data)

$V = Lo/BV = 72/0.5 = 144$ m^3

Theoretical detention time $R_t = V/Q = 144/12 = 12$ h

Air (Pure O_2) Entrance Approaches

Modeling of air or pure O_2 injection in wastewater for decreasing pollution due to detention time at present work (Figures 1.11 and 1.12) is influenced by hydraulic similarity. Two different types of air content models have been proposed in the literature in predicting the transient pressure behavior: the concentrated vaporous cavity model[34] and the discrete air release model.[35]

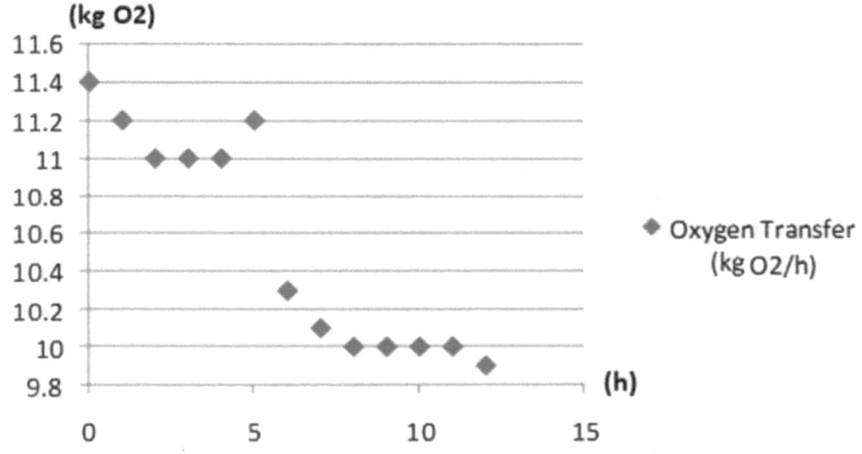

FIGURE 1.11 O_2 required for pollution decreasing due to detention time.

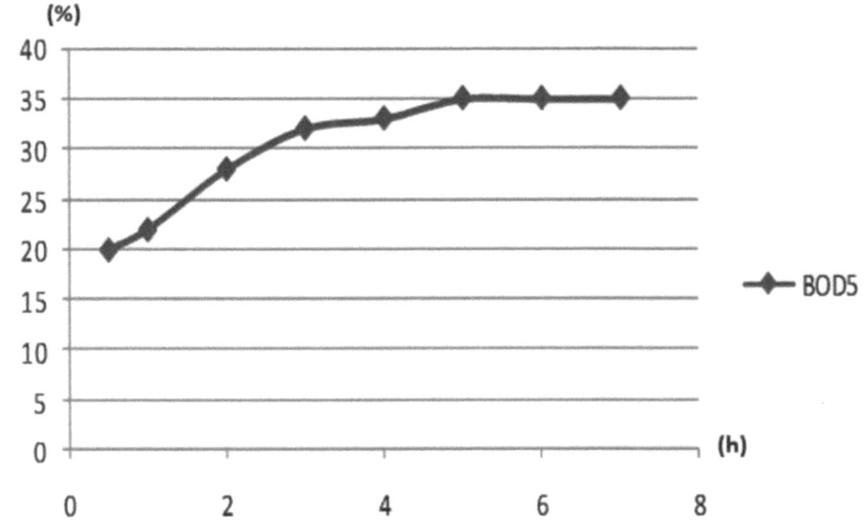

FIGURE 1.12 O_2 required for BOD5 decreasing due to detention time.

The concentrated vaporous cavity model produces satisfactory results in slow transients, but produces unstable solutions for rapid transients (Figures 1.13–1.17), such as pump's stoppage with reflux valve closure. The discrete air release model produces satisfactory results in pump shut down cases, but is susceptible to long-term numerical damping.[36] Typically, in the discrete air release model, the wave speed distribution along a pipeline (with node points $i = 0, 1, \ldots N$) was given by Ref. [37].

Transient analysis results (curve fit) by Regression method (Figures 1.13–1.17) including parameters surge velocity and air injection into wastewater were comparable with actual system measurements (observed).

Behavior of the wastewater fluids flow state as a combination of the diffusing process and remixing process was studied. In this process, high-speed treatment was achieved. Also, CO_2 was released and *O2* was utilized by microorganisms as result of their activity. In this work, the achieved results were coincided with actual system measurements which it was possible by appropriate system data and appropriate assumptions.

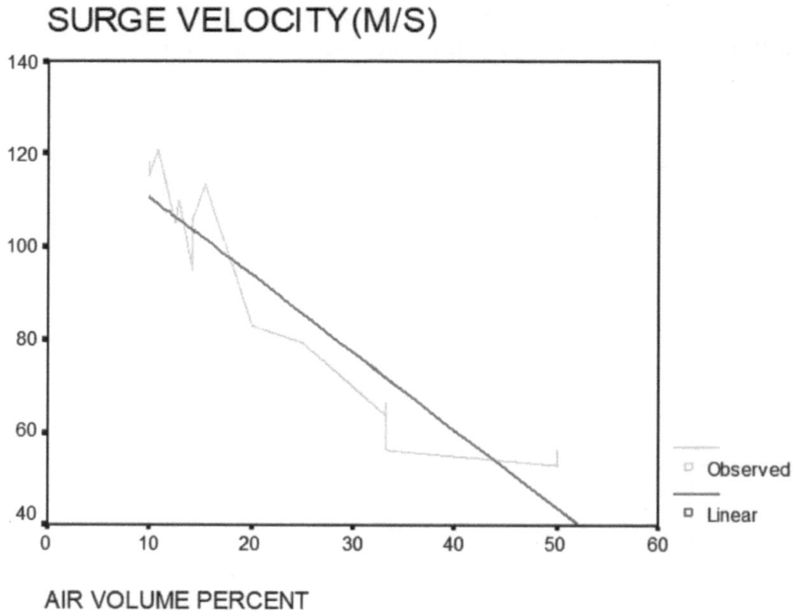

FIGURE 1.13 Transient analysis results by regression method for linear function; including parameters surge velocity and air injection into wastewater.

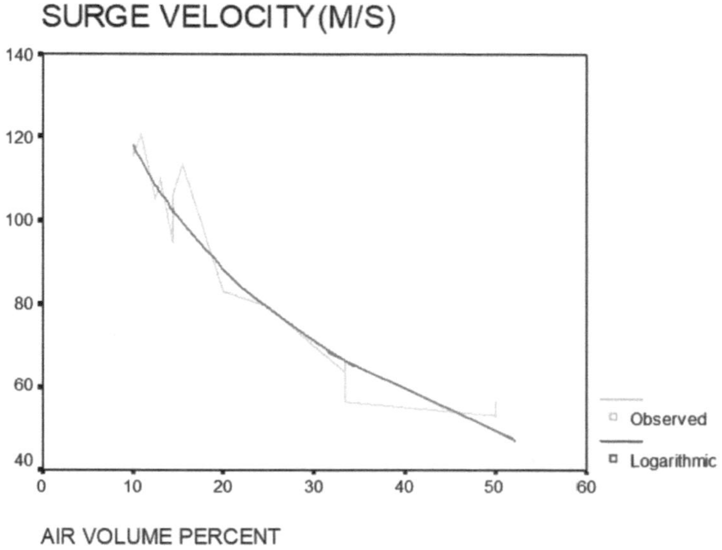

FIGURE 1.14 Transient analysis results by regression method for logarithmic function; including parameters surge velocity and air injection into wastewater.

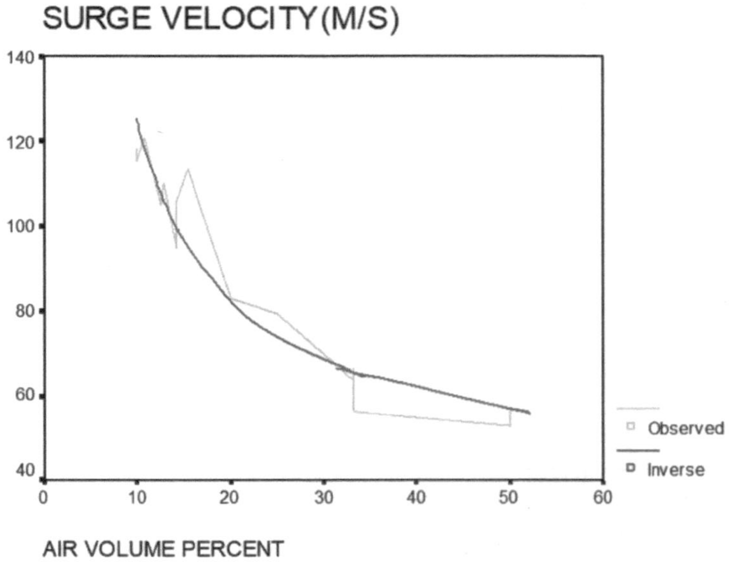

FIGURE 1.15 Transient analysis results by regression method for Inverse function; including parameters surge velocity and air injection into wastewater.

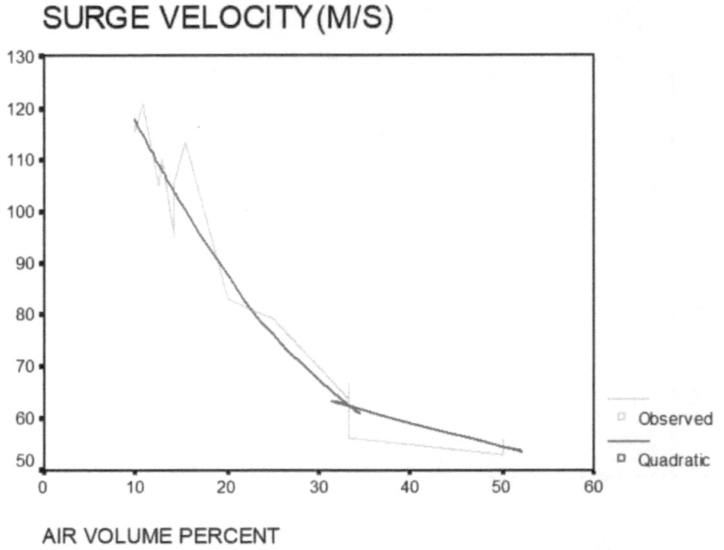

FIGURE 1.16 Transient analysis results by regression method for quadratic function; including parameters surge velocity and air injection into wastewater.

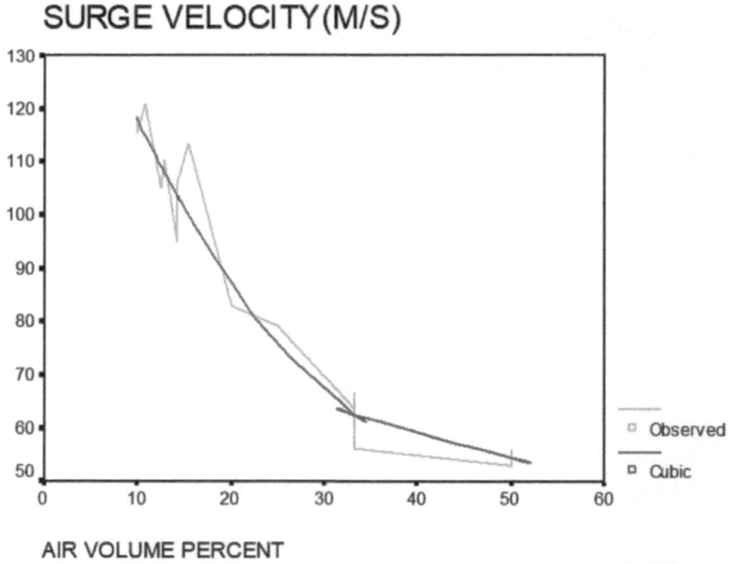

FIGURE 1.17 Transient analysis results by regression method for cubic function; including parameters surge velocity and air injection into wastewater.

In this work, at first step it was a simulated transient pressure in the system due to an emergency power failure without any protective equipment in service. After a careful examination of results, it was selected protective equipment and simulated the system again using modeling to assess the effectiveness of the devices which selected to control transient pressures. The formulated models can accurately describe complex flow features such as wastewater vacuum system-pressurized flow interfaces, interface reversals, and open- pressurized pipeline surges. much of the complex dynamics in unsteady sewer flow is due to the air phase; it is suggested to incorporate the air-phase flow component to the mixed flow model.[38–47,20,48–61]

Due to the parameter limitations, the Logarithmic function had a suitable correlation on the scatter diagram and best-fit curve which was used for surge velocity and air injection into wastewater based on the regression model shown in (Figures 1.13–1.17).

1.3.4 PRESENT WORK COMPARISON WITH OTHER EXPERTS WORKS

Present work comparison with other experts[39–41,62–81] related to laboratory observations on fluid streamlines in pipes for three cases including: laminar flow-transient flow and turbulent flow showed the little error in the prediction of fluid flow software analysis results (Figure 1.18).

1.4 CONCLUSIONS

As it was mentioned in the introduction, a robust and efficient numerical model able to reproduce unsteady gravity flows, unsteady pressurized flows and the simultaneous occurrence of gravity and pressurized flows in sewers was developed by the authors. Transient-state pressures were monitored by high-frequency-response pressure detectors at two locations. By comparing the simulated results using the MOC and the proposed scheme, it can be seen that the pressure traces computed using the MOC are lower than the proposed scheme for all the simulations. This means that the MOC is more dissipative than the proposed scheme. However, the MOC agrees with the experiments slightly better than the proposed scheme, when the physical dissipation is estimated using only a steady friction formulation (as used in present work). The important aim in this research was to decrease of BOD5 to amount of 80–90%. This work was aimed at developing a hydraulic model for real-time simulation of unsteady flows in sewers ranging from gravity flows, to partly gravity-partly surcharged. Numerical tests showed that to achieve a given level of accuracy, a Courant number close to one and a large number of grids were used.

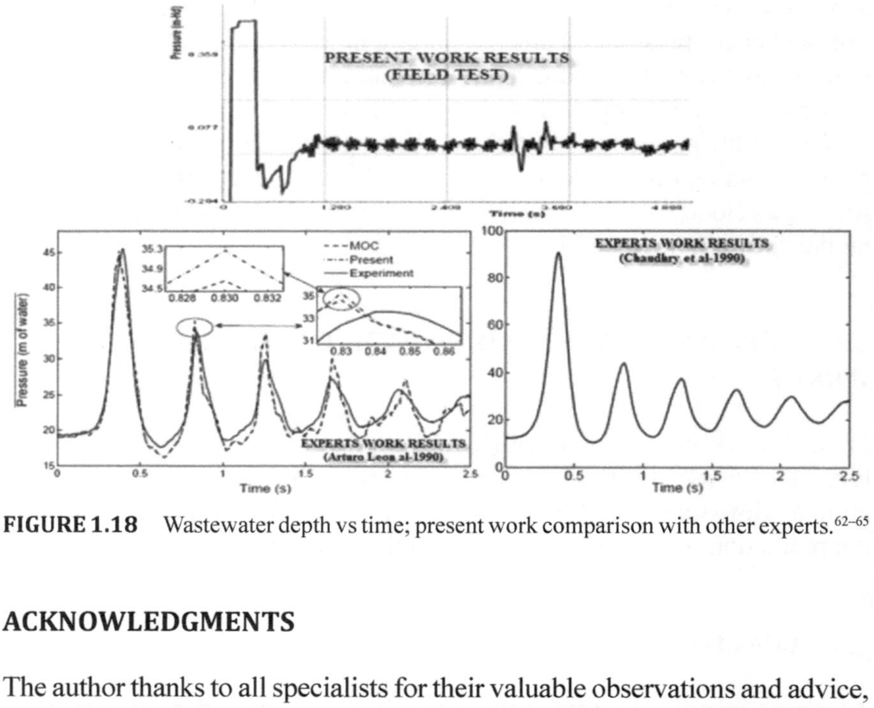

FIGURE 1.18 Wastewater depth vs time; present work comparison with other experts.[62–65]

ACKNOWLEDGMENTS

The author thanks to all specialists for their valuable observations and advice, and the referees for recommendations that improved the quality of this paper.

KEYWORDS

- **numerical methods**
- **method of characteristics**
- **wastewater treatment**
- **aeration**

REFERENCES

1. Kevin A. Internet of Things. *RFID J.*, 22 June **2009**.
2. Parker, D. S.; Merrill, M. S. Oxygen and Air Activated Sludge: Another View. J. Water Pollution Control Federation **1976,** *48* (11), 2511–2528. http://www.jstor.org/stable/25040054
3. Asli, K. H.; Aliyev, S. A. O.; Thomas, S.; Deepu, A. *Handbook of Research for Fluid and Solid Mechanics: Theory, Simulation, and Experiment*, 1st ed.; Apple Academic Press, 2017. https://doi.org/10.1201/9781315365701.
4. Asli, K. H.; Naghiyev, F. B. O.; Haghi, R. K.; Asli, H. H. *Advances in Control and Automation of Water Systems*, 1st ed. Apple Academic Press, 2012. https://doi.org/10.1201/b13114.
5. Asli K. H. Nonlinear Dynamics Modeling by Geospatial Information System (GIS). *Mapping Geospatial Inf. J. Guilan* **2019,** *3* (1). www.mpogl.ir
6. https://en.wikipedia.org/wiki/Artificial_intelligence.
7. Kelly, K. *New Rules for the New Economy*; First published in 1998 by Viking Penguin, 1998.
8. https://fa.wikipedia.org/wiki/ Geographic Information System.
9. Wylie, E. B.; Streeter, V. L. *Fluid Transients*; Feb Press: Ann Arbor, 1983, corrected copy: pp.166–171, 1982.
10. Lee, T. S.; Pejovic, S. Air Influence on Similarity of Hydraulic Transients and Vibrations. *ASME J. Fluid Eng.* **1996,** *4*, 706–709.
11. Yoon, S. H.; Park, C.S. Objective Building Energy Performance Benchmarking Using Data Envelopment Analysis and Monte Carlo Sampling. *Sustainability* **2017,** *9*, 780. DOI: 10.3390/su9050780
12. Asli, K. H.; Nazari, S. Water Age and Leakage in Reservoirs; Some Computational Aspects and Practical Hints. *Larhyss J.* **2021,** *48*, 151–167. http://larhyss.net/ojs/index.php/larhyss/article/view/808/807.
13. Asli, K. H.; Hozori, A. Non-Revenue Water (NRW) and 3d Hierarchical Model for Landslide. *Larhyss J.* **2021,** *48*, 189–210. http://larhyss.net/ojs/index.php/larhyss/article/view/810/810
14. Kensek, K. *Buildings* **2015,** *5*, 899–916. DOI: 10.3390 /buildings5030899
15. Wang, S. K. *Handbook of Air-Conditioning and Refrigeration*; McGraw-Hill, Inc., 1994. ISBN 0-07-068138-4.
16. https://fa.wikipedia.org/wiki/ Internet of Things_IoT
17. Ferras, D.; Manso, A. P. A.; Schleiss, A. J.; Covas, D. I. C. One-Dimensional Fluid-Structure Interaction Models in Pressurized Fluid-Filled Pipes: A Review. *Appl. Sci.***2018,** *8*, 1844–1856. https://doi.org/10.3390/app8101844
18. Vardy, A. E.; Brown, J. M. B.; He, S.; Ariyaratne, C.; Gorji, S. Applicability of Frozen-Viscosity Models of Unsteady Wall Shear Stress. *J. Hydraul. Eng.* **2015,** *141*, 1–13. https://doi.org/10.1061/(ASCE)HY.1943-7900.0000930
19. Lema, M.; Peña, F. L.; Buchlin, J. M.; Rambaud, P.; Steelant, J. Analysis of Fluid Hammer Occurrence with Phase Change and Column Separation Due to Fast Valve Opening by Means of Flow Visualization. *Exp. Therm. Fluid Sci.* **2016,** *79*, 143–153. DOI10.1016/j.expthermflusci.2016.07.008

20. Zecchin, A. C.; Maier, H. R.; Simpson, A. R.; Leonard, M.; Nixon, J. B. Parametric Study for an Ant Algorithm Applied to Water Distribution System Optimization. *IEEE Transitions Evol. Comput.* **2005,** *9,* 175–191. DOI: 10.1109/TEVC.2005.844168
21. Andrade, D. M.; de Freitas Rachid, F. B. A Versatile Friction Model for Newtonian Liquids Flowing Under Unsteady Regimes in Pipes. *Meccanica* **2022,** *57,* 43–72. https://doi.org/10.1007/s11012-021-01458-5
22. Sarkardeh, H.; Zarrati, A. R.; Jabbari, E.; Marosi, M. Numerical Simulation and Analysis of Flow in a Reservoir in the Presence of Vortex. *J. Eng. Appl. Comput. Fluid Mech.* **2014,** *8* (4), 598–608. DOI:10.1080/19942060.2014.11083310
23. Zhao, L.; Yang, Y.; Wang, T.; Han, W.; Wu, R.; Wang, P.; Wang, Q.; Zhou, L. An Experimental Study on the Water Hammer with Cavity Collapse under Multiple Interruptions. *Water* **2020,** *12,* 2566. https://doi.org/10.3390/w12092566
24. Duan, H. F.; Pan, B.; Wang, M.; Chen, L.; Zheng, F.; Zhang, Y. State-of-the-Art Review on the Transient Flow Modeling and Utilization for Urban Water Supply System (UWSS) Management. *J. Water Supply Res. Trans.* **2020,** *69,* 858–893. https://doi.org/10.2166/aqua.2020.048
25. Zhou, L.; Liu, D.; Karney, B. Investigation on Hydraulic Transients of Two Entrapped Air Pockets in a Water Pipeline. *J. Hydraul. Eng.* **2013**. DOI:10.1061/(ASCE)HY.1943-7900.0000750
26. Möller, G.; Detert, M.; Boes, R. M. Vortex-Induced Air Entrainment Rates at Intakes. *J Hydraul Eng.* **2015**. DOI:10.1061/(ASCE)HY.1943-7900.0001036
27. Asli, K. H.; Nagiyev, F. B.; Haghi, A. K. Interpenetration of Two Fluids at Parallel Between Plates and Turbulent Moving in Pipe. In Chapter 7, *Computational Methods in Applied Science and Engineering*; Nova Science Publisher, 2009; pp 115–128. https://www.novapublishers.com/catalog/product_ info.php?products_id=10681
28. Asli, K. H.; Nagiyev, F. B.; Haghi, A. K. Some Aspects of Physical and Numerical Modeling of Water Hammer in Pipelines. *Int. J. Nonlinear Dyn. Chaos Eng. Syst.*, **2009**. ISSN: 0924-090X (print version) ISSN: 1573-269X (electronic version) Journal no. 11071 Springer, USA, http://nody.edmgr.com/
29. Asli, K. H.; Arabani, M.; Golpour, Y. Reclaimed Asphalt Pavement (RAP) Based on a Geospatial Information System (GIS). *Slovak J. Civ. Eng.* **2020,** *28* (2), 36–42. https://doi.org/10.2478/sjce-2020-0013
30. Asli, K. H. Investigation of the Factors Affecting Pedestrian Accidents in Urban Roundabouts. *Trans. Civ. Environ. Eng.* **2022,** *8,* 1–4. Article ID: 2226, Special Issue: NCTT 2021. https://doi.org/10.52547/crpase.8.1.2255
31. Asli, K. H.; Arabani, M. Analysis of Strain and Failure of Asphalt Pavement, *Comput. Res. Progress Appl. Sci. Eng. Trans. Civ. Environ. Eng.* **2022,** *8,* 1–11. Article ID: 2250. https://doi.org/10.52547/crpase.8.1.2250
32. Lee, T. S.; Pejovic, S. Air Influence on Similarity of Hydraulic Transients and Vibrations. *ASME J. Fluid Eng.* **1996,** *118* (4), 706–709.
33. Monzavi, M. T. Wastewater Treatment, Tehran, Iran, 1991; pp 7–162.
34. Filion, Y.; Karney, B. W. A Numerical Exploration of Transient Decay Mechanisms in Water Distribution Systems. In *Proceedings of the ASCE Environmental Water Resources Institute Conference*, American Society of Civil Engineers, Roanoke, Virginia, 2002.

35. Shirzad, S.; Hassan, M. M.; Aguirre, M. A.; Mohammad, L. N.; Daly, W. H. Evaluation of Sunflower Oil as a Rejuvenator and Its Microencapsulation as a Healing Agent. *J. Mater. Civ. Eng.* **2016,** *28* (11), 04016116. https://ascelibrary.org/doi/abs/10.1061/%28ASCE%29MT.1943-5533.0001625
36. Chen, M.; Leng, B.; Wu, S.; Sang, Y. Physical, Chemical, and Rheological Properties of Waste Edible Vegetable Oil Rejuvenated Asphalt Binders. *J. Const. Build. Mater.* 2014. *66*, 286–298. https://doi.org/10.1016/j.conbuildmat.2014.05.033
37. Fini, E. H.; Kalberer, E.W.; Shahbazi, A.; Basti, M.; You, Z.; Ozer, H. Chemical Characterization of Bio-Binder from Swine Manure: Sustainable Modifier for Asphalt Binder. *J. Mater. Civ. Eng.* **2011,** *23* (11), 1506–1513. https://ascelibrary.org/doi/abs/10.1061/%28ASCE%29MT.1943-5533.0000237
38. Hassan, M.; Essam, T.; Yassin, A.S.; Salama, V. Optimization of Rhamnolipid Production by Biodegrading Bacterial Isolates Using Plackett–Burman Design. *Int. J. Biol. Macromol.* **2016,** *82* (4), 573–579. DOI: 10.1016/j.ijbiomac.2015.09.057
39. https://en.wikipedia.org/wiki/Computational_physics
40. NZS 3106. Design of Concrete Structures for the Storage of Liquids. Cement and Concrete Association of New Zealand, Standards Association of New Zealand, 2009.
41. Recommended Standards for Water Works, Great Lakes- Upper Mississippi River Board of State and Provincial Public Health and Environmental Managers, Health Research Inc.; Health Education Services Division, P. O. Box 7126, Albany, NY 12224, 2012 102-SA Water TG.
42. Planning and Infrastructure, Above Ground Circular Concrete Tank, Manager Engineering, 2008. 103- Selected Earthquake Engineering Papers of George W. Housner, American Society of Civil Engineers, 1990.
43. Specification for Construction of Concrete Reservoirs, Standard Specifications and Drawings Specification for Construction of Concrete Reservoirs, Gold Coast, Australia, SS 11, 2004.
44. TM 5-809-10, Seismic Design Guidelines for Essential Buildings, US Department of the ARMY, Washington, DC, 1986.
45. USEPA Finished Water Storage Facilities, U.S. Environmental Protection Agency, 2002.
46. Walski, T. M. Hydraulic Design of Water Distribution Storage Tanks. In Chapter 10, *Water Distribution Systems Handbook*; Mays, L. W.; McGraw-Hill: New York, 2000.
47. Farmani, R.; Walters, G.; Savic, D. Evolutionary Multi-Objective Optimization of the Design and Operation of Water Distribution Network: Total Cost vs. Reliability vs. Water Quality. *J. Hydroinf.* **2006,** *8* (3), 165–179.
48. Farmani, R.; Savic, D. A.; Walters, G. A. The Simultaneous Multi-Objective Optimization of Anytown Pipe Rehabilitation, Tank Sizing, Tank Siting and Pump Operation Schedules, Critical Transitions in Water and Environmental Resources Management, ASCE, American Society of Civil Engineers, Salt Lake City, 2004; pp 4663–4672.
49. Haarhoff, J.; Van Zyl, J. E.; Nel, D.; Van der Walt, J. J. Water Supply and Water Quality in Rural Areas: Stochastic Analysis of Rural Water Supply Systems. *Water Supply* **2000,** *18* (1), 467–470.
50. Maier, H. R.; Simpson, A. R.; Zecchin, A. C.; Foong, W. K.; Phang, K. Y.; Seah, H. Y.; Tan, C. L. Ant Colony Optimization for the Design of Water Distribution Systems.

Proceedings of World Water and Environmental Resources Congress 2001, ASCE, The Rosen Plaza Hotel, Orlando, Florida, United States, 2001.
51. Maier, H. R.; Simpson, A. R.; Zecchin, A. C.; Foong, W. K.; Phang, K. Y.; Seah, H. Y. and Tan, C.L. Ant Colony Optimization for Design of Water Distribution Systems. *J. Water Resour. Plan. Manag.* **2003,** *129* (3), 200–209.
52. Walski, T. M.; Chase, D. V.; Savic, D. A.; Grayman, W.; Beckwith, S.; Koelle, E. *Advanced Water Distribution Modeling and Management*; Haestad Press: Watertown, 2003.
53. Hashemi, S. S. *Optimization of Water Networks with Minimization of Pumping Energy*; Master Thesis, University of Tehran, Iran, 2010.
54. Mau, R. E.; Boulos, P. F.; Clark, R. M.; Grayman, W. M.; Tekippe, R. J.; Trusell, R. R. Explicit Mathematical Models of Distribution Storage Water Quality. *ASCE J. Hydraul. Eng.* **1995,** *121* (10), 669–709.
55. Rossmann, L. A. EPANET 2 User's Manual. Water Supply and Water Resources Division. U.S. 2000. Environmental Protection Agency. Cincinnati, EE.UU. http://www.epa.gov/ORD/NRMRL/wswrd/epanet.html
56. Rossmann, L. A.; Grayman, W. M. Scale-Model Studies of Mixing in Drinking Water Storage Tanks. *J. Environ. Eng.* **2000,** *125* (8), 755–761, 2000.
57. Yeung, H. Modeling of Service Reservoirs. *J. Hydroinf.* **2001,** *3* (3), 165–172.
58. Martínez-Solano, F.; Javier; Rey, Pedro L.; Iglesias; Gualtieri, C.; López-Jiménez, P. A. Modelling Flow and Concentration Field in Rectangular Water Tanks. International Congress on Environmental Modelling and Software. 537. 2010. https://scholarsarchive.byu.edu/iemssconference/2010/all/537
59. Mahmood, F.; Pimblett, J. G.; Grace, N. O.; Grayman, W. M. Evaluation of Water Mixing Characteristics in Distribution System Storage Tanks. *J AWWA* **2005,** 97 (3), 74–88.
60. Marek, M.; Stoesser, T.; Roberts, P. J. W.; Weitbrecht, V.; Jirka, G. H. CFD Modeling of Turbulent Jet Mixing in a Water Storage Tank. *XXXII IAHR Congress*, Venice (Italy), July 1/6, 2007.
61. Van Zyl, J. E.; Haarhoff, J. Reliability Analysis of Municipal Storage Reservoirs Using Stochastic Analysis. *J. South African Inst. Civ. Eng.* **2007,** *49* (3), 27–32.
62. Arturo Leon, S. Improved Modeling of Unsteady Free Surface, Pressurized and Mixed Flows in Storm-Sewer Systems, Submitted in Partial Fulfillment of the Requirements for the Degree of Doctor of Philosophy in Civil Engineering in the Graduate College of the University of Illinois at Urbana-Champaign, 2007; pp 57–58.
63. Asli, K. H.; Nagiyev, F. B.; Haghi. A. K. Physical Modeling of Fluid Movement in Pipelines; Nanomaterials Yearbook, USA. https://www.novapublishers.com/catalog/product_info.php?products_id=11587
64. Arturo, S. L.; Mohamed, S.; Ghidaoui, M.; Arthur R.; Schmidt, M.; Marcelo, H.; Garc, A. M. Godunov Type Solutions for Transient Flows in Sewers, 20–44, Illinois, USA, 2005.
65. Feizbahr, M.; Mirhosseini, S. M.; Joshaghani, A. H. Improving the Performance of Conventional Concrete Using Multi-Walled Carbon Nanotubes. *Express Nano Lett.* **2020,** *1* (1), 1–9.
66. Asli, K. H.; Asli, K. H. Isolated Pressure Zones Based on GIS as a Solution for Water Network Problems. *Water Practice Technol.* **2022,** *17* (10), 2125–2140. https://doi.

org/10.2166/wpt.2022.119. https://iwaponline.com/wpt/article/doi/10.2166/wpt.2022.119/91338/Isolated-pressure-zones-based-on-GIS-as-a-solution

67. Asli, K. H.; Asli, K. H. Minimum Night Flow (MNF) and Corrosion Control in Compliance with Internet of Things (IoT) for Water Systems. *Water Practice Technol.* **2023,** *18* (3), 608–625. https://doi.org/10.2166/wpt.2023.012. https://iwaponline.com/wpt/article/doi/10.2166/wpt.2023.012/93513/Minimum-night-flow-MNF-and-corrosion-control-in
68. Nazeryan, M.; Feizbahr, M. Seismic Evaluation of the Cheng and Chen Modified Model Using Shear Keys in Steel Beam-to-Concrete Column Connections. *Trans. Civ. Environ. Eng.* **2022,** *8* (2), 1–6, Article ID: 2787.
69. Asli, K. H.; Asli, K. H. Smart Water System and Internet of Things. *J. Mod. Ind. Manuf.* **2023,** *2*, 5. DOI: 10.53964/jmim.2023005
70. Shukla, A.; Asli, K. H.; Rawat, N. K.; Abraham, A. R.; Haghi, A. K. *Technological Advancement in Clean Energy Production*; Apple Academic Press, 2024. Hard ISBN: 9781774915585. https://www.appleacademicpress.com/title.php?id=1416
71. Pourhashemi, A.; Asli, K. H.; Abraham, A. R.; Haghi, A. K. *Sustainable Water Engineering; Smart and Emerging Technologies*; Apple Academic Press, 2024. Hard ISBN: 9781774915714. https://appleacademicpress.com/sustainable-water-engineering-smart-and-emerging-technologies/9781774915714
72. Abraham, A. R.; Susanto, H.; Haghi, A. K.; Asli, K. H. *Sustainability in Energy and Environment Engineered Materials and Smart Computational Techniques*; Apple Academic Press, 2024. Hard ISBN: 9781774916209. https://www.appleacademicpress.com/sustainability-in-energy-and-environment-engineered-materials-and-smart-computational-techniques/9781774916209#bios
73. Khanna, S.; Abraham, A. R.; Asli, K. H.; Haghi, A. K. *Sustainable Environmental Engineering Water Security, Energy Conservation, and Green Processes*; Apple Academic Press, 2024. Hard ISBN: 9781774916902. https://www.appleacademicpress.com/sustainable-environmental-engineering-water-security-energy-conservation-and-green-processes/9781774916902
74. Asli, K. H. *Technological Advancement in HVAC&R System Energy Efficiency, Technological Advancement in Clean Energy Production*; Apple Academic Press, 2024. Hard ISBN: 9781774915585. https://www.appleacademicpress.com/title.php?id=1416
75. Feizbahr, M.; Jamshidi, M.; Jayaprakash. Review on Various Types and Failures of Fibre Reinforcement Polymer. *Middle-East J. Sci. Res. 13* (10), 1312–1318.
76. Asli, K. K. *IoT-Based Smart Water Leak Detection System for a Sustainable Future: A Case Study, Sustainable Water Engineering; Smart and Emerging Technologies*; Apple Academic Press, 2024. Hard ISBN: 9781774915714. https://appleacademicpress.com/sustainable-water-engineering-smart-and-emerging-technologies/9781774915714
77. Feizbahr, M.; Bhattarai, S. R. Some Aspects of Research for Reclamation of Asphalt Based on Remote Sensing (RS). *CRPASE: Trans. Civ. Environ. Eng.* **2022,** *8* (3), 1–6, Article ID: 2810.
78. Asli, K. H. *Measurement, Modeling, and Sustainable Control of Environmental Contaminants Using Smart Computational Technologies: A Case Study, Sustainability in Energy and Environment; Engineered Materials and Smart Computational Techniques*; Apple Academic Press, 2024. Hard ISBN: 9781774916209. https://www.appleacademicpress.

com/sustainability-in-energy-and-environment-engineered-materials-and-smart-computational-techniques/9781774916209#bios

79. Nikookar, M.; Brake, N. A.; Adesina, M.; Rahman, A. Selvaratnam, T.; Snyder, H. A.; Günaydın-Sen, O. Reutilization of Oil and Gas Produced Water in Cement Composite Manufacturing. *J. Clean. Product.* **2022,** *381* (Part 1), 135113. ISSN 0959-6526. https://doi.org/10.1016/j.jclepro.2022.135113
80. Asli, K. H. *Fuzzy Logic for Smart Water and Wastewater Systems, Sustainable Environmental Engineering Water Security, Energy Conservation, and Green Processes*; Apple Academic Press, 2024. Hard ISBN: 9781774916902. https://www.appleacademicpress.com/sustainable-environmental-engineering-water-security-energy-conservation-and-green-processes/9781774916902
81. Oruji, S.; Brake, N. A.; Hosseini, S.; Adesina, M.; Nikookar, M. Enhancing Recycled Aggregate Concrete Using a Three-Stage Mixed Coal Bottom Ash Slurry Coating. *J. Mater. Civ. Eng.* **2023,** *35* (5). https://doi.org/10.1061/(ASCE)MT.1943-5533.0004723

CHAPTER 2

Applications of Networked Sensors and Internet of Things (IoT) for Water Treatment

HOSSEIN HARIRI ASLI

Lamar University, Texas State University System, Beaumont, TX, USA

ABSTRACT

The current research was conducted in the direction of tracking and removing manganese by rice husk or Sujal as an organic biosorbent agent. The model of a natural absorbent paddy husk filter is useful in the water distribution facilities of small towns and rural areas. This model, by the rapid circulation of information according to the geographic information system (GIS), was considered. The hydraulic model of the water distribution network was defined during this work. It was defined under the management of ArcGIS-ArcMap software. The laboratory model was run by a bench-scale column 1 m high, diameter 7 cm, and volume 3850 mL. The water flow rate was 1000 mL per minute, and the amount of manganese at the inlet was 1.5 ppm and at the outlet was 0.3 ppm. The results of the work with similar works have also been compared, and the georeferenced model defined in this research was calibrated with the actual conditions. The possibility of manganese removal by the natural adsorbent was shown in the treatment plant by using networked sensors and the Internet of Things (IoT) in the water distribution network.

Sustainable Water Treatment and Ecosystem Protection Strategies. Hossein Hariri Asli, Tamara Tatrishvili, Ann Rose Abraham, & A. K. Haghi (Eds.)

2.1 INTRODUCTION

Safe drinking water is generally not available. Most of the water resources are not suitable for drinking in terms of chemical and biological quality, and they need to be improved before consumption. Such an operation that takes place in order to adapt water for its specific use is called drinking water. Purification of water is very important from the point of view of medicine and health. Among water sources, surface water has more impurities compared to underground water. Such sources need serious improvement, but in terms of microbiological quality, in order to maintain health, the underground water should at least be disinfected, and maybe some of them need more complete improvement operations due to the presence of some chemical elements, such as heavy metals. Heavy metals are among the polluters of water sources, and the best way to detect them is through laboratory and field methods. The nonchemical method of removing heavy metals is the use of natural adsorbents. One of these natural absorbents is rice paddy husk, a byproduct of rice cultivation. Rice paddy husk, a byproduct of rice cultivation (including 20% of the weight of each rice grain), is considered a wonderful product with multipurpose capabilities. The special ability of paddy husk in water and sewage treatment is due to its special structure, the boat shape of paddy husk, and also the product of burning it (ash), giving it the shape of a basket that can trap water-disturbing components. Removal has also been proven in paddy husk, which is due to the presence of antimicrobial agents such as para-hydroxybenzoic acid and trans-hydroxycinnamic acid. In addition to rice husk, the cheap product of rice husk (Rice Husk Char), which is the burnt residue of rice husk and contains more than 5.1% carbon, has one of the capabilities.[1]

2.1.1 NETWORKED SENSORS AND INTERNET OF THINGS (IOT) FOR WATER SYSTEM

This work, at first, investigated the controlling and monitoring of the water systems, including the water treatment plant, water transmission pipeline, water distribution network, and water reservoirs from the monitoring center. In the second stage, this work presented a new way for water treatment using an organic agent. This system must be controlled by networked sensors and IoT based on the geographic information system (GIS). It means that one or more control systems may be added in the future. This system should be open for communication with networked sensors and programming languages, according to the IoT. The controlling system is connected to the monitoring in operation. In this case, operators can monitor all the processes. The water collection system and water treatment plant, including the sedimentation process, flow measurement unit, aeration tank, chlorination unit, ultraviolet (UV) unit, pumping station, and water pipelines, were investigated in present work.[1]

2.1.2 SPECIFICATIONS AND DEFINITIONS

2.1.2.1 WATER PIPELINE

Water transmission lines are the pipelines that collect water from wells or surface sources such as rivers or dams and transfer it to treatment plants, ground reservoirs, or cities. These lines are generally outside the urban area, and with the expansion of the city, a part of them may also be placed inside the urban area (Figure 2.1).

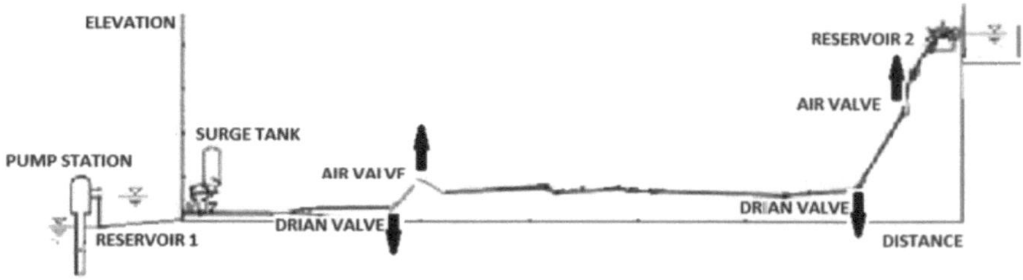

FIGURE 2.1 Water transmission lines facilities.

2.1.2.2 INTRAURBAN WATER TRANSMISSION LINES

It is the pipelines located in the urban area that transfer the required water from the underground reservoirs of the urban service to the distribution network or air sources within the city (Figure 2.2).

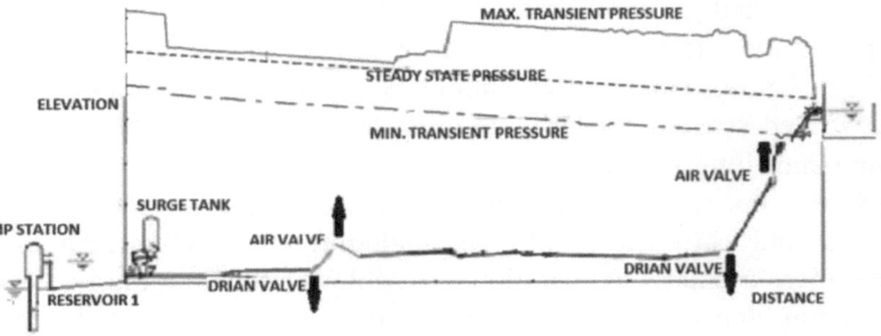

FIGURE 2.2 Pressure changes in water transmission lines.

2.1.2.3 DRINKING WATER DISTRIBUTION NETWORK

It is urban water distribution pipelines within the residential areas of cities or water distribution lines that are built in the preparation projects of settlements before their construction (Figures 2.3 and 2.4).

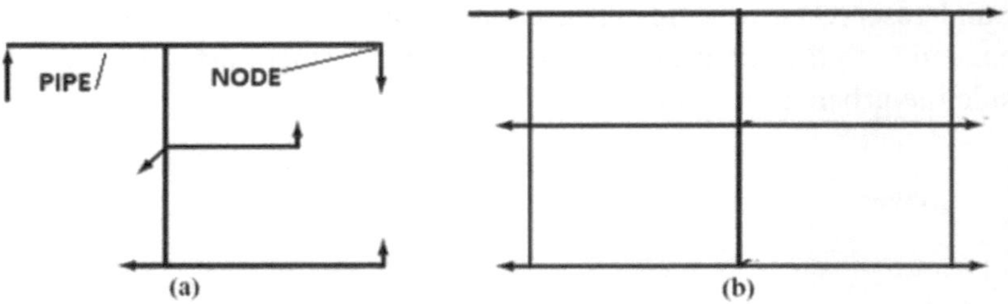

FIGURE 2.3 Water distribution pipelines: (a) branched distribution network and (b) looped distribution network.

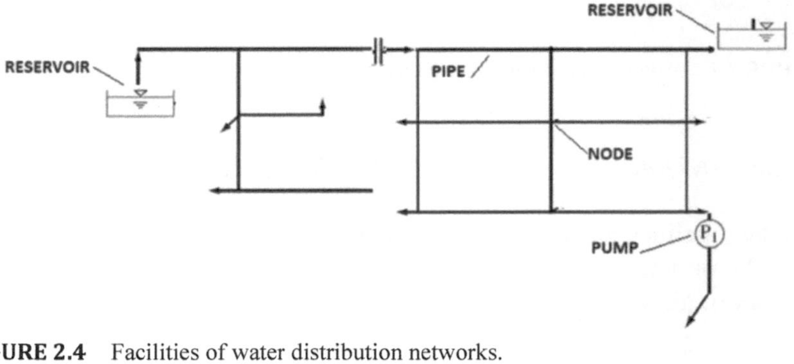

FIGURE 2.4 Facilities of water distribution networks.

Raw water distribution network and green space irrigation: These are urban water distribution pipelines within residential areas that are built for the purpose of urban green space irrigation.

- Water pumping stations
- Normal operation

In operating the pumping station, the following items must be controlled (Figure 2.5):

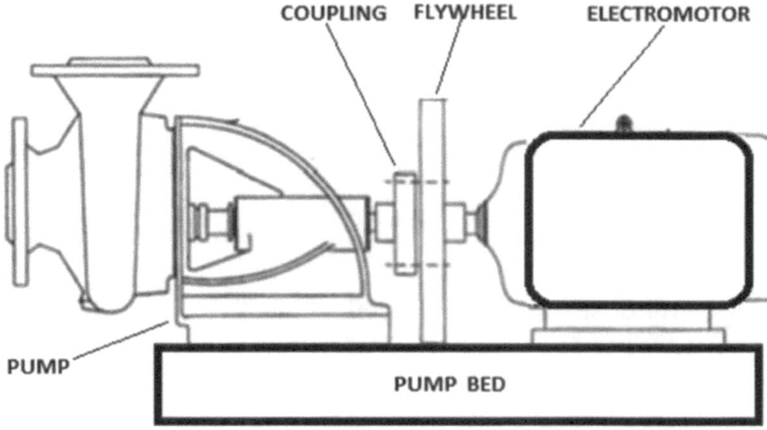

FIGURE 2.5 Water pumping station.

a) Storage tanks

Controlling the water level inside the tanks (Figures 2.6 and 2.7).

– Checking the condition of the inlet and outlet valves

FIGURE 2.6 Schematic of water storage tank.

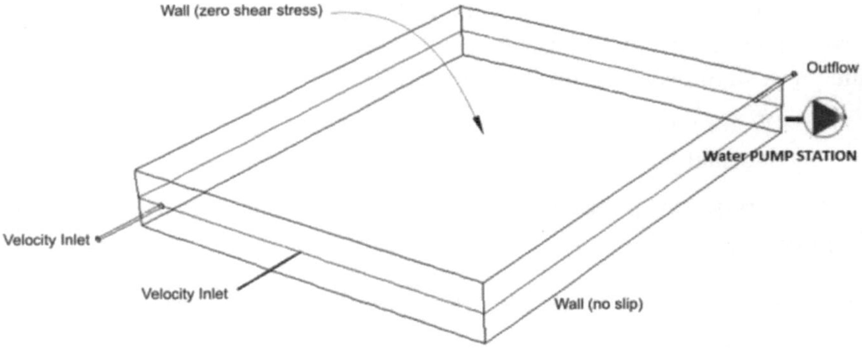

FIGURE 2.7 Reservoirs of water transmission lines.

b) Electric motors

- Engine sound control
- Shell temperature control
- Control of starting ampere, working ampere, voltage and engine speed for surge tank of water transmission lines (Figure 2.8).

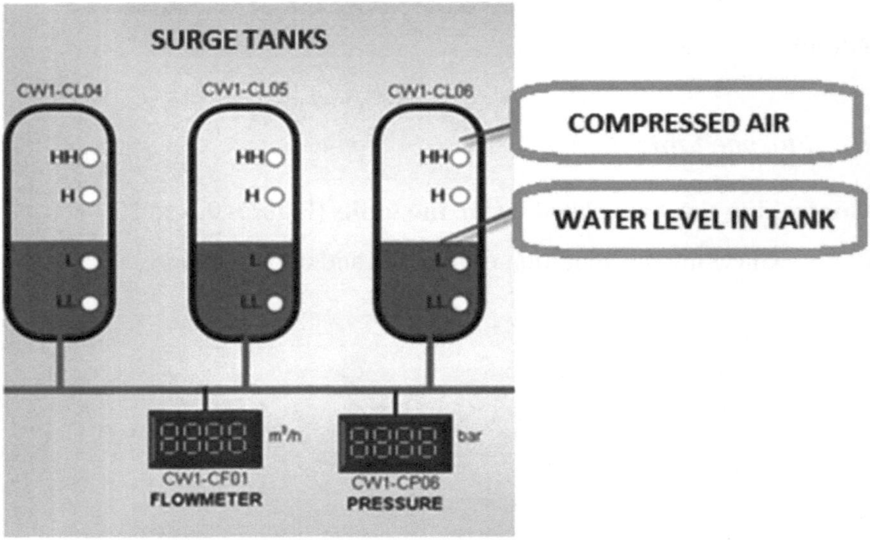

FIGURE 2.8 Surge tank of water transmission lines.

- Electronic consumption registration
- The output signals in cables from the main switches to the start switches of the electro pumps are controlled operationally. If the heat is too high (30°C), let the key be turned off and inform the repair group.
- Cleaning the inside of the electric motor with a dry air compressor with a pressure of about 4 atm

c) Pumps

- Coupling control
- Controlling the sound of the pump and its vibration
- Pump shell temperature control
- Control the circulation of the cooling system, if any
- Ventilation system control
- Control of input and output current intensity

d) Suction pipeline

- Suction line pressure control
- Checking the status of the valves on and off
- Checking the sealing of valves and fittings

e) Thrust pipeline

- Positive line pressure control by one-way impact tanks of water transmission lines (Figure 2.9).
- Checking the status of faucets and connections
- Checking the sealing of faucets and fittings

f) Security and lighting system

- Controlling the status of security systems in the facilities
- Controlling the status of the whole system by setting up daily reports

- The lighting of the pump house during the day, control and if there is a defect, take action to fix it so that there is no problem at night.

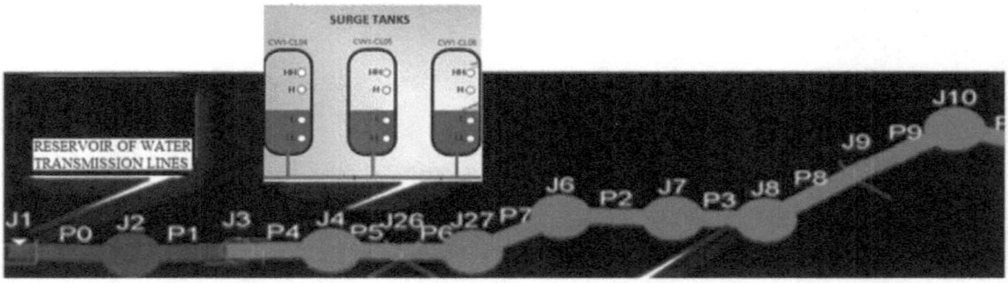

FIGURE 2.9 One-way impact tanks of water transmission lines.

2.1.2.4 FAULT FINDING AND FIXING IT

If periodic services and repairs are carried out according to the schedule, the workload of the repair unit will be greatly reduced. In any case, there are reasons to form an experienced construction group at the pumping station to fix the defects and prevent disruptions in operation. Defects and faults created in the electric motor of the pumps can have hydraulic or mechanical reasons (Table 2.1).

In the first case, the pump does not have the ability to pump, or the pump operates with less than its nominal capacity, or does not provide enough pressure. In the second case, the amount of electricity consumed increases, and problems such as the value of the pump cause noise, corrosion, and breakage of parts. In general, any defects or problems in the operation of pumps can be caused by one or more reasons. By finding the cause, the repair department manager can easily determine the percentage of its resolution. In this connection, Table 2.1 lists the problems and their causes that generally occur in centrifugal pumps.[2]

2.1.2.5 MAINTENANCE AND LUBRICATION

a) Maintenance principles of pumping stations

The maintenance of a pumping station, like any industrial system, is divided into three parts:

- Continuous control of system details
- Periodic services
- Period replacements

*In a pumping station, the details of the station must be continuously controlled by the maintenance group. Find system faults, predict the causes, and fix them.

TABLE 2.1 Problems and Causes of Defects in Centrifugal Pumps.

The reason	Problems	Row
Air has entered the pump	Water is not pumped	1
The pump or suction pipeline is not completely filled with water		
The suction height is high		
The suction pressure is equal to the water vapor pressure		
The presence of air bubbles in the suction pipeline		
The inlet of the suction pipeline is not completely in the water		
The pump speed is very low		
The direction of rotation of the pump is wrong		
The overall system overhead is more than the design overhead		
Operating the pumps in parallel is not correct in the operating conditions		
The presence of foreign materials in the pump impeller		
The pump or suction pipeline is not completely filled with water	Pump with less capacity	2
The suction pressure is equal to the water vapor pressure	Doing	
The height of suction is more than allowed		
There is water, air, or other disturbing gases inside		
Air infiltration into the pump suction pipe		
Air infiltration into the pump chamber		
The one-way valve at the end of the suction pipeline (suction valve) has been pinched or seized		
The rotation speed of the pump is low		
The overall system overhead is more than the design overhead		
The viscosity of the fluid is higher than the design value		
Parallel operation of pumps is not suitable in operating conditions		
The presence of foreign materials in the pump impeller		
Corrosion and damage to the pump impeller		
Deterioration of the sealing gasket		

TABLE 2.1 *(Continued)*

The reason	Problems	Row
The presence of air bubbles in the incoming water	The pump head is insufficient	3
The rotation speed of the pump is low		
The direction of rotation of the pump is wrong		
The overall system overhead is more than the design overhead		
Fluid viscosity is different from the design value		
Corrosion and damage to the pump impeller		
Deterioration of pump sealing washers		
The water supply pipe (water seal) is closed.	Packing soon _	4
The pump is installed flush	It breaks down	
Bending of the pump axis or shaft		
Shaft corrosion		
Installation of wrong packing		
The type and material of packing is not suitable for working conditions		
Engine misalignment		
The flower is very tightly closed		
The distance between the shaft and the body		
The presence of foreign substances in the sealing liquid		
The pump is not aerated	The pump overheats	5
The difference between suction and steam pressure is small		
The pump is operated with less capacity		
The parallel operation of the pump is not suitable in the operating conditions		
Uneven pump installation		
Abrasion of moving parts on fixed parts of the pump		
Axis or shaft bending		
Imbalance of the moving engine		
The speed of the pump has reached the critical speed		
The pump or suction pipeline is not completely filled with water	The pump vibrates or makes noise	6
The suction height is high	It works	
Suction pressure is the degree of water vapor pressure		
The suction line valve is too small or blocked		
The inlet of the suction pipe is insufficiently located in the water		
It is operated with less capacity		
The presence of foreign materials in the pump impeller		

TABLE 2.1 *(Continued)*

The reason	Problems	Row
Uneven installation of pump and drive motor		
The foundation is not strong enough		
Bending of the pump shaft		
The pillow is a moving part to a fixed part		
Damage to the pump impeller		
Lack of lubricant		
There is rotation in the pump axis		
The speed of the era is very high	Power consumption is high	7
The drive motor works in reverse direction		
The head of the whole system is more or less than the design head		
The difference between the specific gravity and the viscosity of the fluid with the design value		
The presence of foreign materials in the pump impeller		
Uneven installation of pump and motor		
Bending and deviation of the pump shaft		
Rubbing a moving part on a fixed part		
Wrong packing installation		
The type and material of packing is not suitable for working conditions		
The flower is very tightly closed		
The pump is not ventilated	After turning on the pump	8
The suction height is high	It takes air	
There is water, air, or other gases inside		
The presence of air in the suction line		
The inlet of the suction pipe is not completely in the water		
The water pipe is flooded		
Misalignment of pump and motor	Bearings break down quickly	9
Bending of the pump shaft	It takes air	
Rubbing a moving part against a fixed part		
Dry working of bearings		
Incorrect installation		
The presence of foreign materials in the bearing		

*The periodical services of a pumping station for each of the constituent parts usually include the following:

1. Control of gate valves and one-way valves
2. Lubrication of the bearings of electric motors and pumps
3. Sealing rims service
4. Opening and closing the pumps and checking the condition of the shaft, impeller, and internal wall of the pump shell
5. For every 2000 working hours of the pump, the bearing grease should be replaced.
6. Starting amp test—working amp—head of electric motors
7. Service of electrical panels and sensitivity testing of safety and protective circuits
8. Service of control systems such as limit switches and flowmeters
9. Service and control of loose couplings

*Periodic replacements of a pumping station for each of the components in a way that can include the following:

1. Bearings and roller bearings (pumps and electric motors)
2. Rings and graphite (every 6 months)
3. Pump impeller
4. Pump shaft
5. Ball shaft of valves
6. Manometers
7. Couplings
8. Water flow measurement meters (if available)
9. Sensitive parts of electrical panels

b) Periodic operation of inspection and maintenance of the pumping station

1. Daily

All defects that are observed during hourly and daily visits in the operation of the pumps, such as changes in the sound of the pump motor, sudden changes in the temperature of the pump shell, leakage from the pump, and pressure and flow readings, are observed by the operators.[3]

2. Monthly

In addition to the daily visits by the operators, the following visits and operations must be performed by the construction team on a monthly basis:

- Investigating the condition of the bolt and gland
- The state of packing and sealing gaskets should be checked and replaced if necessary.
- Adjusting and aligning the pump motor, or in other words, inspecting and measuring the distance of the couplings
- Greasing and lubrication of electric motor bearings with 160° fireproof grease
- Inspection and repair of pipes and connections, if needed
- Inspection and repair or replacement of pressure gauges, if necessary
- Pump electromotor vibration control
- Cleaning the inside of the electric motor of the air compressor belt pump with a pressure of about 4 atm

3. Annually

The measures that should be taken during the annual visits are as follows:

- The pump components should be opened, washed, and cleaned
- Descaling of impeller and pump shaft
- Lubrication of bearings and roller bearings
- Painting faucets, fittings, and pump shell
- Inspect the floor and walls of the pump house and repair it if any defects are found

From the operation of the unit

- Close the valve located on the thrust pipeline of the desired pump
- Turn off the engine
- Close the valve located on the suction pipeline

Safety

- The electropump should never be used outside the specified range in terms of water flow.
- Motor speed
- Pressure and temperature
- Avoid keeping the electropump on for a long time while the outlet valve is closed.
- Under no circumstances should the pump work without a load.

Faucets of the distribution network

The valves in the urban water distribution network are considered key control points of the network. The absence of a suitable plan and method for the maintenance and operation of the faucets of the distribution network often causes them to lose their desired and necessary performance after a period of installation. Instead of being used as executive management tools, they cause problems. They are distributed throughout the system. The first step to repair and maintain these units is to know their function and role in the distribution network. After that, the existing obstacles should be removed and their protection and maintenance should be known, and then, with the schedule of periodic visits and service, while applying the necessary controls, repair and maintenance should be considered.

A. Gate Valve

This valve is used to completely shut off and connect the water flow in the distribution network. These valves are manufactured and marketed in different sizes to withstand different pressures. The material of these faucets is generally cast iron or ductile iron and is widely used in distribution networks and water transmission lines. A gate valve, along with its components and the most common installation points, is:

- In the place of branches and the intersection of pipelines, in order to cut and connect the water flow of the desired branch or the connection between the lines
- At the entrance and exit of tanks
- In the water transmission lines known as taps

There are two types of valves in terms of how they work:

- A valve with a rising stem (RISING STEM VALVE) in which the stem or rod of the valve also rises and water is allowed to pass
- Gate valve with a fixed stem (NON-RISING STEM VALVE) in this group, the valve moves on a rod or stem. These valves are generally used for underground installation.

Advantages of a gate valve

- When it is completely open, the water pressure drop is very small and can be ignored.

- Long life span
- High work pressure
- There is less water leakage in them.

Disadvantages of a gate valve

- In large diameters and at high pressures, it is difficult to open and close it by hand, and the valve must have a passage.
- If the gate valve is not maneuvered (open and close) regularly, due to the accumulation and deposition of particles in the water, opening and closing it will not be done easily and correctly.

Repairs

Valves often need to be repaired due to the following reasons:

- Deterioration of the sealing thread (packing)
- Corrosion and rusting of the tap rod or stem
- Corrosion of the tongue or disc
- Deterioration of sealing rings

Each of the above causes malfunctions and leakage from the valves. In order to fix the defect and repair each of the abovementioned items, it is done in the following way[3]:

Replacing the sealing thread

- Open the gland.
- The sealing thread must be removed from the sealing chamber (SUFFING BOX), and its place should be cleaned.
- The thread of the new seal should be placed in place and secured with a gland.
- Pour a few drops of oil on the sealing thread, and then put the valve gland in its place and close it tightly.

The service of the bar or stem of valves

- Open the gland.
- Take out the sealing thread.
- Open the cap and remove the rod.
- Clean the surface of the rod with sandpaper or sanding oil, and if there is a lot of corrosion and rust, the rod should be replaced.
- Put the rod inside the cap.

- Seal gaskets should be replaced.
- The cap is firmly in place.
- Replace the sealing thread.
- Open and close the valve.

Fixing the corrosion of the disk of valve and sealing ring

- Open the gland and cap.
- Remove the rod.
- Remove the disk and clean it with sandpaper and replace the disk if there is a lot of corrosion.
- Clean the sealing rings and replace them otherwise.
- Mount the valve parts again. Sealing gaskets and sealing thread should be replaced.

Keeping

Gate valves should not be open or closed for a long time, especially in a water transmission line, because the water inside the valve is filled with salts when it is open, the valve is not completely closed and sealed. In the case of being closed for a long time, the solutes in the water accumulate around the valve drawer and cause a lot of pressure on the valve axis when opening the valve, in which case there is a possibility of the axis being cut. It is necessary to maneuver the valve at least once a month (Figure 2.10).

FIGURE 2.10 Connections of surge tank in water transmission line.

B. Check Valve

A one-way valve is actually a valve that is installed in the pipeline, and when water flows, the moving force of the water opens it, and when the water flow changes direction, the plate closes on a tight seat. In general, one-way valves work automatically and prevent reverse flow in pipelines. One-way valves are generally divided into two main groups: hinged one-way valves (LIFE CHECK) and one-way ball checks (BALL CHECK). Hinge one-way valves are often used when the goal is to pass the flow completely, and they are the most common and most used one-way valve. And the one-way valve is used in cases where the valve must operate frequently and in networks with high pressure and flow rates. Among the factors that cause defects and disruptions in the operation of the one-way valve are the corrosion and wear of the thorn holder, cap maintenance, valve deposition, and disc corrosion. In order to fix such defects, proceed as follows (Figure 2.11):

- Open the valve cover.
- The nut or spine of the axis should be removed.
- The valve disc should be sanded or ground.
- In case of damage to the axis, replace the axle pin and nut and their cover.
- Mount the valve components again.
- Valve should be tested.

FIGURE 2.11 Connections of surge tank in water transmission line.

C. Butterfly Valve

The construction of butterfly valves is simpler than other valves. These valves have a part in the shape of a rotating plate or disk that moves around an axis, or a rod, or a stem. Opening and closing these valves is simple and is done only by turning the disk so that it fills the entire width of the pipe. Depending on its type, the plastic lining of the seat of these valves is made of adhesive material, or it is fixed to the disc with mechanical devices, or it is embedded in the disc, or it is made of a mechanical spring seal, or a metal seat that is placed in the form of an "O-ring" around the edge of the disc. Among the advantages of these valves are:

- The coefficient of water passage in it is directly related to the opening percentage of the valve, and therefore it can be used to control the flow and create a pressure drop in the network.
- Opening and closing are done with more ease and speed.
- It needs less space for installation.
- Its repair and maintenance costs are relatively low.

These valves are easy to maintain and generally do not need lubrication until the stem O-ring is replaced. The disc and stem, as well as the O-ring disc and its plastic seats, are easily replaceable. Replacement of the plastic metal lining of the sealing surfaces with adhesive materials cannot be done on site and is done only by the manufacturer.[4]

D. Air Valve

An air valve is a device to let air out of a pipeline or into it. Air valves are divided into three categories:

- Single-hole air valves with a small hole: They are used to release air from the pipeline that is under hydraulic work pressure. The diameter of the opening in these valves will not be more than 4 mm.
- Single-hole air valves with a large hole: This valve is used to ventilate and empty the pipe. When the pipeline is filled with water, air is removed from it, and when the pipe is emptied of water, air enters the pipeline. When the pipeline is under hydraulic pressure, the above valve is not able to release the small amount of collected air, and in this sense, the hydraulic pressure prevents the ball from separating from the hole.
- Two-hole air valve: It performs the work of both of the above valves. In terms of operation, if the valves are installed in the right

place, they will not have any particular problems. Of course, it is necessary to visit it periodically, and if its opening is clogged with sediment, it should be descaled and rusted.

Periodic visits, repair, and replacement of valves

Due to the fact that during the use of the distribution network, there is no justification for cutting off the water for a long time, sometimes it will cause protests from the subscribers. Therefore, it should not be enough to fix problems as corrective maintenance. Rather, it is necessary to optimize the use of the facilities by training and equipping the crews and conducting periodic visits and services (while preventing the creation of crises). Each of the valves used in the network, even if they are made of the best metal materials, will suffer over time. Corrosion, rusting, and sedimentation become clogged. Like other mechanical systems, failure to maintenance the valve on time will gradually make it impossible to maintenance the valve. So, their maintenance is recommended twice a year. On the other hand, the access valve to the line valve due to the spread and various implementation operations in the city such as drilling, surface changes, and asphalt may be invisible, intentionally or unintentionally, and as a result, the possibility of valve maintenance is lost. The more important point is that each faucet is installed for a reason and to connect and disconnect the water in a specific area. If a pipe breaks due to drilling, leakage, or as a result of any kind of accident, by closing the valve of the relevant line, the outflow of water from the accident site is controlled. And therefore, if one of the valves in the network becomes invisible, at least several other line valves should be closed instead, which will cause the water to be cut off in a wider area, and water will be wasted for a long time. In addition to the mentioned points, if any of the valves in the line are not opened and closed continuously, and if it gets corroded, it will automatically lose its water content after a maintenance and will leak from the outside, which results in water wastage and water loss. As a result, the soil settles under the pipe and breaks the pipelines. If there is no sealing valve, the repair of the accident pipe is impossible in some cases, and in some cases, a pump is needed to drain the water from the accident site. In the maps of the water distribution network, which have been updated with a local survey, all the visible and invisible taps in the network have been identified and reflected on the maps. According to the fact that pipes with a diameter of 300 mm and more are considered to be part of the main pipes of the distribution network, the program and budget organization, taking into account the common standards, has recommended that the faucets related to these lines have basins, so to be able to inspect periodically and to ensure the correct operation of the main faucets of the network when there is a break in these pipes, and to reduce water losses caused by this, faucets of 300 mm and more should be placed inside the basin.

Faucets

Faucets that are generally used in urban water distribution networks are valve, butterfly, one-way, and air valves. Since in determining the standard dimensions of the faucet basin, the technical and standard dimensions of these faucets are needed, in this section, while introducing the above valves, their technical dimensions are presented. In connection with the given technical dimensions, it is necessary to mention that the above dimensions are a type, and different manufacturers produce and market their products with slight differences.

A. Gate Valve

This valve is used to completely shut off and connect the water flow in the distribution network. These valves are manufactured and marketed in different sizes to withstand different working pressures. The material of these valves is cast iron or ductile iron.

B. Butterfly Valve

The structure of this valve is simpler than other valves, and they are made in two types: flanged and wafer. This valve is used to control and regulate the amount of flow passing through a pipeline.

C. One-Way Check Valve

This valve is used to prevent backflow in a pipeline. These valves have a simple structure, and their material is cast iron or ductile iron. These valves are made in various forms, such as hinged valved, and ball valves.

D. Air Valve

Air valves are among the valves that facilitate the entry and exit of air inside a pipeline in Max. air entrance into water transmission lines. These valves are manufactured and marketed in different shapes and types, such as single orifices and double orifices. Unlike other valves, the air valves are not located in the pipeline route but are mounted on it. Therefore, air distance and side distances are important for them in the construction of ponds (Figure 2.12).

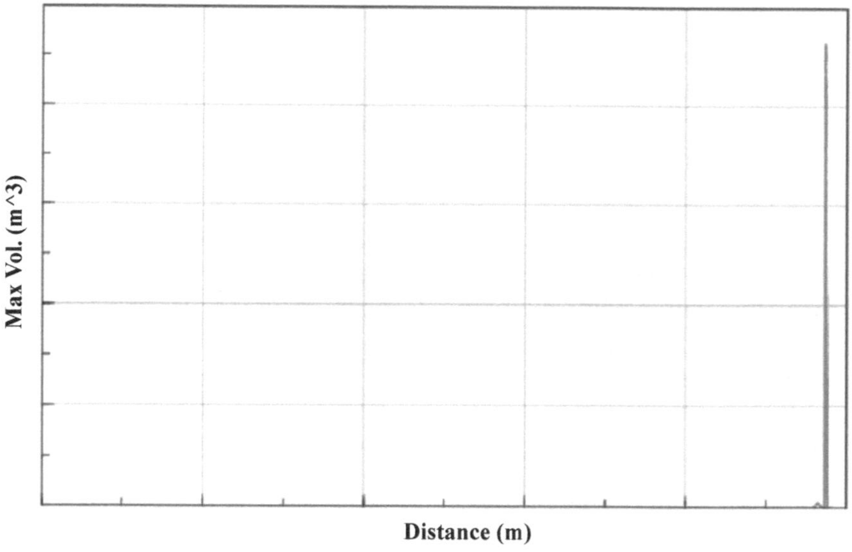

FIGURE 2.12 Max. air entrance into water transmission lines.

Water distribution network pipes

The skeleton of the network is made up of pipes. These pipes can be used in sizes of 60, 80, 100, and 150 mm for subsidiary networks.

Pipes are the only part of the urban water network that are always buried underground, so they are more vulnerable than other parts of the network and will have accidents. Therefore, maintenance of the network in the sense of visiting them is not possible only when accidents occur, but it is better to carry out leak detection by a leak detection device during a regular program and according to the priority of the areas according to the age of the network and its importance. The network should be repaired before causing any massive and costly damage. In this regard, it is necessary to mention that usually the amount of water wasted from small leaks in the network is more than from large leaks that cause accidents.

The reasons for reducing the water transfer capacity of pipes are:

- Foreign objects inside the pipes, such as wood, stone, etc., that enter the pipes during pipe laying or necessary repairs
- Sedimentation of sludge, iron, and manganese
- Creating a coating layer of sedimentary materials on the inner wall of pipes and protrusions caused by rusting, water corrosion, or lime deposits
- Air accumulation at the highest points of the level
- Reducing the cross-sectional area of the pipe due to sedimentation of faucets in semiclosed states or bending of pipes

Therefore, in order to prevent large losses, the accurate implementation of the network, the use of quality materials, and the careful monitoring of its implementation while observing the technical principles of implementation are factors that can have a great effect on the health of the network and increase its life.[5–10]

Otherwise, and in the conditions where it is possible that the water transmission line is reduced, the following measures should be taken:

- Full air evacuation control throughout the length of the pipeline
- Washing pipes
- Visiting the accessible parts of the pipe to control the sediment layer created on the surface of the inner wall of the pipes

Tanks

A) Prelaunch

Before starting the exploitation operation, it must be ensured that the following things are fully fulfilled (Figure 2.13):

- All the problems that caused the failure of the unit are recorded in the operational sheets.

- The repair and replacement of the parts that were involved in the occurrence of problems have been done.
- The tank overflow system is ready for operation.
- The performance of the faucets entering the tanks is guaranteed.
- The performance of the outlet taps from the tanks is guaranteed.
- The performance of the chemical injection and chlorination system is guaranteed.
- Preparing the daily report book (operational sheets)

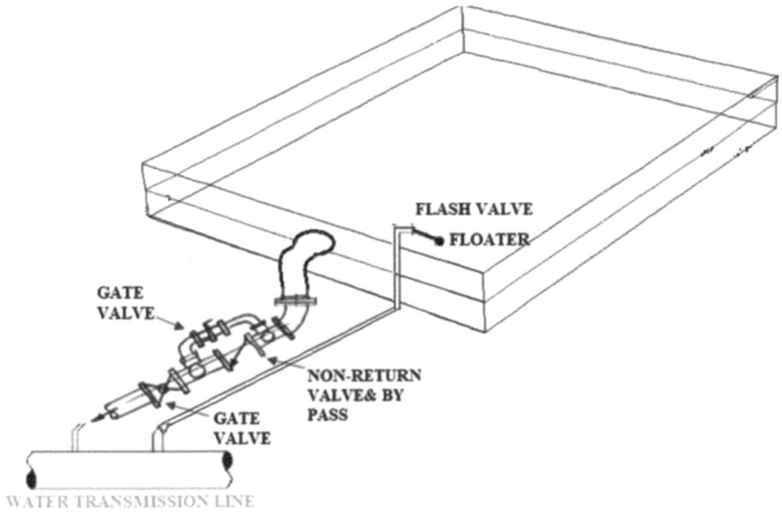

FIGURE 2.13 Inlet and outlet taps of tanks in water transmission line.

B) Normal operation (Figures 2.14–2.16)

- Continuous control of the height of the water level in the tank and recording it in the relevant worksheet
- Ensuring that the inlet and outlet valves are open
- Ensuring the performance of ventilators and ventilation systems
- Inspection and notification of the absence of leaks in the walls and roof
- Ensuring the operation of inspection valves
- Visiting the appearance of water

- Recording the report of each of the above cases in the information sheets
- Continuous control of the chemical injection rate

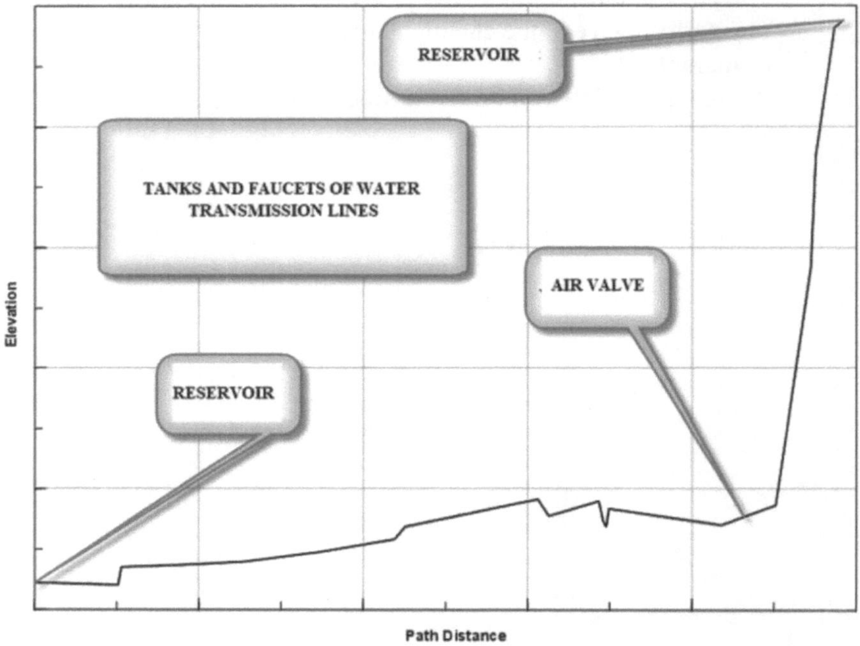

FIGURE 2.14 Tanks and faucets of water transmission lines.

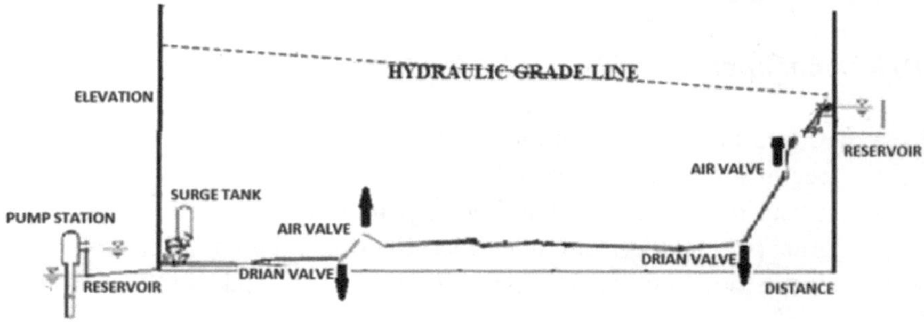

FIGURE 2.15 Tanks and inlet and outlet taps of tanks in water transmission lines.

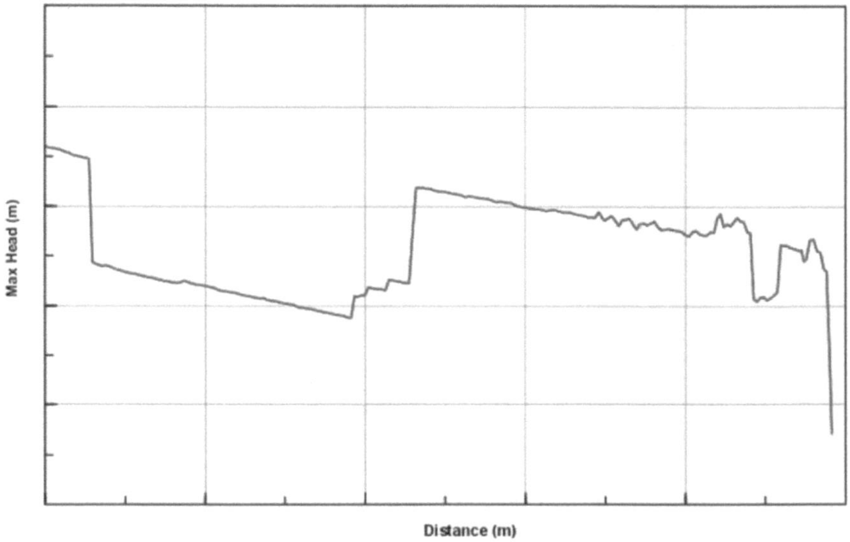

FIGURE 2.16 Max. pressure changes in water transmission lines.

Connections

Connections are generally used in the place of the faucet basin for the operation of various faucets.[11–19]

Valves manholes

The standard technical dimensions of the faucet basin are calculated and presented based on the faucets and equipment used in the urban water distribution network in different situations. In relation to the dimensional details provided, the following points can be mentioned (Figures 2.17–2.22):

- The height of the ponds is considered variable, and this is due to the difference in the depth of the piping.
- The minimum standard dimensions of the number of ponds in which faucets and pipes with a diameter of less than 100 mm are used are 100 × 100 × H cm.

In cases where the height of the basin (H) is more than 1.5 m, it is necessary to install a suitable staircase for the access of the responsible personnel to the internal equipment.

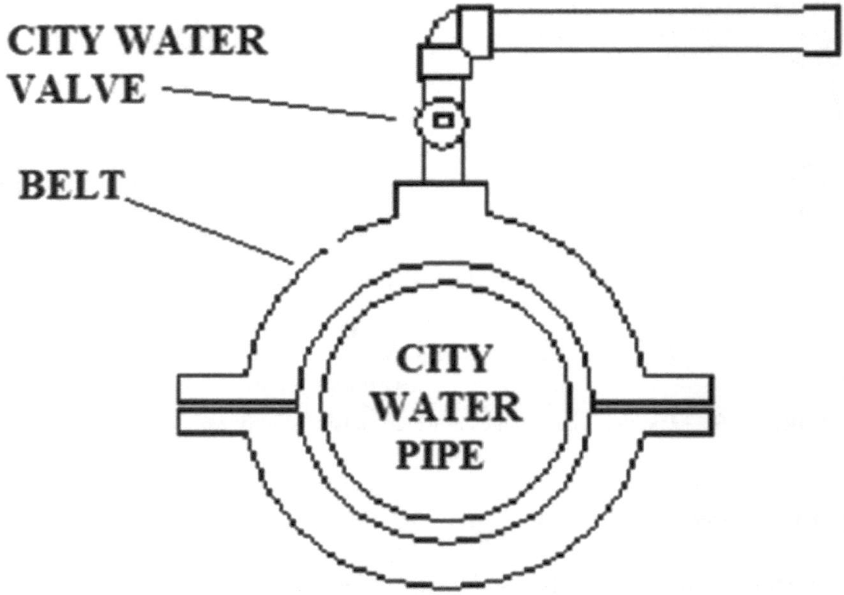

FIGURE 2.17 Belt installation in the urban water distribution network.

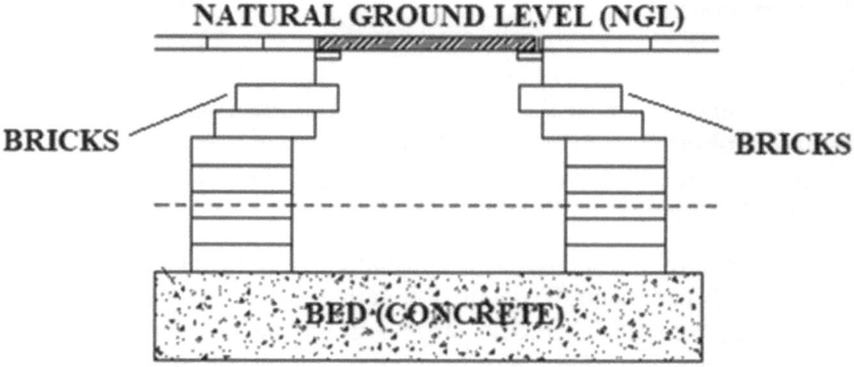

FIGURE 2.18 Installation basin.

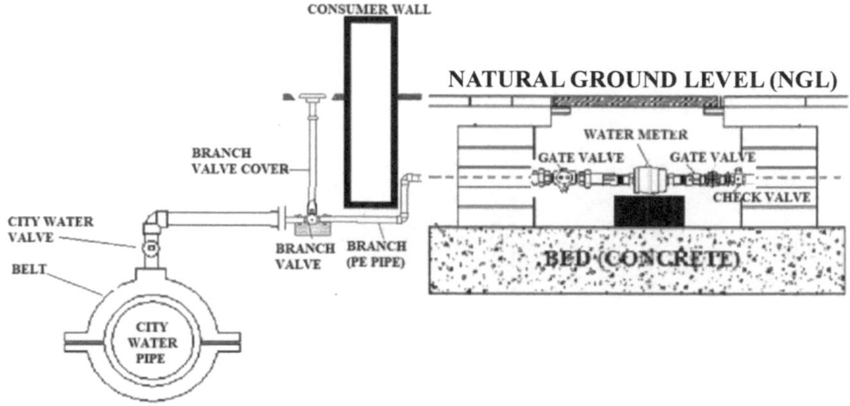

FIGURE 2.19 Installation basin.

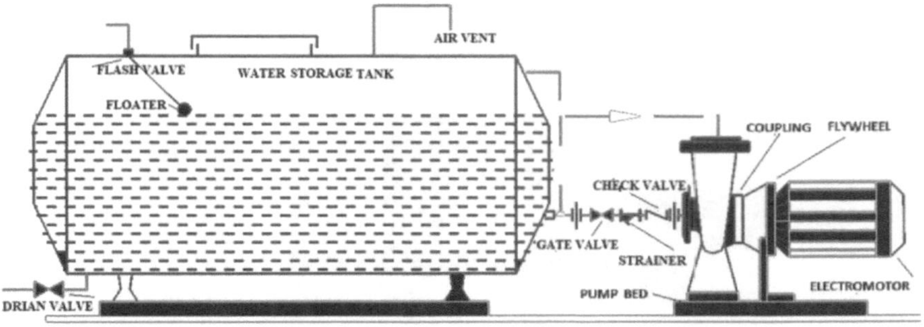

FIGURE 2.20 City water pump station in consumption.

2.1.2.6 A NEW WAY FOR WATER TREATMENT BY ORGANIC AGENT

There are many, such as fine filter medium in water treatment, physical absorption medium, and help in flocculation and coagulation in water treatment. Rice, as a natural absorbent agent, can absorb heavy metals in raw water (Figures 2.23 and 2.24).

The model of a natural absorbent paddy husk filter is useful in the water distribution facilities of small towns and rural areas. This model, by the rapid circulation of information according to the GIS, was considered. The hydraulic model of the water distribution network was defined during this work. It was defined under the management of ArcGIS-ArcMap software.

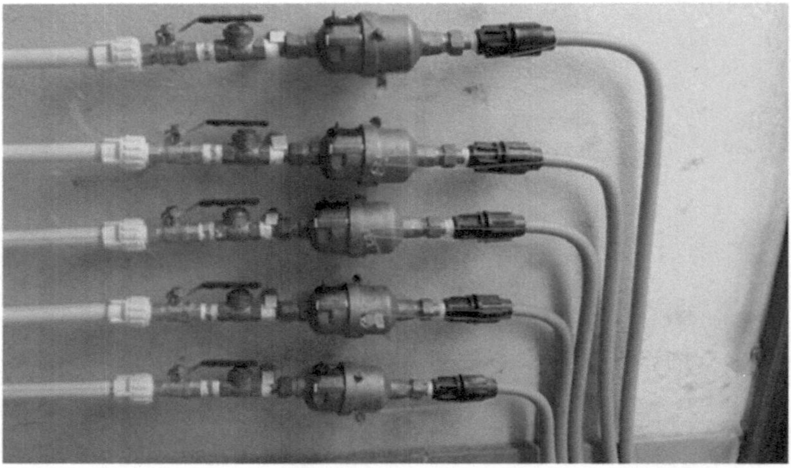

FIGURE 2.21 Installation of the horizontal branch of the wall.

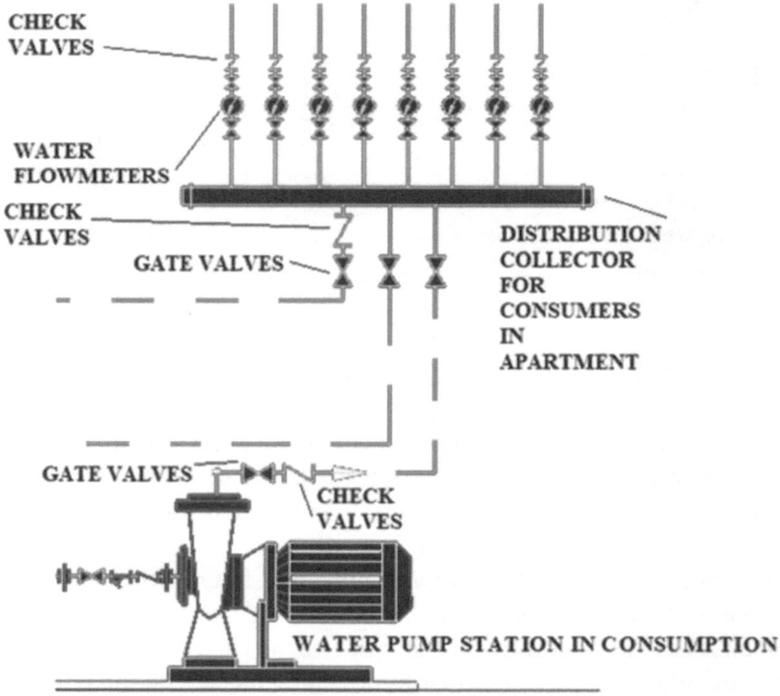

FIGURE 2.22 Water pump station in consumption and distribution collector for consumers in apartments.

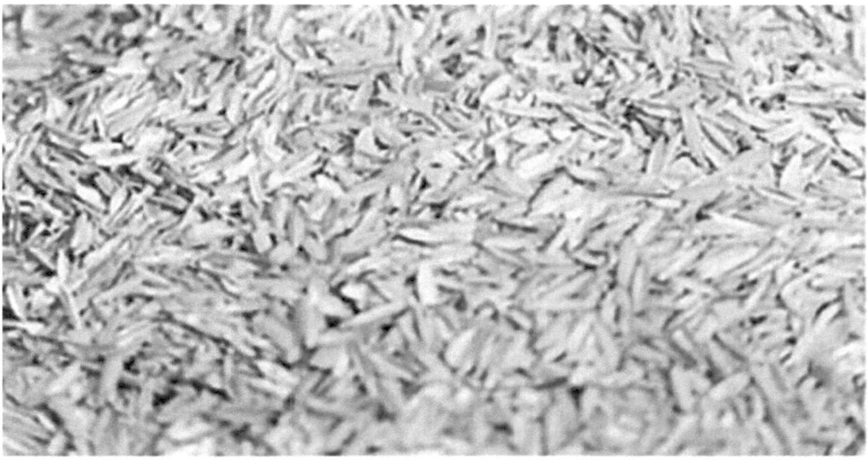

FIGURE 2.23 Rice paddy shell.

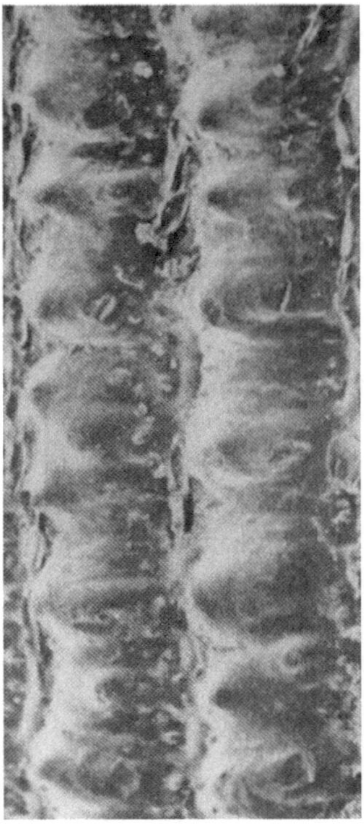

FIGURE 2.24 Micrograph of rice paddy shell.

The laboratory model was run by a bench-scale column 1 m high, diameter 7 cm, and volume 3850 mL. The water flow rate was 1000 mL per minute, and the amount of manganese at the inlet was 1.5 ppm and at the outlet was 0.3 ppm.

In addition to the mentioned chemical antimicrobial agents, the absorption properties of amorphous silica in paddy husk give it a porous filter, hence the use of paddy husk as a biological filter to remove odors and reduce COD. Demand is especially effective in the stage of the physical purification of water. At present, paddy husk is considered a natural and cheap absorbent. Due to its special physical and chemical characteristics and its very low price, paddy husk can be used as one of the most important sources for the production of low-cost absorbents for the removal of manganese. Today, the analytical quality method of reference land laboratory information management system (LIMS) has been accepted as an important and effective technique for understanding the hydraulic behavior and quality of urban water distribution systems and generally overcoming the information gap in these systems. In this regard, there are still many ambiguities in describing the basic relationships that show the effect of different parameters in water distribution systems. Therefore, in order to solve the existing shortcomings of this research, the intelligent pressure management model based on remote sensing in the GIS ready platform is an effective tool. It has been used in tracking and removing manganese with the help of paddy husk as a natural absorbent (biosorbent) in urban water distribution systems.[20–29]

2.2 MATERIALS AND METHODS

In the treatment of water and water resources, the following are considered pollutants:

Heavy metals: These are considered among the pollutants of water sources, and the best way to detect them is through laboratory methods.

Turbidity: Turbidity refers to suspended solid particles and gives the water a cloudy, dark, and unpleasant appearance.

Organic compounds: Water resources are polluted by a number of organic compounds, such as chloroform, gasoline, pesticides, and herbicides. These contaminants are also identified by laboratory methods.

Pathogenic microbes (pathogens): Protozoan systems are the largest pathogenic microbes in drinking water.

Bacteria: Bacteria are smaller than protozoa and are responsible for many diseases such as typhoid, cholera, diarrhea, and dysentery.

Viruses: After protozoa, viruses are the most dangerous pathogenic microbes.

Two physical and chemical methods are effective for water purification.

Boiling is a method to purify water and clean it from pathogenic microbes. In thermal processes, the temperature of boiling water for 10 min or the temperature of 161°C for 15 s and heat exchangers are used.

Reverse osmosis: In reverse osmosis, water is passed under pressure through a membrane that is impermeable to pollutants. The most common use of it is the preparation of freshwater from salt water, and its membrane works better in absorbing weak and nonionized acids and bases and organic molecules (molecular weight lower than 200) compared to salts and heavier organic molecules and microbes. The salts are removed in this way.

Distillation: The distillation process includes the evaporation and condensing of water in order to filter and purify it. Distillation has two major defects in practice:

- A lot of energy is needed.
- Chemical pollutants enter the system with a boiling point lower than the boiling point of water.

Today, distillation is used to separate mineral salts and salts from water, the simplest form of which is solar distillation.

Microfilters: Microfilters are small-size filters that are used to separate floating solid particles, protozoa, and in some cases, bacteria from water. Most of the filters are made of ceramic compounds or fibers, so they can return to their original state after use. Microfilters do not have the ability to separate viruses, and along with charcoal filters, they create the problem of bacteria growth in the environment of the filter. Some experts consider the use of silver to be effective in inoculating filter elements, and some recommend the use of freezers to store filters for a short period of time. Many microfilters include sludge filters, activated charcoal, or iodine resins, and if charcoal filters are used, excess iodine is removed from the water. Charcoal filters separate other natural and artificial polluting compounds from water.

Slow sand filter: Slow sand filters slowly pass water through the sand bed. Pathogenic microbes and turbidity are removed due to natural death, biological actions, and smoothing. The height of the sand in the filter is 24 in, and then a gravel layer is placed that surrounds the discharge pipe. The sand is not in contact with the flat walls, so water cannot easily come down from the flat walls and reach the sand. Building a wall with hard materials helps with this process. The speed of passing through the sieve is 0.2 m per hour of the day. The filter can be cleaned several times before it is replaced with new sand. Slow sand filters can only be used for continuous water purification systems. If a continuous source of water is not available, another method should be substituted. It is important to have low turbidity (suspended solid particles) in the water under treatment. Continuous turbidity is reduced by changing the collection method, allocating more retention time (using a water tank), and preliminary filtering (adding a chemical such as alum to bind the suspended materials together). A sand filter is a large chamber with a height of at least 1.5 m. The walls should be made of hard material to reduce the tendency of water to penetrate them. The flat bottom layer is made of sand bed. The sand in such filters should be clean and uniform.

Activated charcoal filters: In activated charcoal filters, chemicals and some heavy metals are absorbed and attached to the surface of the charcoal. Such filters separate some of the pathogenic microbes, and in some cases, activated charcoal acts as a growth medium for bacteria and algae. To prevent such problems, activated charcoal is inoculated with silver. Activated charcoal filters, along with chemical methods, are effective in water purification. Chemical compounds containing iodine or chlorine destroy pathogenic compounds, while carbon filters separate chemical compounds. If the filter reaches its maximum capacity, you should pay attention to the specific smell of iodine and chlorine. Activated charcoal can be reactivated by heat. It is a recovery. Also, the carbon bed should be deep enough to provide maximum contact with water. Activated carbon can be supplied in two forms: powder and granular.

Water treatment using solar energy (SOL-air water treatment): If the dissolved oxygen is enough, sunlight causes the temporary formation of active forms of oxygen such as hydrogen peroxide and oxygen-free radicals. This type of water purification is called solar photooxidative purification, or water purification through solar flow. This method reduces harmful coliform bacteria. There is also evidence that this method affects other bacteria and viruses. This method does not need advanced technology, and exposure to sunlight for 4.5 h has shown that the maximum activity is at a temperature above 50°C, and therefore the solar method is not suitable for cold weather. Plastic or glass containers are used in this method. Plastic containers allow ultraviolet waves to pass through and increase the rate of destruction of microbes, but they may change color and turn yellow with the passage of time, and as a result, the speed of light transmission decreases, and plasticizers may also enter the water. In this regard, containers made of polyethylene terephthalate are preferable to polyvinyl chloride (PVC). Generally, glass containers are more suitable, and if they are free of turbidity, containers with a larger volume are used. The container is filled up to 3/4 in the morning, and then the lid is put on it and shaken vigorously. At the end of the day, the water is free of bacteria and ready for consumption the next day. The primary goal of water purification is to provide potable water. It is human; in this regard, it is important to improve the microbial condition and control dangerous chemicals and metals. The second purpose of water purification is to protect water distribution systems from corrosion problems and maintain the quality of taste, smell, color, and hardness of water. The choice of water treatment method in small towns and rural areas depends on the type of drinking water source. Therefore, knowing the type of water source and its pollutants will lead to the correct choice of water purification method.

In this research, the modeling of the water distribution network (Figures 2.25 and 2.26) includes the process of computer statistics in the simulation of the rural distribution network.

Measuring the parameters of water pressure and flow entering the water distribution network based on the recommendations of drinking water treatment techniques is recommended by the World Health Organization (WHO) for rural areas with the use of advanced technologies.

Remote reading is done. The hydraulic model of the present research was defined in accordance with the GIS and in the software environment WATERGEMS 8.2 under the management of ArcGIS 9-ArcMap 9.3 software. This hydraulic model was designed in order to remove manganese with the help of natural absorbent paddy husk in the treatment plant and track it with the sensors installed in the network and the reference ground model in order to achieve the following goals:

- Quantitative and qualitative optimization of the water distribution system.
- Quantitative and qualitative updating of existing facilities in compliance with GIS.[30–48]

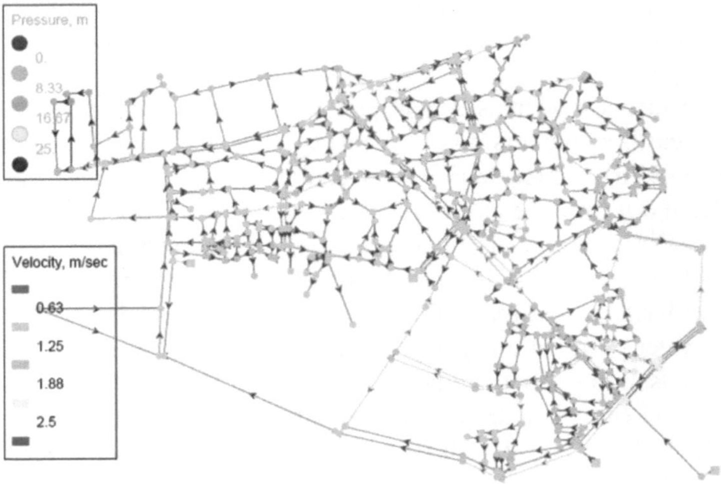

FIGURE 2.25 Distribution network analysis in compliance with GIS.

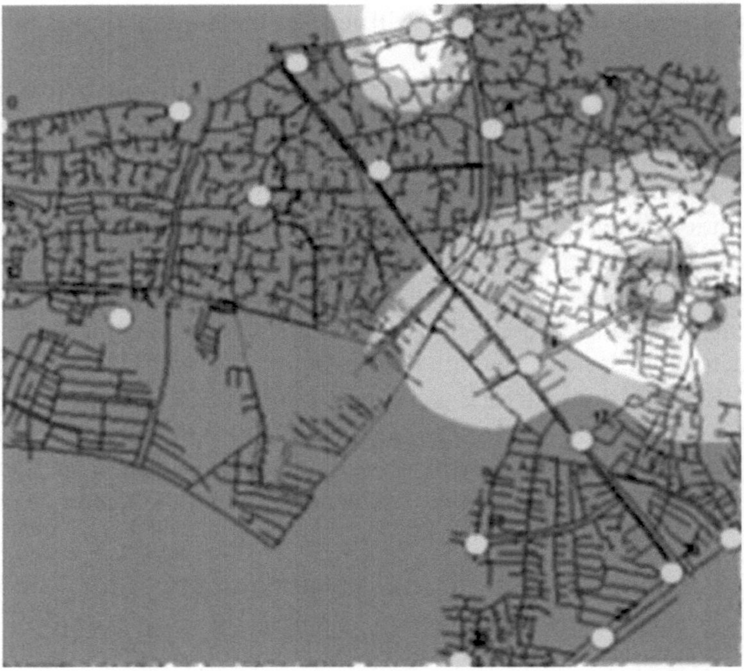

FIGURE 2.26 Pressure detection by networked sensors based on the geographic information system (GIS).

In the current research, the samples were passed through a 1-m column and at a suitable pH (Figure 2.27), and the results of remeasurement of manganese content are given in Tables 2.2–2.4.

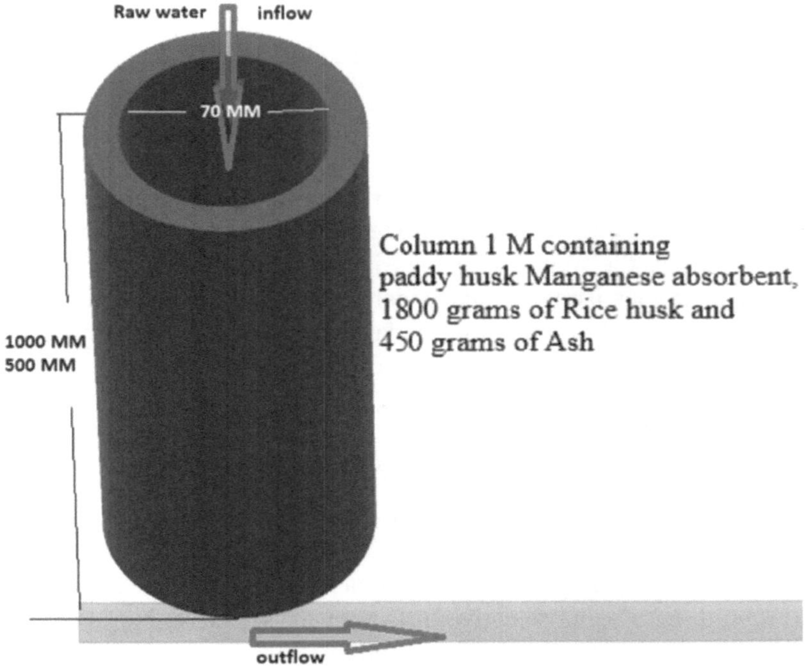

FIGURE 2.27 Column containing paddy husk absorbent.

TABLE 2.2 The Results of Measuring Manganese Absorption by Paddy Husk After Passing Through the Column Containing the Adsorbent.

Sample description	Absorption percentage by paddy husk
0.5 m column with **pH = 3**	77.32
1 m column with **pH = 3**	88.98
0.5 m column with **pH = 6**	89.23
1 m column with **pH = 6**	89.12
0.5 m column with **pH = 8**	90.44
1 m column with **pH = 8**	87.57
0.5 m column with **pH = 10**	96.02
1 m column with **pH = 10**	98.25

Sample with 1.20 dilution prepared from 1 mg/mL manganese standard solution, flow rate 10 mL/min.

TABLE 2.3 Comparison of the Results of Manganese Absorption by Paddy Husk After Passing Through the Column Containing the Adsorbent.

The amount of manganese	Example 1	Example 2	Example 3	Sample 4
Filter input (ppm)	0.50	0.48	0.53	069
Filter output (ppm)	0.53	0.39	0.38	0.30

The spectroscopic method based on the color of the permanganate ion is considered a selective method for measuring manganese concentration. In this method, the manganese ions in the solution are converted into permanganate ions using strong oxidants, which is the basis of the spectroscopic method. Two common oxidants for this purpose are potassium periodate and ammonium persulfate, eqs 2.1 and 2.2:

$$2Mn2+ + 5IO4- + 3H2O\ 2MnO4- + 6H+ \quad (2.1)$$

$$2Mn2+ + 5S2O82- + H2O \longrightarrow 2MnO4- + 10HSO4- + 6H+ \quad (2.2)$$

Oxidation of manganese with periodate or persulfate is carried out in sulfuric acid, nitric acid, or a mixture of the two. The molar absorption coefficient of the permanganate ion at max = 528 nm is equal to 2 × 103; therefore, despite being selective, it is not very sensitive, but in the range of the desired concentration in absorption experiments, its efficiency is very favorable (Figures 2.28–2.30).

TABLE 2.4 Manganese Absorption by Paddy Husk Through the Column Containing the Adsorbent.

Meter column (m)	pH	Absorption percentage by paddy husk (%)
0.5	3	77.32
1	3	88.98
0.5	6	89.23
1	6	89.12
0.5	8	90.44
1	8	87.57
0.5	10	96.02
1	10	98.25

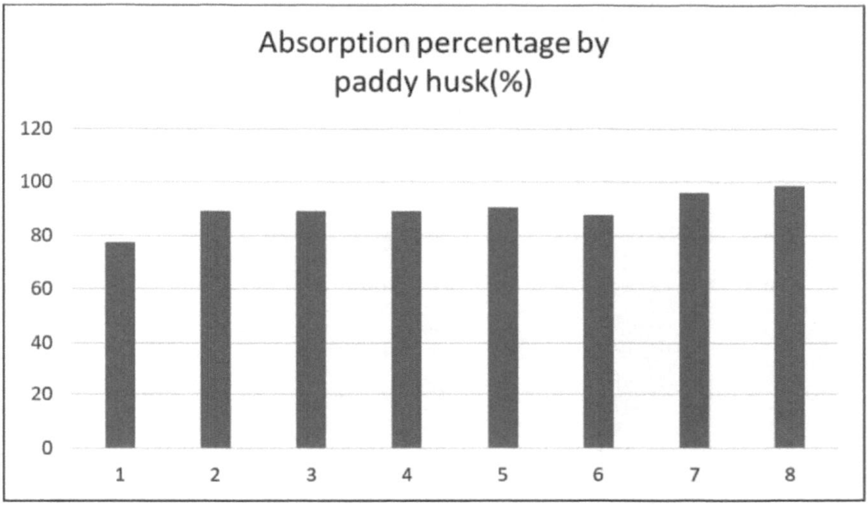

FIGURE 2.28 Column containing paddy husk absorbent.

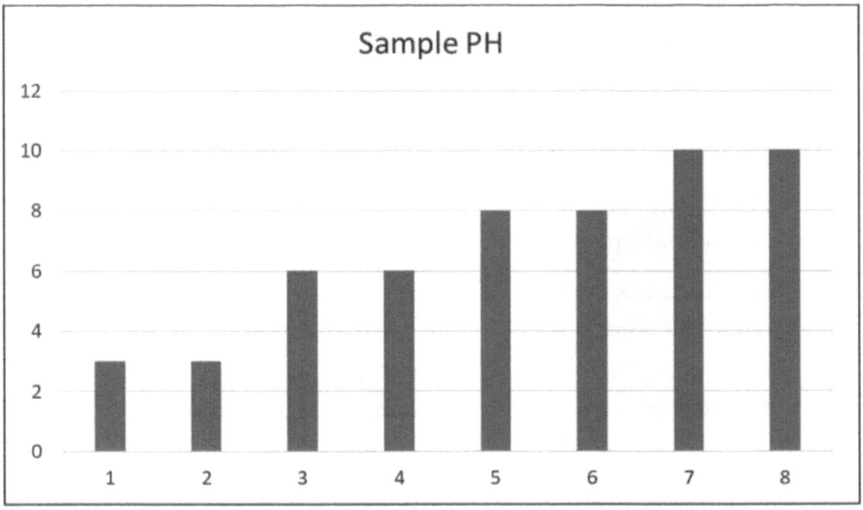

FIGURE 2.29 Column containing paddy husk absorbent against sample pH.

To perform the tests, a standard solution of 1 mg/mL manganese with a dilution of 1.20 was prepared, and after adjusting the pH to 3, 6, 8, and 10, the solution was passed through the half and one meter columns at a speed of 10 mL/min, and then it was absorbed by the method. Permanganate was measured at a wavelength of 528 nm.

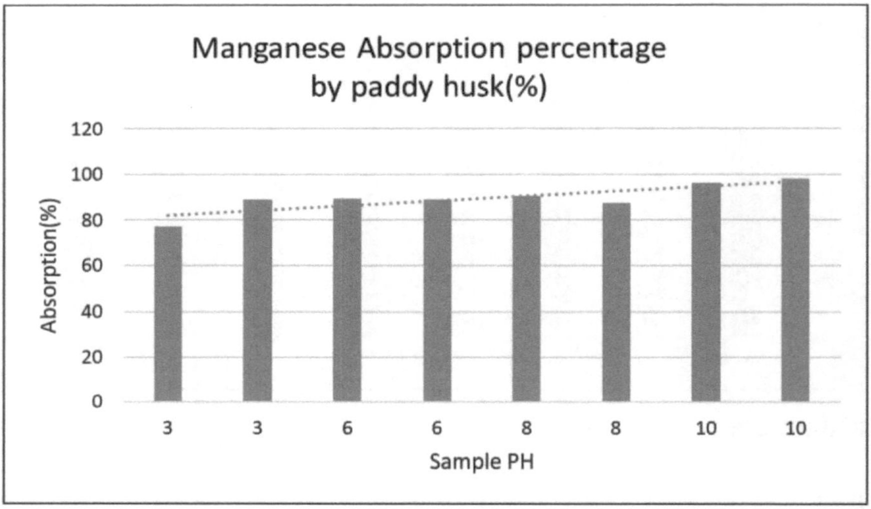

FIGURE 2.30 Column containing paddy husk absorbent of manganese; 1800 g of rice husk and 450 g of ash.

Required materials and reagents:

- Potassium periodate
- Ammonium persulfate

Silver nitrate, about 0.100 M: Dissolve 85 mg of silver nitrate in 500 mL of water, and store the solution in a brown bottle in the refrigerator.

- Manganese standard solution 1 mg/mL: Dissolve 2.873 g of $KMnO_4$ in water containing 2 mL of concentrated sulfuric acid and then dilute it with water to a volume of 1 L.

In the method of measuring manganese and oxidation with potassium periodate:

A sample solution containing up to 300 mg of manganese (free from chloride and other reducing agents) was added in a beaker to about 15 mL with diluted water, then 2 mL of concentrated sulfuric acid, 0.5 mL of concentrated nitric acid, 1 mL of concentrated phosphoric acid and 0.12

g of KIO4 were added and stirred. The solution was heated to boiling temperature and kept at 90°C for 10 min, and then after cooling, it was transferred to a 25 mL volumetric flask and made up to volume. Then the absorbance of the solution was measured at Max. = 528 nm.

In the oxidation method with ammonium persulfate:

Half a milliliter of concentrated sulfuric acid and concentrated nitric acid, 1 mL of concentrated phosphoric acid, 0.001 M silver nitrate, and 0.25 g of ammonium persulfate were added to the sample containing manganese and heated to boiling temperature. Then, it was kept in a boiling state for 5 min and after cooling, it was made up to volume with water in a 25 mL volumetric flask, and then its absorbance was measured at max = 528 nm.[50–73]

2.3 RESULTS AND DISCUSSION

In the current research, the filter made of washed and dried rice husk with a dry weight of 500 g has the following results (Figure 2.7 and Tables 2.1 and 2.2):

Column conditions = height 1 m, diameter 7 cm
Column volume = 3850 mL
1000 mL per minute = Flow rate:
The amount of iron in the input is 3 and in the output = 0.1 (in ppm)
The amount of manganese in the input 1.5 and in the output = 0.3 (in ppm)
Temporary hardness at the entrance is 415 and at the exit = 282.5 (in ppm)

The filter made of washed and dried rice husk with a dry weight of 600 g has the following results:

Column conditions = height 1 m, diameter 7 cm
Column volume = 3850 mL
Flow rate = 1000 mL per minute
Amount of iron in the input 3 and in the output = 0.1 (in ppm)
Amount of manganese in the input 1.5 and in the output = 1 (in ppm)
Temporary hardness at the entrance is 415 and at the exit is 19 = 9.2 (in ppm)

2.3.1 COMPARISON OF THE RESEARCH RESULTS WITH OTHER RESEARCHES

The filter made of washed and dried almond husk with a dry weight of 450 g has the following results:

- Column conditions: height 1 m, diameter 7 cm
- Column volume: 3850 mL
- Flow rate: 1000 mL per minute
- The amount of iron in the input is 3 and in the output is 2.6 (in ppm)
- The amount of manganese in the input 1.5 and in the output 1.5 (in ppm)
- Temporary hardness at the entrance 415 and at the exit 380 (in ppm)

2.4 CONCLUSION

The effectiveness of the rice husk filter in removing manganese and the best efficiency in terms of hardness removal are related to the filter made of rice husk ash. In the next step, to check the efficiency of the filter, a combined filter with a ratio of 30% ash and 70% rice husk was made, and this filter was installed at the research pilot site. The filter specifications are as follows:

Filter height: 1.5 m and diameter: 20 cm. Paddy husk: 1800 g and paddy husk ash: 450 g.

As a final conclusion, considering that this research was conducted on a laboratory and field scale for small towns and rural areas in a limited manner and according to the available facilities in a short period of time, it is suggested that an urban pilot scale plan be used to determine the quality. The completeness of the filter and the determination of other parameters, such as the duration of the filter's exact performance, are defined, and other parameters that were not possible in this research should be determined accurately.

AKNOWLEDGMENT

The author thanks to all specialists for their valuable observations and advice, and the referees for recommendations that improved the quality of this paper.

KEYWORDS

- **rice**
- **water treatment**
- **manganese**
- **networked sensors**
- **Internet of Things**
- **geographic information system**

REFERENCES

1. Dashti-Khavidaki, S.; Khalili, H.; Vahedi, S. M.; Lessan, M. Serum Zinc Concentrations in Patients on Maintenance Hemodialysis and Its Relationship with Anemia, Parathyroid Hormone Concentrations and Pruritus Severity. *Saudi J. Kidney Dis. Transpl.* **2010,** *21* (4), 641–645.
2. Khalili, H.; Vahedi, S. M.; Lessan, M. Serum Zinc Concentrations in Patients on Maintenance Hemodialysis and Its Relationship with Anemia, Parathyroid Hormone Concentrations and Pruritus Severity. *Saudi J. Kidney Dis. Transpl.* **2010,** *21* (4), 641.
3. Gibbon, G. A. A Brief History of LIMS. *Lab. Autom. Inform. Manag.* **1996,** *32* (1), 1–5.
4. Simon W. *Comprehensive Laboratory Informatics: A Multilayer Approach*. American Laboratory, 2007; p 1.
5. Jones, J. *The LIMS Book & Buyer's Guide*. Laboratory Informatics Institute, Inc., 2012.
6. Volesky, B. *Biosorption of Heavy Metals*. CRC Press: Florida. 1990.
7. Hariri Asli, K.; Aliyev S. A. O. *Fluid Mechanics and Heat Transfer; Advances Research in Nonlinear Dynamics Modeling*. CRC Press: U.S.A., Canada, 2015.
8. Kevin, A. Internet of Things. *RFID J.* 2009.
9. Asli, K. H.; Aliyev, S. A. O.; Thomas, S.; Deepu, A. *Handbook of Research for Fluid and Solid Mechanics: Theory, Simulation, and Experiment*, 1st ed.; Apple Academic Press, 2017.
10. Asli, K. H.; Naghiyev, F. B. O.; Haghi, R. K.; Asli, H. H. *Advances in Control and Automation of Water Systems*, 1st ed.; Apple Academic Press, 2012. DOI: 10.1201/b13114
11. Hariri Asli, K.; Nonlinear Dynamics Modeling by Geospatial Information System (GIS). *Map. Geosp. Inform. J. Guilan.* **2019,** *3* (1), Iran. www.mpogl.ir.
12. https://en.wikipedia.org/wiki/Artificial_intelligence
13. Kelly, K. *New Rules for the New Economy*. First published in 1998 by Viking Penguin, 1998.
14. https://fa.wikipedia.org/wiki/ Geographic Information System.
15. Wylie, E. B.; Streeter, V. L. *Fluid Transients*. Feb Press: Ann Arbor, MI, 1983, corrected copy: pp 166–171, 1982.

16. Lee, T. S.; Pejovic, S. Air Influence on Similarity of Hydraulic Transients and Vibrations. *ASME J. Fluid Eng.* **1996,** *118* (4), 706–709.
17. Yoon, S. H.; Park, C. S.; Objective Building Energy Performance Benchmarking Using Data Envelopment Analysis and Monte Carlo Sampling. *Sustainability*, **2017,** *9,* 780.
18. Hariri Asli, H.; Nazari, S. Water Age and Leakage in Reservoirs; Some Computational Aspects and Practical Hints. *Larhyss J.* **2021,** *48,* 151–167.
19. Hariri Asli, H.; Hozori, A. Non-Revenue Water (NRW) and 3d Hierarchical Model for Landslide. *Larhyss J.* **2021,** *48,* 189–210.
20. Kensek, K.; *Buildings* **2015,** *5,* 899–916. DOI: 10.3390 /buildings5030899
21. Wang, S. K. *Handbook of Air-Conditioning and Refrigeration.* McGraw-Hill, Inc., 1994.
22. https://fa.wikipedia.org/wiki/ Internet of Things _IoT
23. Ferras, D, Manso, A. P. A.; Schleiss, A. J.; Covas, D. I. C. One-Dimensional Fluid-Structure Interaction Models in Pressurized Fluid-Filled Pipes: A Review. *Appl. Sci.* **2018,** *8,* 1844–1856.
24. Vardy, A. E.; Brown, J. M. B.; He, S.; Ariyaratne, C.; Gorji, S. Applicability of Frozen-Viscosity Models of Unsteady Wall Shear Stress. *J. Hydraul. Eng.* **2015,** *141,* 1–13.
25. Asli, K. H.; Aliyev, S. A. O.; Thomas, S.; Deepu, A. *Handbook of Research for Fluid and Solid Mechanics: Theory, Simulation, and Experiment*, 1st ed.; Apple Academic Press, 2017.
26. Asli, K. H.; Naghiyev, F. B. O.; Haghi, R. K.; Asli, H. H. *Advances in Control and Automation of Water Systems*, 1st ed.; Apple Academic Press, 2012.
27. Lema, M.; Peña, F. L.; Buchlin, J. M.; Rambaud, P.; Steelant, J. Analysis of Fluid Hammer Occurrence with Phase Change and Column Separation Due to Fast Valve Opening by Means of Flow Visualization. *Exp. Therm. Fluid Sci.* **2016,** *79,* 143–153.
28. Zecchin, A. C.; Maier, H. R.; Simpson, A. R.; Leonard, M.; Nixon, J. B. Parametric Study for an Ant Algorithm Applied to Water Distribution System Optimization. *IEEE Trans. Evol. Comput.* **2005,** *9,* 175–191.
29. Andrade, D. M.; de Freitas Rachid, F. B. A Versatile Friction Model for Newtonian Liquids Flowing Under Unsteady Regimes in Pipes. *Meccanica* **2022,** *57,* 43–72.
30. Sarkardeh, H.; Zarrati, A. R.; Jabbari, E.; Marosi M Numerical Simulation and Analysis of Flow in a Reservoir in the Presence of Vortex. *J. Eng. Appl. Comput. Fluid Mech.* **2014,** *8* (4), 598–608.
31. Zhao, L.; Yang, Y.; Wang, T.; Han, W.; Wu, R.; Wang, P.; Wang, Q.; Zhou, L. An Experimental Study on the Water Hammer with Cavity Collapse under Multiple Interruptions. *Water* **2020,** *12*, 2566.
32. Duan, H. F.; Pan, B.; Wang, M.; Chen, L.; Zheng, F.; Zhang Y. State-of-the-Art Review on the Transient Flow Modeling and Utilization for Urban Water Supply System (UWSS) Management. *J. Water Supply Res. Trans.* **2020,** *69,* 858–893.
33. Zhou, L.; Liu, D.; Karney, B. Investigation on Hydraulic Transients of Two Entrapped Air Pockets in a Water Pipeline. *J. Hydraul. Eng.* **2013,** DOI: 10.1061/(ASCE) HY.1943-7900.0000750
34. Möller, G.; Detert, M.; Boes, R. M. Vortex-Induced Air Entrainment Rates at Intakes. *J. Hydraul. Eng.* **2015,** DOI: 10.1061/(ASCE)HY.1943-7900.0001036
35. Hariri Asli, K.; Nagiyev, F. B.; Haghi. A. K. Interpenetration of Two Fluids at Parallel Between Plates and Turbulent Moving in Pipe. In *Chapter 7, Computational*

Methods in Applied Science and Engineering. Nova Science publisher: USA, 2009; pp. 115–128.

36. Hariri Asli, K.; Nagiyev, F. B. O.; Haghi. A. K. Some Aspects of Physical and Numerical Modeling of water hammer in pipelines. *Int. J. Nonlinear Dyn. Chaos Eng. Syst.* **2009**.
37. Hariri Asli, K.; Nagiyev, F. B.; Haghi. A. K. Physical Modeling of Fluid Movement in Pipelines. In *Nanomaterials Yearbook*. USA, 2009.
38. Arturo, S. L.; Mohamed, S.; Ghidaoui, M.; Arthur, R.; Schmidt, M.; Marcelo, H.; Garc, A. M. Godunov Type Solutions for Transient Flows in Sewers. 20–44, Illinois 61801, USA. 2005.
39. Arturo Leon, S. Improved Modeling of Unsteady Free Surface, Pressurized and Mixed Flows in Storm-Sewer Systems. In *Submitted in Partial Fulfillment of the Requirements for the Degree of Doctor of Philosophy in Civil Engineering in the Graduate College of the University of Illinois at Urbana-Champaign*, 2007, pp 57–58.
40. Hariri Asli, H.; Arabani, M.; Golpour, Y. Reclaimed Asphalt Pavement (RAP) Based on a Geospatial Information System (GIS). *Slovak J. Civil Eng.* **2020,** *28* (2), 36–42.
41. Hariri Asli, H. Investigation of the Factors Affecting Pedestrian Accidents in Urban Roundabouts. *Transac. Civil Environ. Eng.* **2022,** *8* Article ID: 2226, 1–4, Special Issue: NCTT 2021. DOI: 10.52547/crpase.8.1.2255
42. Hariri Asli, H. Arabani, M. Analysis of Strain and Failure of Asphalt Pavement, Computational Research Progress in Applied Science & Engineering. *Transac. Civil Environ. Eng.* **2022,** *8,* 1–11.
43. Lee, T. S.; Pejovic, S. Air Influence on Similarity of Hydraulic Transients and Vibrations. *ASME J. Fluid Eng.* **1996,** *118* (4), 706–709.
44. Monzavi, M. T. *Eastewater Treatment*. Tehran, Iran, 1991; pp 7–162.
45. Filion, Y.; Karney, B. W. In *A Numerical Exploration of Transient Decay Mechanisms in Water Distribution Systems*, Proceedings of the ASCE Environmental Water Resources Institute Conference. American Society of Civil Engineers: Roanoke, Virginia, 2002.
46. Chen, M.; Leng, B.; Wu, S.; Sang, Y. Physical, Chemical, and Rheological Properties of Waste Edible Vegetable Oil Rejuvenated Asphalt Binders. *J. Constr. Buil. Mater.* **2014,** *66* (15 Sep), 286–298.
47. Fini, E. H.; Kalberer, E. W.; Shahbazi, A.; Basti, M.; You, Z.; Ozer, H. Chemical Characterization of Bio-Binder from Swine Manure: Sustainable Modifier for Asphalt Binder. *J. Mater. Civil Eng.* **2011,** *23* (11), 1506–1513.
48. Hassan, M.; Essam, T.; Yassin, A. S.; Salama, V. Optimization of Rhamnolipid Production by Biodegrading Bacterial isolates using Plackett–Burman Design. *Int. J. Biol. Macromol.* **2016,** *82* (4), 573–579.
49. Shirzad, S.; Hassan, M. M.; Aguirre, M. A.; Mohammad, L. N.; Daly, W. H. Evaluation of Sunflower Oil as a Rejuvenator and Its Microencapsulation as a Healing Agent. *J. Mater. Civil Eng.* **2016,** *28* (11), 04016116.
50. https://en.wikipedia.org/wiki/Computational_physics
51. NZS 3106. Design of Concrete Structures for the Storage of Liquids, Cement and Concrete Association of New Zealand, Standards Association of New Zealand, 2009.
52. Recommended Standards for Water Works, Great Lakes-Upper Mississippi River Board of State and Provincial Public Health and Environmental Managers, Health

Research Inc.; Health Education Services Division, PO Box 7126, Albany, NY 12224, 2012 102-SA Water TG.
53. Planning and Infrastructure, Above Ground Circular Concrete Tank, Manager Engineering, 2008. 103-Selected Earthquake Engineering Papers of George W. Housner, American Society of Civil Engineers, 1990.
54. Specification for Construction of Concrete Reservoirs, Standard Specifications and Drawings Specification for Construction of Concrete Reservoirs, Gold Coast, Australia, SS 11, 2004.
55. TM 5-809-10. *Seismic Design Guidelines for Essential Buildings*. US Department of the ARMY: Washington, D.C., 1986.
56. USEPA Finished Water Storage Facilities.U.S. Environmental Protection Agency, 2002.
57. Zecchin, A. C.; Maier, H. R.; Simpson, A. R.; Leonard, M.; Nixon, J. B. Parametric Study for an Ant Algorithm Applied to Water Distribution System Optimization. *IEEE Trans. Evol. Comput.* **2005,** *9* (2), 175–191.
58. Walski, T. M. Hydraulic Design of Water Distribution Storage Tanks. In *Water Distribution Systems Handbook*; Mays, L. W., Eds.; McGraw-Hill: New York, 2000; Chapter 10.
59. Farmani, R.; Walters, G.; Savic, D. Evolutionary Multi-Objective Optimization of the Design and Operation of Water Distribution Network: Total Cost vs. Reliability vs. Water Quality. *J. Hydroinform.* **2006,** *8* (3), 165–179.
60. Farmani, R.; Savic, D. A.; Walters, G. A. The Simultaneous Multi-Objective Optimization of Anytown Pipe Rehabilitation, Tank Sizing, Tank Siting and Pump Operation Schedules, Critical Transitions in Water and Environmental Resources Management. *ASCE, Am. Soc. Civil Eng.* **2004,** 4663–4672.
61. Haarhoff, J.; Van Zyl, J. E.; Nel, D.; Van der Walt, J. J. Water Supply and Water Quality in Rural Areas: Stochastic Analysis of Rural Water Supply Systems. *Water Supply* **2000,** *18* (1), 467–470.
62. Maier, H. R.; Simpson, A. R.; Zecchin, A. C.; Foong, W. K.; Phang, K. Y.; Seah, H. Y.; Tan, C. L. In Ant Colony optimization for the Design of Water Distribution Systems, *Proceedings of World Water and Environmental Resources Congress 2001, ASCE*, The Rosen Plaza Hotel: Orlando, Florida, United States.2001.
63. Maier, H. R.; Simpson, A. R.; Zecchin, A. C.; Foong, W. K.; Phang, K. Y.; Seah, H. Y.; Tan, C. L. Ant Colony Optimization for Design of Water Distribution Systems. *J. Water Resour. Plann. Manag.* **2003**, *129* (3), 200–209.
64. Walski, T. M.; Chase, D. V.; Savic, D. A.; Grayman, W.; Beckwith, S.; Koelle, E. *Advanced Water Distribution Modeling and Management*. Watertown, Conn: Haestad Press, 2003.
65. Hashemi, S. S. Optimization of Water Networks with Minimization of Pumping Energy. Master Thesis, University of Tehran, Iran, 2010.
66. Mau, R. E.; Boulos, P. F.; Clark, R. M.; Grayman, W. M.; Tekippe, R. J.; Trusell, R. R. Explicit Mathematical Models of Distribution Storage Water Quality. *ASCE J. Hydraul. Eng.* **1995**, *121* (10), 669–709.
67. Rossmann, L. A. EPANET 2 User's Manual. In *Water Supply and Water Resources Divison*. U.S. 2000. Environmental Protection Agency: Cincinnati, EE. UU. Available at: http://www.epa.gov/ORD/NRMRL/wswrd/epanet.html

68. Rossmann, L. A.; Grayman, W. M. Scale-Model Studies of Mixing in Drinking Water Storage Tanks. *J. Environ. Eng.* **2000,** *125* (8), 755–761.
69. Yeung, H. Modeling of Service Reservoirs. *J. Hydroinformatics* **2001,** *3* (3), 165–172.
70. Martínez-Solano, F. J.; Lopez-Jimenez. P. A.; Iglesias-Rey, P.; Perez, R. Utilization of Fluid Dynamics Co. **2006**.
71. Mahmood, F.; Pimblett, J. G.; Grace, N. O.; Grayman, W. M. Evaluation of Water Mixing Characteristics in Distribution System Storage Tanks. *J. Am. Water Works Assoc.* **2005,** *97* (3), 74–88.
72. Marek, M.; Stoesser, T.; Roberts, P. J. W.; Weitbrecht, V.; Jirka, G. H. CFD Modeling of Turbulent Jet Mixing in a Water Storage Tank. XXXII IAHR Congress, Venice (Italy), July 1/6, 2007.
73. Van Zyl, J. E.; Haarhoff, J. Reliability Analysis of Municipal Storage Reservoirs Using Stochastic Analysis. *J. S. Afr. Inst. Civ. Eng.* **2007,** *49* (3), 27–32.

68. [illegible]

69. [illegible] Scientific [illegible]

70. [illegible] 2008, [illegible]

71. [illegible]

72. [illegible]

CHAPTER 3

Modeling of Corrosion for Water System by Networked Sensors and Internet of Things (IoT) in Compliance with Geographic Information System (GIS)

HOSSEIN HARIRI ASLI

Lamar University, Texas State University System, Beaumont, Texas, USA

ABSTRACT

In the new world view, water is considered as an economic-social commodity and as a basic human need. Although water is one of the renewable resources, its quantity is limited. This work investigated the leakage phenomena in the water system due to the corrosion. The real case of non-revenue water (NRW) was modeled by the definition of relationship classes between the spatial and non-spatial data for water system. The relationship classes was assumed as a tool for communication between coordination of water system elements related to NRW data. Therefore, it was affected according to the type of connection and this method provided the access to the corrosion zones and leakage locations. The hydraulic model of NRW for analysis of water loss was defined in compliance with geographic information system (GIS). The result of this work led to reduction of the NRW.

3.1 INTRODUCTION

Due to population growth, expansion of industry, rising level of public health and well-being, per capita renewable resources are decreasing.

Sustainable Water Treatment and Ecosystem Protection Strategies. Hossein Hariri Asli, Tamara Tatrishvili, Ann Rose Abraham, & A. K. Haghi (Eds.)

Therefore, the subject of this work was to investigate the corrosion in water facilities.

Corrosion phenomenon of water facilities in flow transmission pipes and in water supply networks is affected by contact with the soil in terms of chemistry, physics, and biology, as well as the chemical properties of the water inside the pipe and the composition of various types of pipes. In addition to causing leakage in water supply networks, this phenomenon also changes the quality of water. In the operation of water facilities, it is necessary to pay attention to the corrosion of water supply facilities. The purpose of operation is the proper use of equipment within the range of their useful life. The equipment should be tested according to the instructions of the manufacturing plant so that they have the proper function and the useful production with the minimum cost in the appropriate period of time.

Periodic tests according to the accepted standards and instructions of the manufacturer to check the types of corrosion and ways to deal with it, including soil tests, corrosion tests, and deposits in pipes are as follows:

- Soil moisture level of subsurface water
- Preventive maintenance (PM)
- Soil chloride and sulfate ions
- All dissolved solids in soil extract and its oxygen
- Granulation and hydrometric test
- Soil porosity
- Determining electrical resistance or electrical conductivity of soil
- Determination of soil sulfate reducing bacteria in the sample
- Total count of soil bacteria of sample

Corrosion in facilities leads to waste of water and financial resources. Corrosion is considered as the biggest problem in water and sewage facilities. Corrosion is a chemical reaction between the components of the facility equipment and its surrounding environment. For example, the result of this reaction in the case of metal pipes and equipment is metal oxidation, which will lead to water leakage. The effects of corrosion in facilities do not happen suddenly and it is a gradual process that in the long term, its destructive effects can be seen. Corrosion phenomenon does not always happen due to reaction with water or air. The special phenomena of fluids such as very high and very low working pressure of fluids, in unstable conditions of fluid work, such as water hammer, cavitation, and pulsation, also lead to peeling of the walls of pipes, valves, and pumps. The presence of suspended materials in water, such as sand, also causes wear and tear in facility equipment and can cause corrosion. Two adjacent non-homogeneous metal pipes causes sacrificial anode mechanism and causes corrosion. As a result, the pipes have water leakage after corrosion, which will lead to water loss in water facilities.

3.1.1 CORROSION PHENOMENON AND ITS REASONS

Substances soluble in water inside pipes cause corrosion. Oxygen and solutes that are dissolved in water have an effect on the corrosion of facilities. If the pipes of the facility are not well protected, they will suffer from internal corrosion due to the production of sulfuric acid.

Pipes that are buried underground, such as steel and cast iron, will definitely corrode. Pipes can be corroded due to contact with soil and its organic substances. Bacteria that exist in the soil or other biological activities that are carried out in the soil are other factors that cause the phenomenon of corrosion.

3.1.2 WATER HAMMER AND CORROSION PHENOMENON

Corrosion or erosion from cavitation implosion can lead to upsurge, down surge, cracking, or rupture of pipe. Water hammer and corrosion destructive phenomenon can be dumped by surge tank (Figures 3.1–3.6).

3.1.3 RAW WATER

Raw water comes from rivers, springs, lakes, etc. Raw water has the ability to dissolve substances in nature. As a result, due to the dissolution of the substances in the water, some impurity will appear in it, which is created as a result of sediments in the transmission pipes in the facility.

Water that is obtained in nature has a very high corrosion potential. Surface water is not drinkable before treatment. It can be purified by three methods: mechanical, chemical, and biological. In the purification of drinking water, several types of purification methods are combined.

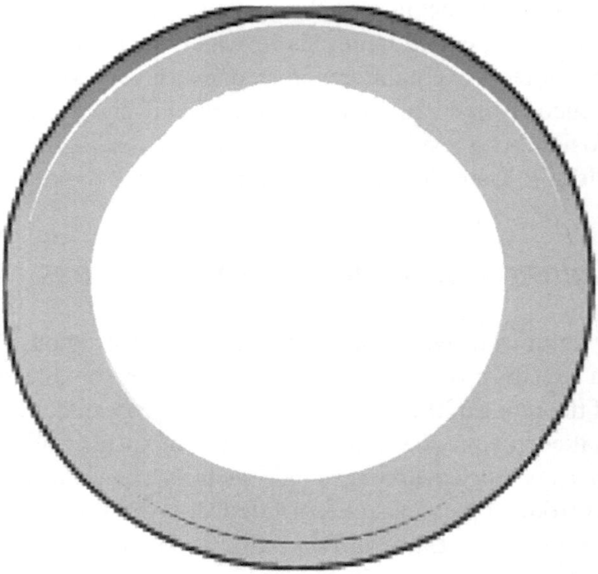

FIGURE 3.1 Water hammer and corrosion phenomenon (weakened section).

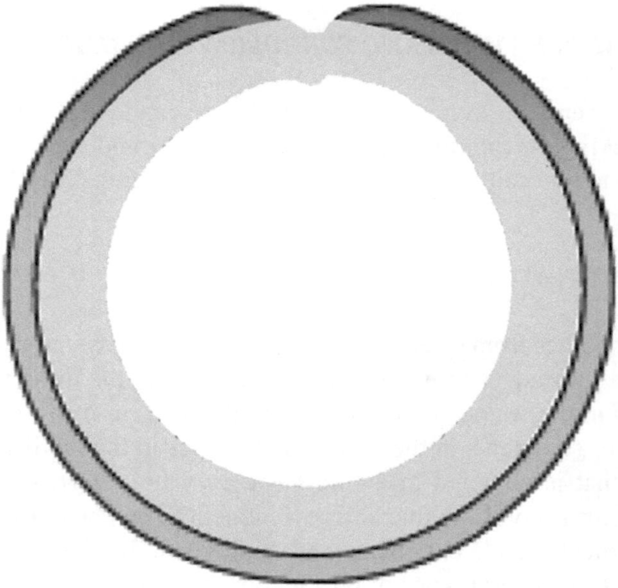

FIGURE 3.2 Water hammer and corrosion phenomenon (downsurge).

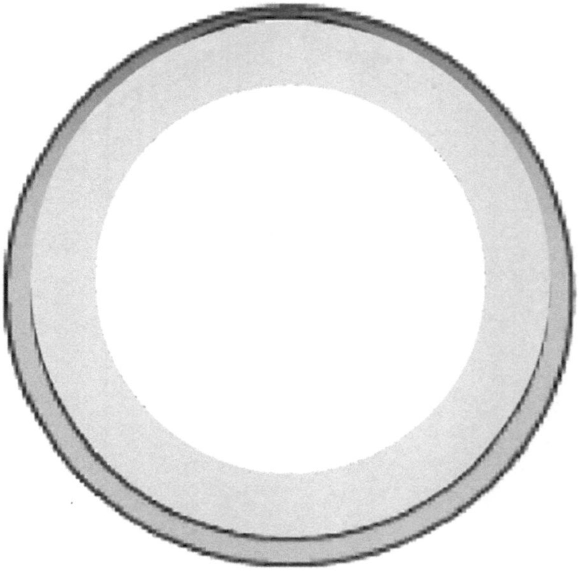

FIGURE 3.3 Water hammer and corrosion phenomenon (upsurge).

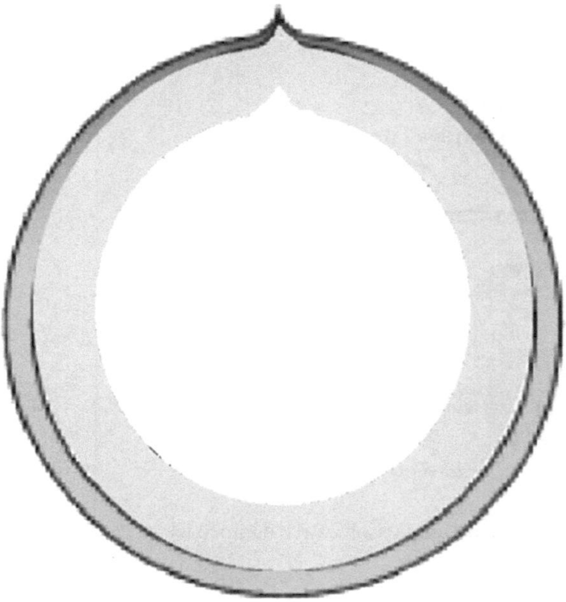

FIGURE 3.4 Water hammer and corrosion phenomenon (cracking).

FIGURE 3.5 Water hammer and corrosion phenomenon (rupture).

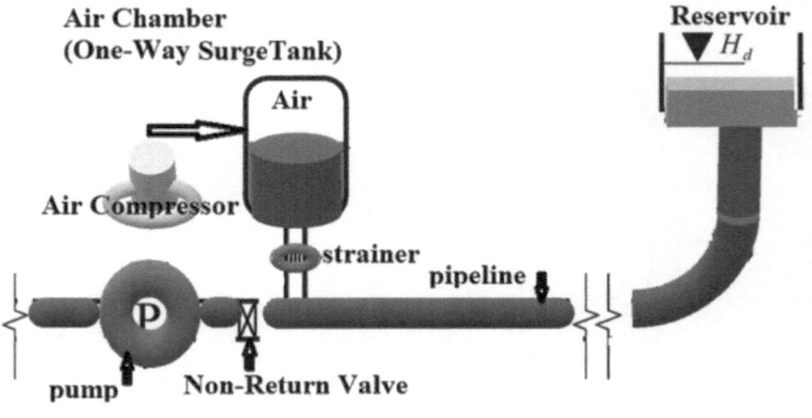

FIGURE 3.6 Water hammer pressure wave dumped by surge tank.

The most important common methods of drinking water purification are:

3.1.4 MESH AGAINST FLOATING MATERIALS

Before the water enters the treatment plant, with the help of these metal nets, large floating materials are captured in the water; therefore, they are only used for harvesting from sources on the ground such as rivers. If there is a lot of floating material in the river water and to prevent the mentioned nets from closing, they are placed in two or three rows with different holes. Holes of 2–5 mm and very fine grids, which are rarely used, are made with holes of 0.2–1 mm.

3.1.4.1 THE METHOD OF SETTLING FOREIGN SUBSTANCES IN WATER

Sedimentation without the use of chemicals: Treatment in sedimentation ponds is usually called preliminary water treatment.

Because the water in nature contains suspended substances, in the first stage, the suspended substances must be removed from the water, therefore, they enter the water into large pools and take it out from the other side. After entering the pool, the water slows down and reaches 2–10 millimeters per second. This speed is enough to allow suspended particles to settle.

3.1.4.2 SEDIMENTATION USING CHEMICALS

To shorten the sedimentation period and also to reduce the amount of soluble salts in water, simple sedimentation ponds are often not enough and the use of chemicals for water purification is necessary. The type and amount of necessary chemicals should be determined in the laboratory according to the type of water and its materials.

3.1.4.3 FILTER

The final purification of water is done by filters. If the water does not contain many foreign and suspended substances, filters can be the only means of water purification.

The types of filters are:

1. Slow smoothness
2. Spicy smooths
3. Fast, open, and unstressed smoothness
4. Sharp and covered head and under pressure

3.1.4.4 AERATION METHOD

The aeration method is very useful in drinking water purification. This method is used for the oxidation of organic and mineral materials and especially for capturing additional gases in water. After receiving enough oxygen and carrying out the biological process by bacteria and oxidation of organic materials and chemical reactions on mineral materials, drinking water should be passed through a filter and through sedimentation pools to remove suspended materials from the water.

Various methods of ventilation are:

- Creation of artificial rain
- Dripping water
- Creating a waterfall
- Pulverizing water
- Ventilation in a closed environment

3.1.4.5 DRINKING WATER DISINFECTION

Disinfecting and killing bacteria in water, which may be done chemically or physically, is the minimum level of purification process that should be implemented in the water supply network of every city.

Disinfection can be done in the following ways:

3.1.4.6 CHLORINATION

Chlorination is the cheapest way to disinfect water. Chlorine can be used in gaseous form or as a combination of calcium hypochlorite or sodium hypochlorite. Chlorine gas has more effect, but it is very toxic and deadly.

3.1.4.7 USE OF OZONE

With the help of alternating electric current with voltage of 8000–10,000 volts and frequency of 500, ozone can be produced and used to disinfect drinking water. The amount of ozone used for water disinfection is 0.5–1 g of ozone per cubic meter of water and the duration of disinfection is 10–15 min. The advantage of ozone is that it does not remain in the water after disinfection, and it does not change the smell and taste of water unlike chlorine, but it is more expensive and has explosion property.

3.1.5 DISINFECTION WITH THE HELP OF ULTRAVIOLET RAYS

This physical method can be used in small houses; the necessary wavelengths are 200–300 mm; the disinfection time is 3 s; and the electricity consumption is 40 watts per cubic meter of water.

3.1.6 THE USE OF IODINE AND BROMIDE

Small amounts of iodine and bromide can be used to disinfect water. It is easier to work with them and today there are tablets of these substances in the market for water disinfection.

3.1.7 BOILING WATER

Practically, bacteria die in water above 75°C, but to be sure, you should boil the water. Disinfection of water by boiling is very expensive and can be used for a small amount in homes.

3.1.8 CORROSION PHENOMENON IN WATER FACILITIES

In nature, there is a large amount of dissolved oxygen gas in water, and it is known as the main constituent elements of water. Due to the reactivity of oxygen with metals and the ability to cause corrosion, dissolved oxygen in the water inside the pipes is one of the important factors in corrosion. The water that flows inside the pipes is not the only cause of corrosion, and rain also has a great effect on corrosion. Because when water is in contact with air particles, it will be more corrosive.

3.1.9 WAYS TO DEAL WITH CORROSION

Performing periodical corrosion tests, periodical repairs of equipment are usually done during production interruptions and due to equipment failure. This practice is still prevalent in some industries. For this reason, the repair time is not certain for the maintenance personnel. In addition to that, the repairs should be done in such a way that as soon as the production line is ready to restart, the repairs are also completed.

It is obvious that this procedure cannot guarantee the satisfactory performance of the equipment against corrosion, and therefore, in the long run, it causes a severe decrease in the life of this equipment. Premature failure of equipment causes an unwanted stop in the system and causes more financial losses.

3.1.10 CORROSION AND NON-REVENUE WATER (NRW)

If the facility equipment is not well protected, the corrosion of the facilities will have a significant impact on the amount of non-revenue water (NRW) in the water system. Therefore, at the same time as the remote reading sensors by remote sensing (RS) are installed, the facility personnel should monitor the water supply process, the electrical network, and the control and start-up circuits. They get to know each other completely. Obviously, the complete map of all remote reading systems should be in the data base.

During a training course, the precision instrument personnel should also gain complete familiarity with the water supply process—the precision instruments installed in the units, the purpose of their installation and operation, and all the existing control circuits and precision instruments. It is obvious that the complete map of control circuits and precision instruments of all parts should be provided to them and they should be familiar with the following:

- Fundamentals of fluid mechanics engineering
- Networked sensors, gauges, and precision instruments

- Remote control systems and RS
- Calibrating the water system and precision instruments
- District metering area (DMA)

At first, all the map characteristics of the studied pilot area for corrosion and NRW and changes in the network and places are reflected on the maps. Then, with local survey, the exact location of the facilities equipment (faucets, pipes, pressure gauges, etc.) are updated and recorded in GIS maps.

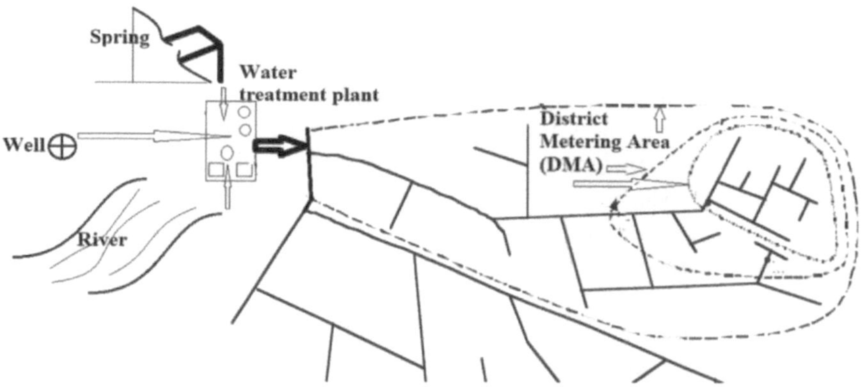

FIGURE 3.7 Facilities equipment for district metering area (DMA).

3.1.11 GIS MAPS OF THE DISTRIBUTION NETWORK

3.1.11.1 IDENTITY CARD OF PIPES AND FAUCETS

In order to hydraulically analyze the distribution network and obtain the pressure in the consumption nodes as well as the amount of leakage through pressure and repair the distribution network, it is necessary to have complete information about the network and the way of connections and the diameter of the pipes, as well as their type and life.

All basins of open faucets and the diameter of the pipes, the complete information of the basins and faucets are recorded in the form of data base.

Along the route, the locations of the buried faucets are identified and recorded so that the faucets can be made visible in the next stages.

During the survey, the places with visible leakage and the places suspected of leaking are identified and recorded in the form. The faucet specifications form includes complete information such as the specifications of the pond, the dimensions of the pond, the type of valve in terms of use, the condition of the pond, the condition of the soil around the pond, the condition inside the pond, the number of valves and how they are placed, the location, the diameter of the inlet and outlet pipes, and the correctness of the valve installation. Also, the location specifications include the replacement of the valve, the sketch of the place and the details of the basin, the characteristics of the valve, the type of connection, the type of valve, the types of replacement, knee and brain, and the condition of the valve.

In the valve information collection form, in addition to general information about the survey equipment; controller; date and specific information page of each valve, including valve code; diameter; type of valve; usage; status (unburied, buried, leaky) diameter, the chamber, the depth of the valve, and the specific description of each valve are included.

In the form to collect information related to distribution network pipelines, general information as well as specific information of each pipe, including the first node code, end node code, first valve code, end valve code, diameter, gender, pressure class, thickness, depth and age are included. The occurrence of the accident (including the number and type of the accident), the number of branches, length, address and considerations are included.

3.1.12 OPERATION

Improper operation of devices and equipment in an industry greatly reduces their useful life, and for this reason, it is necessary to operate and repair a set based on the specific instructions of that set.

All personnel operating and maintaining and repairing a system should have sufficient mastery of the theoretical and practical aspects of this task, at least to the extent of the assigned task. Although this is costly, its usefulness is determined in the long run. In education, more attention should be paid to its practical aspects. The goals that can be achieved as a result of training are:

- Mastery of assigned duties and as a result optimal use of devices
- Reduction of construction costs as a result of optimal operation
- Increasing the probability of reaching the useful life of the system

- Creating confidence in the use of the system, especially in the vital institutions of the society
- Reducing side risks as a result of awareness and correct use of devices in the system

Maintenance of electrical systems and precision instruments should be done by people who have received sufficient training in the desired fields. Although practical training while working is very necessary, it cannot replace specialized professional technical training courses. Therefore, the personnel of electrical units and precision instruments must be familiar with the following during certain training courses:

- Safety regulations and rules for working with electrical equipment
- Basics of electricity and familiarity with production, distribution, and consumption systems of electric energy
- Different consumers and the characteristics of each of them
- Various equipment, including transformers, motors, keys, fuses, etc.
- Different circuits for starting and controlling motors and other electrical consumers
- Measuring equipment in general
- All tools and devices used in installation, start-up, and repair and maintenance of electrical equipment

In addition to the above, the electrical personnel must be fully familiar with the water supply process, the electrical network, the details of how the panels are connected to each other and the transformer substation, how consumers connect with the panels and control circuits and their start-up during a training course inside the facilities. Obviously, the complete map of all electrical systems should be available to all of them.

During a training course, the precision instrument personnel should also gain complete familiarity with the water supply process, the precision instruments installed in the units, the purpose of their installation and operation, and all the existing control circuits and precision instruments. It is obvious that the complete map of control circuits and precision instruments of all parts should be provided to them and they should be familiar with the following:

- Basics of heat and fluids
- Measuring equipment and precision instruments, their working methods and internal structure

- Types of control systems, purpose, and application of each
- Methods of calibrating precision instruments and devices

3.1.13 CONTROL IN OPERATION

In the operation of facilities, there are always important parameters that must be carefully controlled, such as the control of safety and security systems, the control of transmission and pumping line systems, and the control of electrical systems and precision instruments.

3.1.14 DAILY REPORT

A daily report of the operating system should contain the complete status of the systems and include parameters such as height, speed, volume, working hours, work status, incidents and events, statistics, etc., during 24 working hours.

Based on these foundations and principles, it is possible to reach the goals that were mentioned at the beginning and make optimal use of the system according to the existing instructions.

- The criteria for the necessary work for the periodic services of each device is primarily the repair and maintenance manual of that device, which is printed by the manufacturer.
- In the works required for periodic services mentioned in this program, the mention of services and visits that should be done in shorter time intervals has been avoided, while their performance is necessary and the repair personnel should pay attention to this issue. For example, during monthly services.

For urban facilities, maps should be scanned and edited and layered. In addition, it is necessary to reflect the subsequent changes in the network and places on these maps so that they do not lose their effectiveness over time. Considering the vast amount of information and the need to update map information in the form of a GIS system, which is a dynamic and floating system, different data can be extracted from it. Having a strong and efficient database can solve many problems in water systems. Usually, most of the information related to the network and facilities are in the memory of experts and experienced workers, which can easily be out of reach of the system with the passage of time.

The first step in the scientific exploitation of the facilities is to update the map information of the facilities in the form of geographic information system (GIS). Considering the vast amount of operational information, while updating the map information of facilities in GIS format, different data can be extracted from it in minimum time. A strong and efficient database as the main server can solve many problems. If the information about the facilities is in the memory of experts and experienced workers, it will be out of reach of the system with the passage of time. Therefore, at first, facility maps should be scanned, edited, and layered. In addition, it is necessary to reflect the subsequent changes in the network and places on these maps so that they do not lose their effectiveness over time. The components of facilities and equipment mainly include the following:[1]:

- Pipes: In the prepared computer maps, the pipes in the network are carefully updated and reflected on the maps by mentioning the type and diameter in the GIS format.
- Faucets: The location of faucets, including visible faucets, invisible faucets, and all new faucets, are reflected in the GIS format on the distribution network maps.

Today, the analysis methods of urban water distribution systems have been accepted as a well-known and efficient technique in understanding hydraulic behavior, evaluating improvement and development plans, exploitation, and generally overcoming the information gap. In this regard, there are still many ambiguities in describing the basic relationships that show the effect of different parameters in urban water distribution systems. The intelligent pressure management model is used as an effective and suitable tool in hydraulic estimation and evaluation of urban water distribution system.[2] The components of the intelligent pressure management model are as follows:

- Pressure gauges: The location of the pressure gauges is reflected in the GIS format on the distribution network maps.
- Wells: The technical specifications of the wells are reflected in the GIS format on the facility maps.
- Chlorination station: The technical specifications of chlorination stations are reflected on the facility maps.

- The criterion for choosing the pipe was based on the pipe diameter, the placement of the pipe in the ring network, the uniform expansion, and the requirement in different areas.
- Due to the effects of out-of-pilot pipes on the hydraulic conditions of the pilot distribution network, the main out-of-pilot lines are also included in the model.
- According to the local survey and the preparation of updated maps of the distribution network, the computer model is updated based on this information.

3.1.15 TECHNICAL SPECIFICATIONS OF FACILITIES AND EQUIPMENT

Pilot facilities and equipment include the following:

- Pipes: In the computer maps prepared from the pilot area, the pipes in the network are carefully updated and reflected on the maps by mentioning the type and diameter.
- Faucets: The location of these faucets, including visible faucets, invisible faucets, and all new faucets, should be reflected on the distribution network maps according to existing standards.
- Pressure gauges: The location of pressure gauges in the pilot area should be reflected on the distribution network maps.
- Wells: The technical specifications of the wells should be reflected on the facility maps (Table 3.1).
- Chlorination station: The technical specifications of chlorination stations should be reflected on the facility maps (Tables 3.1 and 3.2).

3.1.16 REPAIR AND MAINTENANCE

In the last one or two decades, the concept of repair and maintenance has changed significantly and it includes the following:

- Emergency repair and maintenance (EM)
- Emergency maintenance (EM)
- Preventive maintenance and repair (PM)
- Preventive maintenance (PM)

- Predictive or condition base maintenance and repair (CM)
- Condition base maintenance (CM)

TABLE 3.1 Technical Specifications of the Wells.

No.	Equipment	
		Well number
		Location
		Wall pipe diameter (mm)
		Depth (m)
1	The well	Static level
		Rate (L/s)
		Amount of extractable water (L/s)
		Type of well
		Current situation
		Year of operation
		Pump number
		Purpose
		Number
		Model
		Manufacturer
		Type
		Number of floors
2	Pump	Capacity (m 3/hr.)
		Suction pipe diameter (mm)
		Thrust pipe diameter (mm)
		Operating discharge (m 3/hr.)
		Diameter of bilge pipe (mm)
		Dynamic level (m)
		Type
		Rpm
3	Engine	Power (kw)
		Ampere (A)
		Voltage (V)

TABLE 3.2 Technical Specifications of the Chlorination System.

Description	Specifications
System type	
Number	
Capacity (kg h)	
Year of operation	
Pump	
Number	
Model	
Manufacturer	
Pump type	
Number of floors	
Discharge (m^3/h)	
Head (m)	
Engine	
Round (rpm)	
power (kw)	
Manufacturer	
Suction pipe diameter (mm)	
Diameter of thrust tube (mm)	
Date of start of operation (year of construction)	

3.1.17 EMERGENCY REPAIR AND MAINTENANCE

Emergency maintenance and repair are more or less the same works that are usually known as repairs. This type of repair is done only when the device in question is out of order or at least has problems and the relevant personnel have noticed this.

3.1.18 PREVENTIVE MAINTENANCE AND REPAIR

Preventive maintenance is completely different from emergency repairs. In this regard, the devices are periodically serviced in such a way as to prevent unexpected events and production stoppages due to device breakdowns. To achieve this goal, the following factors should be considered:

- Correct choice of electrical equipment
- Choosing the right time interval for services and visits
- Choosing the right time for basic repairs
- The availability of materials and spare parts in the warehouse
- Establishing the annual maintenance budget
- Recording all the repair and maintenance work done in the device file

3.1.19 PREDICTIVE MAINTENANCE AND REPAIR

This type of maintenance is actually a special form of preventive maintenance. In this type of repairs, by using precise tools such as vibration meter, amplitude meter, pressure meter, thermometer, resistance meter, etc., they predict the problems before they occur and act to solve them.

Periodic corrosion tests, visits, and services are mainly related to preventive maintenance and to some extent predictive maintenance. Corrective maintenance should be done when the equipment is out of order and the personnel did not notice the defect.

3.1.20 TARGETS

The following are among the reasons and goals that make it necessary to prepare and use the system management and maintenance program in urban water facilities:

- Creating integrated management
- Organizing, forming, and coordinating activities in the form of a coordinated system according to the needs of the distribution system
- Increasing the efficiency and capabilities of the personnel by improving their scientific and practical awareness along with the targeted control and guidance of the personnel
- Optimum use of existing capacities and facilities
- Improving and upgrading the exploitation methods by increasing the exploitation power and economic maintenance of water supply networks and facilities, and maximum use of the investments made by preventing their premature consumption

- Preparation of regular and systematic methods of recording the information of the performed activities and improving the methods of obtaining and recording information in order to develop future policies related to maintenance and repairs
- Creating a reliable system in providing services
- Preventing the occurrence of problems caused by incomplete methods of inspection and maintenance and repairs, due to the traditional nature of the existing methods.
- Lack of inspection, repair, and maintenance programs of urban water facilities
- Preventing sudden breakdowns in facilities, stopping services, reducing water loss, and time loss
- Using special recipes for each device to use it
- Modifying and replacing more appropriate methods, recipes and, tools to prevent similar defects from occurring
- Drawing the curves and preparing the tables and comparing them with the operating curves of each device to identify the optimal condition, increase the quality of operation and profitability of the set of facilities in compliance with the standards.[3–6]

3.1.21 OPERATION AND MAINTENANCE

3.1.21.1 PERSONNEL TRAINING IN MAINTENANCE AND REPAIRS

Training of maintenance and repairman is important. It is worth mentioning that the relationship between the maintenance and repair group is mutual and two-way, so that the correct operation directly affects the work of the maintenance group, and the correct repair and maintenance of the system increases the efficiency of the operation group.

3.1.21.2 CONTROL IN REPAIR AND MAINTENANCE

The maintenance of each system is done in the following three ways:

1. Continuous control of system details
2. Periodic services
3. Periodic replacements

In general, the time interval of each service or replacement is determined by the instructions of the manufacturer, but the maintenance manager has interfered in these instructions by changing the operating conditions in such a way that the efficiency of the system is maintained at the optimal level.

Regarding the continuous control of the details of the system, the work of the maintenance and repair group is somewhat similar to the work of the operation group, with the difference that the maintenance and repair group, after observing the problems of the system, follow up the investigations related to fault finding and fix them with more accuracy, and usually comment on these cases with speculation.

Periodic services should be performed in each of the components of the system according to the manufacturer's instructions. In this way, in case of problems in any of the components of the system, before these cases cause serious problems in service or interruption of services, they are investigated and fixed; periodic replacements are also done based on the useful life of each device or part.

The useful life of each part is determined by taking into account the conditions of operation and use of the part, so that in some cases we see that some parts work more or less than the expected useful life, however, although periodic replacements are possible in the first place. They are expensive, but in order to prevent consequences that are sometimes much more expensive, the reasonable and economical method is to replace each part at the end of its useful life.

The totality of the principles mentioned in operation, repairs, and maintenance includes the following steps for each of the components of urban facilities.

3.1.22 PUMPING STATIONS

Normal operation

In operating the pumping station, the following items must be controlled.

Storage tanks

- Water level control inside the tanks
- Checking the condition of the inlet and outlet valves

Electric motors

- Engine sound control
- Shell temperature control
- Control of starting ampere, working ampere, voltage, and engine speed
- Electronic consumption registration
- The output cables from the main switches and the starting switches of the electro-pumps are touched by hand; if the heat is more than reasonable (30°C), the switch should be turned off and the repair group should be informed.
- Cleaning the inside of the electric motor with a dry air compressor with a pressure of about 4 atm.

Pumps

- Control of couplings
- Controlling the sound of the pump and its vibration
- Pump shell temperature control
- Control the circulation of the cooling system, if any
- Ventilation system control
- Control of input and output current intensity

Suction pipeline

- Line negative pressure control
- Checking the status of the valves on and off
- Checking the sealing of valves and fittings

Thrust pipeline

- Positive line pressure control
- Checking the status of faucets and connections
- Sealing control of faucets and fittings

Security and lighting system

- Controlling the status of security systems in the facilities
- Controlling the status of the whole system by setting up a daily report
- The lighting of the house pump during the day, control, and, if there is a defect, take action to fix it so that there is no problem at night.

3.1.23 FAULT FINDING AND FIXING IT

If periodic services and repairs are carried out according to the schedule, the workload of the repair unit will be greatly reduced. In any case, there are reasons to form an experienced construction group at the pumping station, to fix the defects and prevent disruptions in operation.

Defects and faults created in the electric motor of the pumps can have hydraulic or mechanical reasons. In the first case, the pump does not have the ability to pump, or the pump operates with less than its nominal capacity, or does not provide enough pressure, in the second case, the amount of electricity consumed increases and problems such as the value of the pump cause noise, corrosion, and breakage of parts fall in general, any defects and problems in the operation of pumps can be caused by one or more reasons. The person in charge of the repair department can easily find the cause and try to fix it. In this connection, in Table 3.3, the problems and causes of their occurrence that are generally created in centrifugal pumps are given.[7–11]

3.1.24 MAINTENANCE PRINCIPLES OF PUMPING STATIONS

The maintenance of a pumping station, like any industrial system, is divided into three parts:

- Continuous control of system details
- Periodic services
- Period replacements

TABLE 3.3 Problems and Causes of Defects in Centrifugal Pumps.

No.	Problems	The reason
1	Water is not pumped	Air has entered the pump
		The pump or suction pipeline is not completely filled with water
		The suction height is high
		The suction pressure is equal to the water vapor pressure
		The presence of air bubbles in the suction pipeline
		The inlet of the suction pipeline is not completely in the water
		The pump speed is very low
		The direction of rotation of the pump is wrong
		The overall system overhead is more than the design overhead
		Operating the pumps in parallel is not correct in the operating conditions
		The presence of foreign materials in the pump impeller
2	Pump with less capacity	The pump or suction pipeline is not completely filled with water
	doing	The suction pressure is equal to the water vapor pressure
		The height of suction is more than allowed
		There is water, air, or other disturbing gases inside
		Air infiltration into the pump suction pipe
		Air infiltration into the pump chamber
		The one-way valve at the end of the suction pipeline (suction valve) has been pinched or seized
		The rotation speed of the pump is low
		The overall system overhead is more than the design overhead
		The viscosity of the fluid is higher than the design value
		Parallel operation of pumps is not suitable in operating conditions
		The presence of foreign materials in the pump impeller
		Corrosion and damage to the pump impeller
		Deterioration of the sealing gasket
3	The pump head is insufficient	The presence of air bubbles in the incoming water
		The rotation speed of the pump is low
		The direction of rotation of the pump is wrong
		The overall system overhead is more than the design overhead
		Fluid viscosity is different from the design value
		Corrosion and damage to the pump impeller
		Deterioration of pump sealing washers

TABLE 3.3 *(Continued)*

No.	Problems	The reason
4	Packing soon _	The water supply pipe (water seal) is closed
	It breaks down	The pump is installed flush
		Bending of the pump axis or shaft
		Shaft corrosion
		Installation of wrong packing
		The type and material of packing is not suitable for working conditions
		Engine misalignment
		The flow discharge line is very tightly closed
		The distance between the shaft and the body
		The presence of foreign substances in the sealing liquid
5	The pump overheats	The pump is not aerated
		The difference between suction and steam pressure is small
		The pump is operated with less capacity
		The parallel operation of the pump is not suitable in the operating condition
		Uneven pump installation and operation with less capacity
		Abrasion of moving parts on fixed parts of the pump
		Axis or shaft bending
		Imbalance of the moving engine
		The speed of the pump has reached the critical speed
6	The pump vibrates or makes noise	The pump or suction pipeline is not completely filled with water
		The suction height is high
		Suction pressure is the degree of water vapor pressure
		The suction line valve is too small or blocked
		The inlet of the suction pipe is insufficiently located in the water
		The presence of foreign materials in the pump impelled
		Uneven installation of pump and drive motor
		The foundation is not strong enough
		Bending of the pump shaft
		The pillow is a moving part to a fixed part
		Damage to the pump impeller
		Lack of lubricant
		There is rotation in the pump axis

TABLE 3.3 *(Continued)*

No.	Problems	The reason
7	Power consumption is high	The speed of the era is very high
		The drive motor works in reverse direction
		The head of the whole system is more or less than the design head
		The difference between the specific gravity and the viscosity of the fluid with the design value
		The presence of foreign materials in the pump impeller
		Uneven installation of pump and motor
		Bending and deviation of the pump shaft
		Rubbing a moving part on a fixed part
		Wrong packing installation
		The type and material of packing is not suitable for working conditions
		The flower is very tightly closed
8	After turning on the pump	The pump is not ventilated
	It takes air	The suction height is high
		There is water, air, or other gases inside
		The presence of air in the suction line
		The inlet of the suction pipe is not completely in the water
		Turbulence in water pipe
9	Bearings break down quickly	Misalignment of pump and motor
		Bending of the pump shaft
		Rubbing a moving part against a fixed part

In a pumping station, the details of the station must be continuously controlled by the maintenance group—find system faults and predict the causes and fix them.

The periodic services of a pumping station for each of the constituent parts can usually include the following.

- Operation valves and controlling one-way valves
- Greasing the bearings of electric motors and pumps
- Service of sealing rings
- Opening and closing the pumps and checking the condition of the shaft, impeller, and internal wall of the pump shell

- For every 2000 working hours of the pump, the grease of the bearings should be replaced
- Testing the start-up head of electric motors
- Service of electrical panels and sensitivity testing of safety and protective circuits
- Service of control systems such as limit–switches–flowmeters
- Service and control of loose couplings

Periodic replacements of the components of a pumping station includes the following procedure:

- Bearings and roller bearings (pumps and electric motors)
- Rims and sealing graphite (to be replaced every 6 months)
- Pump propellers
- Shaft of pumps
- Ball shaft of valves
- Manometers
- Couplings
- Water flow measurement contours (if available)
- Sensitive parts of electrical panels

3.1.25 PERIODIC OPERATION OF INSPECTION AND MAINTENANCE OF THE PUMPING STATION

3.1.25.1 DAILY

All defects that are observed during hourly and daily visits in the operation of the pumps, such as the change in the sound of the pump motor, sudden change in the temperature of the pump shell, leakage from the pump, and pressure and flow readings are observed by the operators.

3.1.25.2 MONTHLY

In addition to the daily visits by the operators, the following visits and operations must be performed by the construction team on a monthly basis.

- Investigating the condition of the bolt and gland
- Checking the state of packing and sealing gaskets and replace them if necessary

- Adjusting and aligning the pump motor, or in other words, inspecting and measuring the distance of the couplings
- Greasing and lubrication of electric motor bearings with 160-degree fireproof grease
- Inspecting and repairing pipes and connections, if needed
- Inspecting and repairing or replacement of pressure gauges, if necessary
- Pump electromotor vibration control
- Cleaning the inside of the electric motor of the air compressor belt pump with a pressure of about 4 atm.

3.1.25.3 YEARLY

The measures that should be taken during the annual visits are as follows:

- Pump components should be opened, washed, and cleaned
- Degreasing the impeller and pump shaft
- Lubrication of bearings and roller bearings
- Painting faucets, fittings, and pump shell
- Inspect the floor and walls of the pump house and repair it if any defects are found

3.1.25.4 FROM THE OPERATION OF THE UNIT

- Close the valve located on the thrust pipeline of the desired pump
- Turn off the engine
- Close the valve located on the suction pipeline
- Safety
- The electro-pump should never be used outside the specified range in terms of water flow; motor speed; pressure and temperature
- Avoid keeping the electro-pump on for a long time while the outlet valve is closed
- Under no circumstances should the pump work without load

3.1.26 DISTRIBUTION NETWORK FAUCETS

The valves in the urban water distribution network are considered as key and control points of the network. The absence of a suitable plan and method for the maintenance and operation of the faucets of the distribution network often causes them to lose their desired and necessary performance after a period of installation. Instead of being used as executive management tools, it causes problems. They are distributed in the system. The first step to repair and maintain these units is to know their function and role in the distribution network. After that, the existing obstacles should be removed and their protection and maintenance should be known, and then with the schedule of periodic visits and service, while applying the necessary controls, repair, and maintenance should be considered.

3.1.27 GATE VALVE

This valve is used to completely shut off and connect the water flow in the distribution network. These faucets are manufactured and marketed in different sizes and to withstand different pressures. The material of these faucets is generally cast iron or ductile iron and is widely used in distribution networks and water transmission lines. The most common installation points of valve valves are:

- In the place of branch branches and the intersection of pipelines in order to disconnect and connect the water flow of the desired branch or the connection between the lines
- At the entrance and exit of tanks
- In the water transmission lines that go to (tap line)

There are two types of valve in terms of how it works:

- A valve with a rising stem (rising stem valve) in which the stem or rod of the valve also rises and water is allowed to pass
- Gate valve with a fixed stem (non-rising stem valve), where the valve moves on a rod or stem. These valves are generally used for underground installation

3.1.28 ADVANTAGES OF GATE VALVE

- When it is completely open, the water pressure drop is very small and can be ignored
- Long lifespan

- High work pressure
- There is less water leakage in them

3.1.29 DISADVANTAGES OF GATE VALVES

- In large diameters and for high pressures, it is difficult to open and close it by hand, and the valve must have a passage.
- If the gate valve is not maneuvered (open and close) regularly, due to the accumulation and deposition of particles in the water, opening and closing them will not be done easily and correctly.

3.1.30 REPAIRS

Valve faucets often need to be repaired due to the following reasons:

- Deterioration of the sealing thread (PACKING)
- Corrosion and rusting of the tap rod or stem
- Corrosion of tongue or disc
- Deterioration of sealing rings

Each of the above causes malfunctions and leaks from valve faucets. In order to fix the defect and repair each of the above-mentioned items, it is done in the following ways:

3.1.31 REPLACING THE SEALING THREAD

- Open the valve gland.
- The sealing thread must be removed from the sealing chamber and its place must be cleaned.
- The thread of the new seal should be placed in its place and secured in place with a gland.
- Pour a few drops of oil on the sealing thread and then put the valve gland in its place and close it tightly.

- *The service of the bar or stem of valve*
    - Open the valve gland.
    - Take out the sealing thread.

- Open the cap and remove its rod.
- Clean the surface of the rod, and if there is a lot of corrosion and rust, the rod should be replaced.
- Put the rod inside the cap.
- Seal gaskets should be replaced.
- The cap should be firmly in place.
- Replace the water seal thread.
- Open and close the valve.

- *Fixing the corrosion of the disk or valve tongue and sealing ring*
  - Open the gland and cap.
  - Remove the rod.
  - Take out the disk and clean it with sandpaper and replace the disk if there is a lot of corrosion.
  - Clean the sealing rings and replace them otherwise.
  - Mount the valve parts again. Sealing gaskets and sealing thread should be replaced.[12–18]

Gate valves should not be left open or closed for a long time, because the seat is filled with water salts when it is open, and the valve is not completely closed and sealed. In the case of being closed for a long time, the solutes in the water accumulate around the valve drawer and cause a lot of pressure on the valve axis when opening the valve, in which case there will be a possibility of the axis being cut.

3.1.32 ONE-WAY CHECK VALVE

A one-way valve is actually a valve that is installed in the pipeline and when water flows, the moving force of the water opens it, and when the water flow changes its direction, the plate closes on a tight seat. In general, one-way valves work automatically and prevent the return of flow in pipelines. One-way valves are generally divided into two main groups, hinged one-way valve (LIFE CHECK). One-way ball checks (BALL CHECK) are divided.

Hinged one-way valves are often used when the goal is to completely pass the flow rate, and it is the most common and most used one-way valve. The one-way ball valve is used in cases where the valve must operate frequently and in networks with high pressure and high flow rate.

Among the factors that cause defects and disturbances in the performance of the one-way valve are corrosion and wear of the retaining pin, retaining cap, sedimentation of the valve, and corrosion of the disc. In order to fix such defects, it is done in the following ways:

- Open the valve cover.
- The nut or spine of the axis should be removed.
- The valve disc should be sanded or ground.
- In case of damage to the axle, the hub and nut of the axle and their cover should be replaced.
- Reassemble valve components.
- Valve should be tested.

3.1.33 BUTTERFLY VALVE

The construction of butterfly valves is simpler than other valves, especially gate valves. These valves have a part in the shape of a rotating plate or disk that moves around an axis, rod, or a stem. Opening and closing these valves is simple and is done only by turning the disk so that it fills the entire width of the pipe. Depending on its type, the plastic lining of the seat of these valves is either made of adhesive material, it is fixed to the disc with mechanical devices, embedded in the disc, or is made of a mechanical spring seal or a metal seat that is placed in the form of an O-RING (0-RING) around the edge of the disc. The advantages of these faucets are:

- The coefficient of water passage in it is directly related to the opening percentage of the valve and therefore it can be used to control the flow and create a pressure drop in the network.
- Opening and closing it is done with more ease and speed.
- It needs less space for installation.
- Its repair and maintenance cost is relatively low.

Maintenance of these faucets is easy and generally does not need oiling until the stem O-ring is replaced. The disc and stem as well as the O-ring disc and its plastic seats are easily replaceable. Replacement of the plastic metal lining of the sealing surfaces with adhesive materials cannot be done on site and is done only by the manufacturer.

3.1.34 AIR VALVE

An air valve is a device to let air out of a pipeline or to let air into it. Air valves are divided into three categories:

- Single hole air valves, with a small hole: It is used to release air from the pipeline that is under hydraulic work pressure. The diameter of the opening in these valves will not be more than 4 millimeters.
- Single orifice air valves, with a large orifice: This valve is used to ventilate and empty the pipe. When the pipeline is filled with water, air is removed from it, and when the pipe is emptied of water, air enters the pipeline. When the pipeline is under hydraulic pressure, the above valve is not able to release the small amount of collected air, and in this sense, the hydraulic pressure prevents the ball from separating from the hole.
- Two-hole air valve: It performs the work of both the above valves.
- In terms of operation, if the faucets are installed in the right place, they will not have any particular problem. It is necessary to visit it periodically, and if its opening is clogged with sediment, it should be descaled and rusted.

3.1.35 PERIODIC VISITS, REPAIR, AND REPLACEMENT OF VALVES

Due to the fact that during the operation of the distribution network, the water cut for a long time is not justified and will cause protests from the subscribers. Therefore, it should not be enough to fix the problems as an emergency repair and maintenance. Rather, it is necessary to optimize the use of the facilities by training and equipping the crews and conducting periodic visits and services.

Each of the faucets used in the network, even if they are made of the best metal materials, will suffer corrosion, rust, and sedimentation over time.

Like other mechanical systems, failure to maneuver the valve on time will gradually make it impossible to maneuver the valve. So that their maneuver is recommended twice a year.

On the other hand, the access valve to the line valve due to the spread and various implementation operations in the city such as drilling, surface changes, and asphalt may be invisible, intentionally or unintentionally, and as a result, the possibility of valve maneuvering is lost. The more important point is that each faucet is installed for a reason and to connect and disconnect the water in a specific area. If a pipe breaks due to drilling, leakage, or as a result of any kind of accident, by closing the valve of the relevant line, the outflow of water from the accident site is controlled. Therefore, if one of the valves in the network becomes invisible or is maneuverable, at least several other line valves should be closed instead, which will cause the water to be cut off in a wider area, and water will be wasted for a long time.

In addition to the mentioned points, if any of the faucets in the line are not opened and closed continuously, and if it gets corroded, it will automatically lose its water capacity after a maneuver and will leak from the outside, which results in water wastage and water penetration into the ground. As a result, the soil settles under the pipe and breaks the pipelines. If there is no watertight valve, the repair of the accident pipe is impossible in some cases and in some cases a pump is needed to drain the water from the accident site.

The maps of the pilot water distribution network are updated with local survey and all the visible and invisible taps are identified in the network and reflected on the maps. The periodic visit schedule of the water distribution network is given in Table 3.4. Considering the fact that pipes with a diameter of 300 mm and more are considered as the main pipes of the distribution network, the management and planning organization has recommended that the faucets related to these lines have basins, so in order to be possible Periodic inspection and assurance of the correct operation of the main faucets of the network when there is a break in these pipes, and in order to reduce the water losses caused by this, the faucets of 300 mm and more should be placed inside the basin.[19–22]

3.1.36 WATER DISTRIBUTION NETWORK PIPES

Pipes are the only part of the urban water network that are always buried underground, so they are more vulnerable than other parts of the network and will have accidents. Maintenance of the network, in the sense of visiting them, is not possible only when accidents occur, but it is better to carry out the work of leak detection by a leak detection device during a regular program and according to the priority of the areas according to the age of the network and its importance to restore the network from causing any massive and costly losses and damages. In this regard, it is necessary to mention that usually the amount of water wasted from small leaks in the network is more than large leaks that cause accidents.

The reasons for reducing the water transfer capacity of pipes are:

- Foreign objects inside the pipes, such as wood, stone, etc., which enter the pipes during pipe laying or necessary repairs.
- Sedimentation of sludge, iron, manganese, and algae growth
- Creating a coating layer of sedimentary materials on the inner wall of pipes and protrusions caused by rusting or water corrosion or lime deposits.
- Air accumulation in the highest points of the level.
- Reducing the cross section of the pipe due to sedimentation of faucets in semi-closed states or bending of pipes.

Therefore, in order to prevent large losses, the accurate implementation of the network and the use of quality materials and the careful monitoring of its implementation while observing the technical principles of implementation are factors that can have a great effect on the health of the network and increase its life.

Otherwise, where it is possible that the pipe water transmission is reduced, the following measures should be taken:

- Full air evacuation control throughout the length of the pipeline
- Washing pipes.
- Visiting the accessible parts of the pipe to control the sediment layer created on the surface of the inner wall of the pipes.

Reservoirs

Before starting the exploitation operation, it must be ensured that the following things are fully fulfilled:

- All the problems that caused the unit to fail are recorded in the operational sheets.
- The repair and replacement of the parts that were involved in the occurrence of problems has been done.
- The tank overflow system is ready for operation.
- The performance of the taps entering the tanks is guaranteed.
- The performance of the outlet taps from the tanks is guaranteed.
- The performance of the chemical injection and chlorination system is guaranteed.

- Preparing the daily report book (operational sheets).

Normal operation

- Continuous control of the height of the water level in the tank and record it in the relevant worksheet.
- Ensuring that the inlet and outlet valves are open.
- Ensuring the operation of ventilators and ventilation system.
- Inspection and notification of the absence of leaks in the walls and ceiling.
- Ensuring the operation of the inspection valves.
- Visiting the appearance of the water.
- Record the report of each of the above cases in information sheets.
- Continuous control of chemical injection rate.

3.1.37 MAINTENANCE AND LUBRICATION OF RESERVOIRS

The schedule of inspection and maintenance of tanks reservoirs and related faucets is given in Table 3.4.

TABLE 3.4 Schedule of Periodic Visits to Reservoirs.

No.	Operation period	Description of operations
1	Seasonal	Inspection of internal and external surfaces
		General inspection of faucets and ensuring their correct operation
		Inspection of manholes and repair of possible damage to them
		Inspecting the roof vents and making sure they are healthy
2	Yearly	Taking the tank out of service
		Full inspection of internal and external surfaces in terms of leaks and possible damages
		General service of faucets and necessary repairs
		Washing and disinfecting the tank
		Maneuver of existing lions
		The outer coating of faucets and connections is painted to protect against corrosion
		Inspection of the gearbox, the butterfly valves should be replaced and adjusted if necessary

Repairs

- Examining faucets and other connections and repairing them if necessary.
- Tanks should be checked for leaks and if necessary, isolation should be done.
- Investigate the installation locations of valves and ventilators, and in case of infiltration of surface water and runoff through the wall of the valves and ventilators, necessary repairs should be made.

Safety

- Preventing the entry of surface water, as a result of atmospheric spills that have accumulated on the tank as waste water.
- Full sealing of inspection valves.
- Ensuring that all the outlet taps of the tank are left handed.
- Preventing and hindering the entry of different people.

Leak detection

Leak detection (Figures 3.8 and 3.9) is performed with a leak detection device in the pilot.[23–28]

FIGURE 3.8 Leak detection and repair of distribution network leakage.

FIGURE 3.9 Leak detection and repair of water pipeline leakage.

3.1.38 INTELLIGENT PRESSURE MANAGEMENT MODEL

3.1.38.1 INVESTIGATING THE WORK OF THE WATER DISTRIBUTION NETWORK

According to the preparation of computer hydraulic model and model calibration using the results of pressure measurement, problems in the water distribution network of pilot are examined and comments are made.

By examining the results of the readings of pressure gauges and related charts, it is clear that the pilot distribution network system at the maximum (peak) consumption has optimal efficiency or does not meet the current needs of the pilot. At the peak of consumption, if the pressures are much lower than the standard, considering the drops caused by the branches of the subscribers, including the drop of the contour, cut-off valve, and one-way valve, which is a significant figure, these pressures are not desirable at all.

The main problem of pilot water distribution network can be seen in the following three sections:

1. The low capacity of the main pipes of the network (transmission lines).

2. The existence of inappropriate branches from the main lines of the network, so that the branches taken from these lines were made after high pressure drops.
3. The presence of pumps with low capacity in the network, which can be seen especially from the pressure measurement results, causes very low pressures in the network.

– Examining the problems of the pilot distribution network and providing suggestions to improve the situation:

   A. Low capacity of water supply pipelines
   According to the results of the network analysis, it is suggested to make changes in the hydraulic model.
   B. Investigating the reduction of leakage by controlling the water pressure in the network:

Pilot water supply sources can be ground reservoirs and deep and semi-deep wells inside and outside the pilot. These wells directly or indirectly (pumping from the storage tank next to the wells) pump water into the distribution network. Determining the scope of the operational pilot and measuring the water entering the water distribution network of the cities of the province can be done by using advanced technologies of remote reading of pressure and flow parameters at the entrance of facilities and distribution networks in the form of implementing intelligent pressure management and in order to analyze water loss in the software environment. WATER GEMS should be done under the management of ArcGIS-ArcMap software (Figure 3.10).

The low water pressure in the pilot during peak consumption indicates the low efficiency of the distribution system (problems in pipelines and low power of pumps). The capacity of distribution network pumps should be such that during peak consumption, the pressures remain at the standard level, and this problem increases the pressure in the distribution network when there is no consumption (from 2 to 4 in the morning), which is unavoidable. This problem can be seen in the results of pressure measurement, and the use of the usual methods of pressure control during low consumption times (use of pressure relief valves or pumps with lower power during these hours) due to the large number of these wells and lack of It will not be cost-effective to concentrate them in one place, and therefore, it is not possible to control the pressure in the wells during low consumption hours in order to reduce leakage.

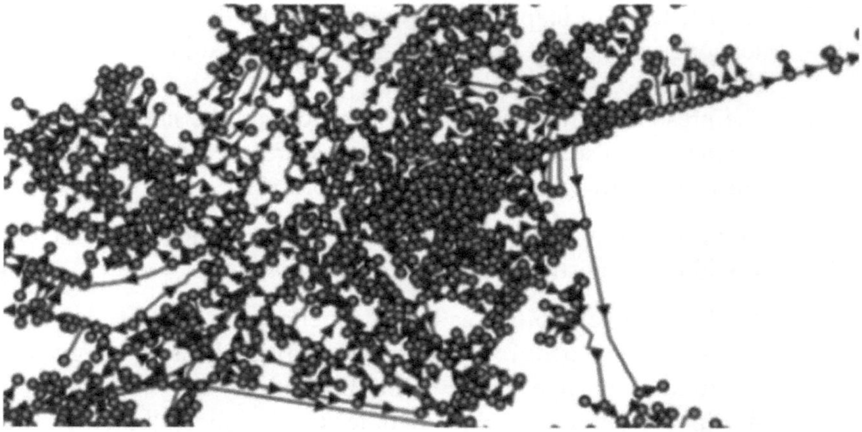

FIGURE 3.10 Pressure management by analyzing water loss (WATER GEMS/ArcGIS-ArcMap).

There are different types of pressure relief valves. Some of them are set for a minimum pressure in the lower hand and some work on the basis of reducing the constant pressure between the upper hand and the lower hand.

The design and installation of pressure relief valves in the main route should be done in such a way that they do not reduce the pressure in the water distribution system during peak consumption.

This issue greatly reduces water losses and increases network pressure during peak consumption times, and in low consumption seasons such as winter, it will be very effective in reducing losses caused by accidents and breakages.[8,29,30,31,32]

3.1.39 PREPARATION OF DATABASE

3.1.39.1 COMPUTER MAPS AND EVALUATION OF FACILITIES INFORMATION

Preparing a computer map of water supply, transmission, storage, and distribution facilities in pilot supply, transmission, storage maps, as well as distribution network maps in the pilot area, were updated during local survey operations, and new pipelines along with all new faucets and pressure gauges are reflected on these maps.

Since the maps of the existing situation are usually manual and related to the past years, first the mentioned maps are scanned and edited, and then according to the information obtained during the local survey operations regarding the technical specifications of the network facilities and equipment. The actions after scanning and editing maps are as follows:

- Integrating and replicating maps
- Aligning the maps in terms of font, title, contractual signs of urban facilities (water faucets, pressure gauges, conversions, etc.)
- Updating the computer maps of the water distribution network and control of places (specifications of places)
- During the place control activity, all the places that had a wrong code in the pilot area were identified and these cases are corrected on the computer maps. These items need to be corrected in the subscriber affairs software.

3.1.40 EXAMINING THE PRESSURE MEASUREMENT RESULTS

After the completion of the work of installing pressure gauges in the pilot area, the activity of reading the pressure gauges was carried out during the provided schedule, and the results of this activity are presented in the relevant forms along with the equivalent pressure chart at different hours of the day and night (Figure 3.11).

Based on the investigations, the main reasons for the low pressure in the network are as follows:

- The low capacity of the pipeline entering the network.
- Existence of inappropriate branches from the pipeline entering the network.

If a branch is made at a shorter distance from this pipeline, the flow through this pipeline will be reduced and the amount of pressure drop will also be reduced.

- Low power of network pumps.

FIGURE 3.11 Pressure detection for analyze of water loss (WATER GEMS /ArcGIS-ArcMap).

3.1.41 PREPARATION OF COMPUTER MAPS OF EQUAL PRESSURE LINES (ISO-PRESSURE LINES)

According to the pressure measurement and the pressure gauge readings (Figure 3.12–3.14) and the calibration of the hydraulic model prepared using the results of the pressure gauge readings, the computer map of the iso-pressure lines was prepared and the map of the iso-pressure lines in the water distribution network shows the pilot area.

3.1.42 REVIEWING STATISTICS AND INFORMATION RELATED TO ACCIDENTS

Examining the statistics and information related to accidents according to the information available in the employer's incident offices.

FIGURE 3.12 Pressure measurement by networked sensors.

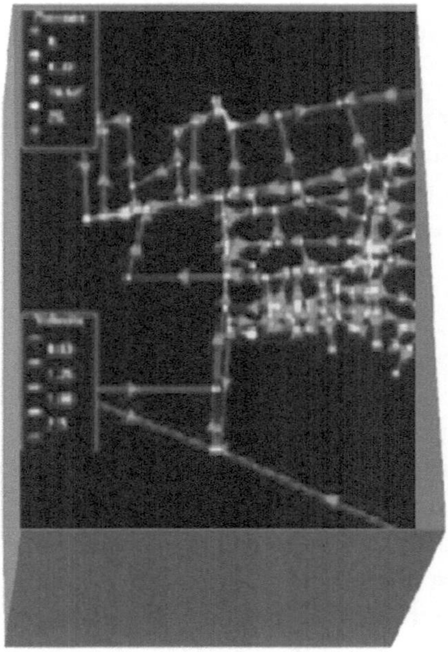

FIGURE 3.13 Analysis of water loss (WATER GEMS /ArcGIS-ArcMap).

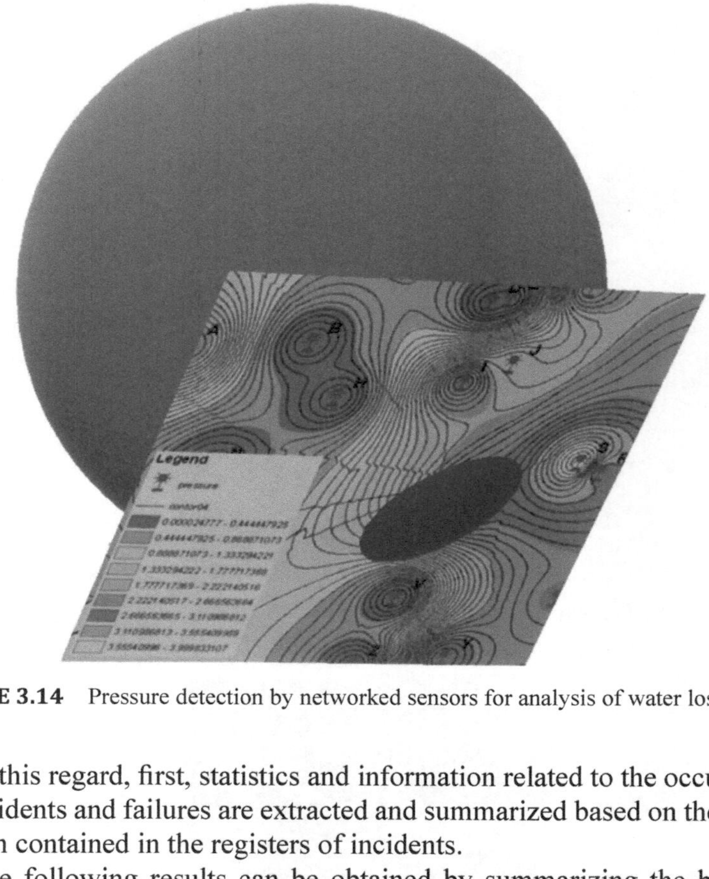

FIGURE 3.14 Pressure detection by networked sensors for analysis of water loss.

In this regard, first, statistics and information related to the occurrence of accidents and failures are extracted and summarized based on the information contained in the registers of incidents.

The following results can be obtained by summarizing the book of events:

- More than 80% of accidents and failures are related to the branching of subscribers.

  As it can be seen from the summary of the events book, the only results that can be obtained are the above row with different pipe diameters, which is manual and very time-consuming. However, in order to fully and comprehensively understand, other factors need to be raised, examined and processed by the events software.
- Examining statistics and information related to accidents

  Due to the fact that providing any kind of quantitative and qualitative analysis requires the presence of more accurate information about the occurrence of incidents, completion of the incident work-sheet according to the relevant form, the topic under discussion is evaluated and reviewed from various aspects as described below.

- Determining the appropriate strategy in dealing with incidents
- Evaluating the efficiency of different incident response teams
- Evaluation of the method adopted to carry out system maintenance and repair operations
- Calculation of water losses
- Calculation and estimation of costs caused by accidents
- Planning to correct matters and rebuild sensitive points of the network
- Modifying and updating water distribution network maps
- Determining the weaknesses of the water distribution system
- Investigating the causes of accidents

The investigation of incidents is summarized as follows:

- The table of the number of monthly incidents according to the location of the incident
- The table of the number of monthly occurrences on the same pipe by gender and diameter
- The table of incidents by separating the location and type of incident, the type of incident and the type of pipes, fittings, and faucets
- The table of the number of incidents on the pipes by the cause of leakage
- The table of the number of incidents by the cause of the incident
- Street surface destruction table

Investigations have shown that worn pipes are one of the main causes of leakage.

3.1.43 PREPARING A COMPUTER MAP RELATED TO DIFFERENT TYPES OF INCIDENTS IN THE REGION

The statistics and information related to the incidents in the pilot distribution network were examined in detail. As mentioned, you should be present at the place of occurrence to complete the incident worksheet day and night.

3.1.44 DETERMINING THE AGE OF NETWORK PIPES

In the absence of as-built maps of the water distribution network, the age of the network can be determined based on the information contained in the subscriber's file. In this regard, by examining the files of the subscribers of the pilot area, the age of the pipes of the distribution network is determined.

3.1.45 PREPARATION OF COMPUTER MAPS OF NETWORK AGE

In the absence of integrated computer maps of the pilot area, all scanned and edited maps of the distribution network are integrated and the age map of the pipes of the distribution network of the pilot area is determined.

3.1.46 DETERMINING THE ENTRY POINTS AND EXIT POINTS OF THE PILOT

3.1.46.1 EXAMINING THE HYDRAULIC CONDITIONS IN THE SAMPLE AREA AND COMMENTING ON IT

Today, the analysis methods of urban water distribution systems have been accepted as a well-known and effective technique in understanding hydraulic behavior, evaluating improvement and development plans, exploitation, and generally overcoming information gaps. In this regard, despite the fact that there are still many unknowns in describing the basic relationships that show the effect of different parameters in urban water distribution systems, models can be used as an effective and appropriate tool in hydraulic estimations and evaluations of distribution systems (Figures 3.15 and 3.16).[33–38]

Water distribution network modeling or network analysis is a computer statistical process in simulating a real water distribution network. Using numerical methods, computer models solve a large number of equations iteratively (Hardy cross method) or simultaneously (Newton-Raphson and linear theory methods). The hydraulic model is carried out in compliance with the GIS in the WATERGEMS software environment under the management of ArcGIS-ArcMap software and in the form of waste analysis.

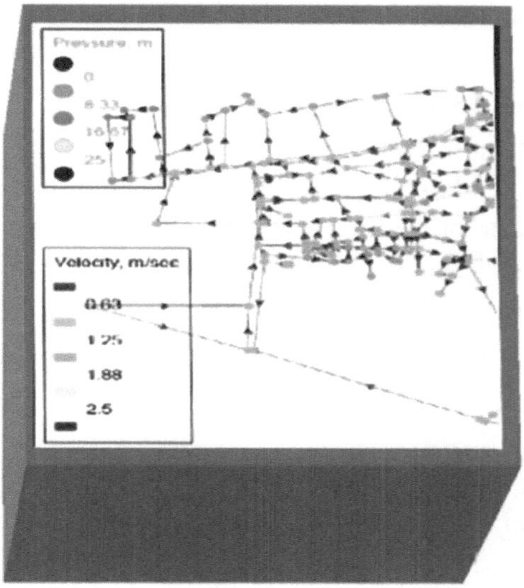

FIGURE 3.15 Hydraulic estimations and evaluations of distribut:on systems by networked sensors.

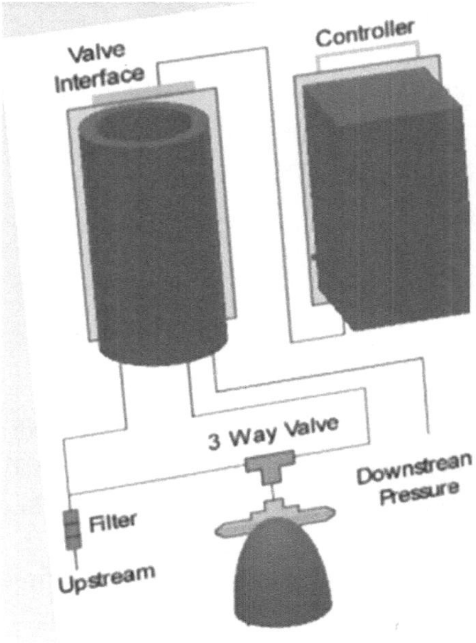

FIGURE 3.16 Advanced control by networked sensors for distribution network.

The hydraulic model of a distribution network is made in order to achieve various goals, including:

- Water comprehensive plans
- Water quality studies
- Estimation and evaluation of water distribution system
- Energy waste prevention studies
- Improvement and development of water distribution system
- Connecting with nearby networks
- Updating existing facilities

3.1.47 HYDRAULIC MODEL

3.1.47.1 MODEL SELECTION

In choosing the model and type of software, the following are used:

- Hydraulic analysis
- Water quality analysis
- Simulation features and specifications
- Graphic features
- Hardware requirements
- Software requirements
- Software capabilities
- Required data
- Cost

3.1.48 REQUIRED INFORMATION

The information needed in the preparation of the hydraulic model of the water distribution system includes information related to pipes, faucets, pumps, nodes, operation status, water consumption and needs, consumption fluctuations, and water consumption patterns in the pilot and loss coefficient. Therefore, quantitative information required for modeling and hydraulic analysis of the network has been prepared, studied, and used under three branches as described below.

Information about the network

- Information related to exploitation
- Information about consumption

Pipes information, including

- Pipe code
- Flow direction in the pipe
- Node code of the beginning and end of the pipe
- Diameter
- Material
- The length
- Roughness coefficient
- Hazen–Williams coefficient
- Partial loss coefficients
- Working pressure and design pressure

Information about pumps

- Start and end node code
- The diameter of the vane suction pipe
- Pump head
- Pump capacity
- Altitude number of the place
- Pump installation height figure
- Pump impeller diameter
- Pump characteristic curve

Information about nodes

- Node type
- Reference node or fixed slope

Consumption node or constant flow

- Node code
- Altitude number
- Node location

Connecting node

- Pipe diameter changes at the node location

- Gender changes
- Areas covered by each nodes
- Information related to exploitation

Tanks

Pumps

- Thrust pressure
- Suction pressure
- Flow rate
- Performance curve
- Pipes
- The amount of flow in the pipes

Information related to consumption

Reading statistics

Water production statistics

- Changes caused by climate changes
- Seasonal changes
- Daily changes
    - Min
    - Average
    - Maximum
- Hourly changes
    - Min
    - Majority limit

Consumption classes

- Residential
    - Household use
    - Irrigation expenses
- Industrial
- Commercial
- official
- General
- Entertainment

Water losses

Water needs

Per capita consumption

Model preparation

After completing the above activities, the following steps are taken to prepare the model:

- Preparation of the coded map of the computer model
- Entering the collected information in the selected software
- Output
- Collecting information for model calibration

After preparing the distribution network model, in order to calibrate the model, or in other words, adapt the theoretical parameters as much as possible to the existing situation, the following actions must be taken:

- Information about the maximum daily, hourly consumption, etc.
- 24-hour flow measurement
- Off and on times of the pumps
- The flow of the pumps and the discharge head of the pumps
- Pressure measurement at different points of the network

3.1.49 HYDRAULIC MODEL CALIBRATION

In order to analyze and make decisions, after building the model, it should be calibrated so that the model matches the existing reality of the system. The purpose of doing this is:

- In order to analyze and make decisions, after building the model, it should be calibrated so that the model matches the existing reality of the system. The purpose of doing this is:
    - Increasing the credibility of the model
    - Making it possible to predict the future situation
    - Obtaining accurate results compared to the current state of the system
    - Discovering problems and unknowns in the system

For this purpose, the information collected in the previous paragraph is used. It should be mentioned that the model should be continuously calibrated according to the changes made.[39–42]

The purpose of performing network analysis calculations in the water reduction plan is not taken into account.

In the water reduction plan, network analysis is done to adapt the theoretical parameters to the existing reality and to achieve the following goals.

- Determine the amount and direction of the incoming and outgoing currents from the pilot.
- Determine the amount of flow passing through the sections, the maximum pressure in the pipes, and the speed and direction of the flow.
- How to solve issues and problems such as pressure drop, lack of sufficient discharge, and failure of faucets to function properly should be studied.
- The results of network analysis are analyzed and compared with any future changes in the network.

3.1.50 MODEL AND SOFTWARE SELECTION

There are many issues in choosing a model and the ability of models to meet these needs will depend on them. In choosing a model, there are various questions that can be answered easily by reviewing the capabilities of the available software, and the desired software can be selected and used.

The hydraulic model of the distribution system is defined by the software in accordance with the GIS in the WATER GEMS 8.2 software environment under the management of the ArcGIS 9-ArcMap 9.3 software and in the form of waste analysis, which is one of the strongest software for the hydraulic analysis of water distribution networks.

Pipes

Due to the fact that the function of the pipes and the effects of the fluid at this time are effective in depositing and changing the diameter of the pipe. Therefore, the pressure drop coefficients should be selected in such a way that the effect of the above parameters is considered in it. For example, if the pipes used in the main water transmission lines of the city are made of pre-stressed concrete and the pipes related to the water distribution network are often made of cement. to be Manufacturing factories suggest the Hazen-Williams formula coefficient for cement pipes 120-140 (Tables 3.5 and 3.6).

TABLE 3.5 Roughness Coefficient of New Pipes; Manning and Darcy-Wiesbach Coefficients.

No.	Manning N	Darcy-Wiesbach E (mm)
1	0.012–0.015	0.26
2	0.012–0.017	0.3–3.05
3	0.015–0.017	0.152
4	0.011–0.015	0.00152
5	0.015–0.017	0.046

TABLE 3.6 Roughness Coefficient of New Pipes; Hazen-Williams Coefficient.

No.	Hazen-Williams C	Material
1	130–140	Ductile iron
2	120–140	Concrete or pre-stressed
3	120	Galvanized iron
4	140–150	plastic
5	140–150	Steel

Obviously, in every water distribution network, there are accessories such as sliding valves, butterfly valves. Air intake valve, discharge valve, three-way valve, knee, switch, etc. are used as controllers of the transmission line and network, or depending on the nature and location of the land and design. By using any of the above, there will be a pressure drop in the network. Calculations related to such partial losses are time-consuming and error-prone due to their extensiveness and complexity. Therefore, in the analysis of the pilot network, the equivalent length method has been used to calculate the local losses. In this way, due to the extent of the network and the number of equipment, a percentage has been added to the length of the pipe to account for local drops. This value for water supply networks is between 5 and 10%, which is considered 7% in the analysis of the pilot network.[43–46]

3.1.51 NODES INFORMATION

- In order to distribute the flow in the network, it is assumed that the consumption flow of the network is concentrated in the nodes, and also the consumption flow of each pipe is equally divided between the two nodes at the beginning and end of the pipe.

- Topographical maps have been used to obtain the height figure of different points of the network.
- The amount of water consumption is variable during different days of the year, even the amount of consumption during the day and night has fluctuations.

The intensity and weakness of hourly fluctuations in pilot depends on the size of the pilot and the daily schedule of most of its people. As the city grows, the inconsistency in people's life schedule increases, so the hourly fluctuations become less. Therefore, due to the special importance of these fluctuations, two coefficients of maximum daily consumption and maximum hourly consumption are defined to analyze the water supply network according to the maximum amount of consumption; in this analysis, to calculate the two coefficients of the maximum daily consumption and the maximum hourly consumption (Tables 3.7 and 3.8).

TABLE 3.7 The Coefficient of Maximum Daily Consumption in Different Regions. The weather of the country.

No.	Climate zone	The coefficient of maximum daily consumption
1	Cold mountain	0.2–1.3
2	Caspian temperate and humid	5.1–3.1
3	Mediterranean/semi-desert	6.1–3.1
4	Hot semi-desert/dry desert	8.1–5.1
5	Hot coastal dry/coastal dry	6.1–3.1

TABLE 3.8 The Coefficient of Maximum Hourly Consumption.

No.	People (population)	The coefficient of maximum hourly consumption
1	5 ≥ thousand people	5.2–2
2	5–20 thousand people	5.2
3	20–150 thousand people	8.1–4.1
4	100–500 thousand people	6.1 4.1
5	500 ≤ thousand people	1.4–1.2

According to the above tables and weather conditions, and by examining consumption statistics during different months of the year, the maximum daily consumption coefficient is 1.8 and the maximum hourly coefficient is 1.6. And therefore the coefficient of maximum consumption is equal to [2.8 = 1.6 * 1.8].

- The maximum pressure in the water supply network should be enough for the pipes to withstand it, especially at the joints. The risk of pipe bursting in low-lying areas of the pilot increases in the middle of the night when the water flow in the network is small and as a result the pressure drop is the lowest and the pressure is the highest.
- The minimum pressure in the urban network should be enough to have the necessary pressure for the consumer in the pipe at the beginning of each branch. This pressure should meet the following needs.
    - Compensate the pressure drop in the branch pipe, water inlet valve, and one-way valve.
    - Compensate the pressure drop in the water meter. It should be noted that the pressure drop in the water meter, unlike other local pressure drops, is usually high and around 2–6 meters.
    - Compensate the pressure drop of plumbing in each floor of the building.
    - Compensate the height of the building floors. (The first floor is 4 m and the other floors are 3 m each.)
    - At least 3 m of pressure remains in the milk located at a height of 2 m above the last floor.

According to the mentioned needs, the lowest suitable pressure depending on the number of floors of buildings in the pilot is given in Table 3.9.

TABLE 3.9 The Lowest Suitable Pressure in the Network in Meters of Water Height Against Each Building.

No.	The number of building floor	Building height (m)
1	1	5
2	2	8
3	3	11
4	4	14
2	5	17

The minimum pressure against a one-story building is 14 m, and 4 m is added for each additional floor in the building.

3.1.52 INFORMATION ABOUT PUMPS AND STORAGE PLACES (WELLS AND TANKS)

- The information related to the technical specifications of the well pumps and tanks and the production statistics of these units have been prepared in two stages each for one month. This information is used in the network analysis. An example of this information is given in Table 3.10.

TABLE 3.10 Specifications of Distribution Network Tank Pumps.

The well no.	Number of pumps	How to work
101	2	They work with each other and at the same time, and the way to connect them to the network is in parallel.
102	1	Due to the presence of iron in the water of this well, the water first enters the tank and then is pumped to the network.
103	2	One of the pumps is used as a spare.
104	3	Two pumps work together and in parallel (connected to the network) and the third one is spare.

- Information related to well pumps such as well depth, pump installation depth and pump head and pump discharge, active and inactive wells have also been collected and included in the hydraulic model. An example of this information is given in (Table 3.11).
- The head of the pumps has also been extracted using the operating method and technical specifications provided by the factories that manufacture the pumps.

3.1.53 CALIBRATION OF THE PREPARED HYDRAULIC MODEL

After reading the pressure gauges, the hydraulic model of the distribution network is calibrated using the results. The results of the hydraulic model are the same with the barometer results in many cases and it can be stated that the prepared hydraulic model is completely consistent with the environmental and hydraulic conditions in the pilot's distribution network.

3.1.54 ANALYSIS OF WATER AND SOIL TESTS

Evaluation of the used equipment according to the results of the tests.

In evaluating the equipment used according to the results of the water and soil tests in the region, it is necessary to first carefully examine the characteristics of each of the equipment used and then evaluate their performance according to the results of the tests.

In this regard, due to the fact that the pipes used in the pilot water distribution network assumed in this book are generally made of asbestos cement, their characteristics are given as follows:

3.1.55 INSPECTION OF DISTRIBUTION NETWORK PIPES

3.1.55.1 CEMENT PIPES

Asbestos cement pipes were produced for the first time in Italy (1920) by mixing a mineral called asbestos with mountain cotton and Portland cement. Asbestos cement pipes of two types, PR and PN (according to international ISO standards, are produced in classes A, B, D, and E from 80 to 1200 mm in diameter). According to the pressure, the work of asbestos cement pipes of different classes according to the type of use, weight per unit length in two types PR, PN, and its physical characteristics are given (Tables 3.11–3.13).

TABLE 3.11 Working Pressure of Asbestos Cement Pipes of Different Classes.

No.	Brigade	Working pressure kg/cm^2
1	A	2
2	B	6
3	c	9
4	d	12
5	E	15

TABLE 3.12 Weight Per Unit Length of Cement Pipes in Two Types—PR and PN.

Diameter (mm)	Asbestos cement—PN type			Asbestos cement—PR type		
	B	c	d	B	C	D
80	667	6.76	8.18	6.76	7.39	8.52
100	881	8.99	10.73	8.99	10.25	11.02
150	12.09	15.95	20.85	14.27	18.22	22
200	18.52	23.89	37.62	21.47	28.85	39
250	27.52	36.12	51.04	29.41	39.67	53.08
300	37.16	49.69	75.95	39.41	54.34	77.18
350	48.19	69.43	101.46	50.79	77.62	101.46
400	62.09	92.46	130.66	67.07	97.79	132.29
450	76.08	113.53	167	82.77	123	174.15
500	93.31	136.73	213.84	104.47	151.91	223.84
600	124	188.74	290.45	144.17	215.23	315.46

TABLE 3.13 Physical Characteristics of Asbestos Cement Pipes.

No.	Physical characteristics	Unit	Amount
1	Tensile strength	kg/mm^2	2.5–2
2	Bending strength	kg/mm^2	2.5
3	Modulus of elasticity	kg/mm^2	$2.7*10^3$–2.1
4	Longitudinal expansion percentage	%	–
5	Brinell hardness coefficient	HB	35–40
6	–	kg/mm^2	2.08
7	Thermal expansion coefficient	1/C	$0.6–0.8*10^2$
8	Poisson's ratio		0.25–0.18
9	Impact resistance	kg/mm^2	–

The advantages of asbestos cement pipes are:

– Medium resistance against external loads (tolerance of external load resistance up to 15,000 kg per meter length)
– Relative ease of length

Relative ease of repair and replacement

- It does not react with solutes and dissolved oxygen in water and does not form sediment

- It has good resistance against the abrasion of fast water flow
- Insulation against cold and heat
- Due to relatively low roughness, the amount of sediment and pressure drop is low
- Lifespan of 30–35 years

The disadvantages of asbestos pipes are as follows.

- Transportation problem due to fragility
- The limitation of manufacturing with a diameter of more than 700 mm, especially in pressure pipes

It is weak against sulfuric and chloride corrosion.

3.1.56 HOW TO CONNECT

Sleeve joints are used to connect cement pipes to each other. In the installation of cement pipes, it is necessary to mention, although these pipes are not considered as flexible pipes, but due to the fact that its connections have very little flexibility, therefore, the deviation in the piping route is somewhat limited. And therefore, it tolerates incongruous and little meetings. The amount of permissible deviation in the installation of cement pipes (Figures 3.17–3.19) in Table 3.14 were given.[47–49]

TABLE 3.14 Physical Characteristics of Asbestos Cement Pipes.

No.	Pipe diameter (mm)	Degree	f (cm)
1	50–80	8	69.6
2	175–100	7	61.4
3	300–200	5	43.7
4	350–500	4	35
5	1000–600	2	17.4

3.1.57 FIELD OPERATIONS IN THE DMA

A detailed examination of the network's weaknesses and strengths in each region in terms of leakage and measuring the amount of unaccounted water is one of the goals of separating the regions. In other words, with this measure, in addition to being aware of the amount of casualties in different parts of the pilot, which is caused by physical and non-physical casualties, the condition of the network and the sensitive points of the network in terms of leakage have been identified, thus drawing the attention of the relevant officials to the problems in each region. Policies and control measures in the field of leakage reduction will be carried out in the right direction.

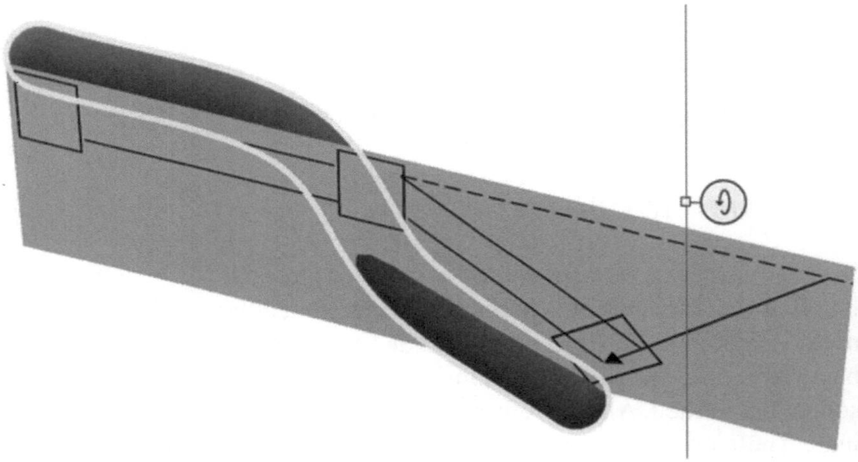

FIGURE 3.17 Degree of freedom in the installation of cement pipes.

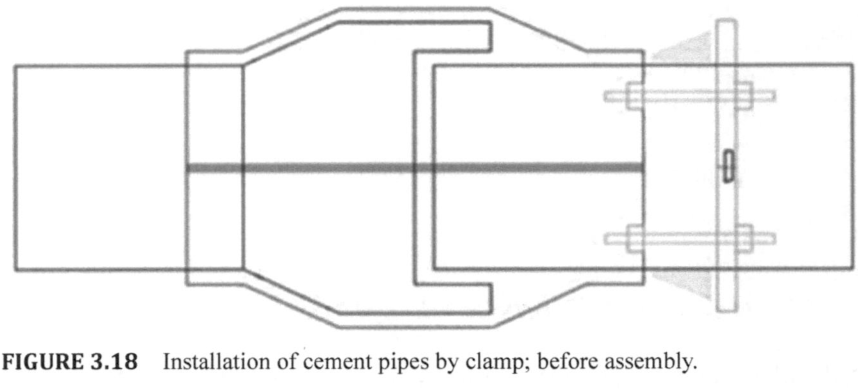

FIGURE 3.18 Installation of cement pipes by clamp; before assembly.

Usually, the amount of water delivered and discharged from areas under different pressures is rarely measured, and estimating the amount of flow entering each area is a difficult task, but in order to control and calculate the unaccounted water, we have to measure the amount of water delivered to each area with the same accuracy. Let's measure what is offered to customers. This requires the use of equipment such as meters, taps, connectors, etc., in certain points of the network.

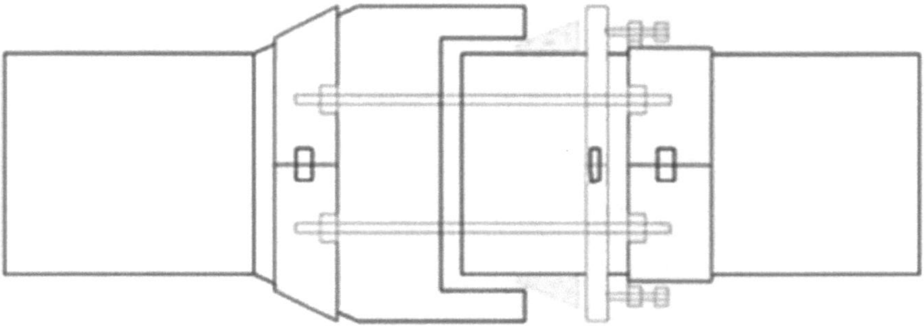

FIGURE 3.19 Installation of cement pipes by clamp; after assembly.

Determining the installation location of the above equipment is important according to the general points below.

- The flow passing through the contour should be within the measuring range of the contour
- Installed meters should have sufficient accuracy even in minimum currents
- The variety of contours used in all regions should be minimized

As mentioned, the separation of each area is done by installing a number of meters at the entry and exit points of each area, but due to the high cost that the installation of each meter and related valves imposes on the project, it is possible to use disconnecting and connecting valves in some places. did obviously, before making this decision, it is necessary to study the effects of the disconnection of connecting pipes in the separate area on the hydraulic parameters of the flow in all points by using a computer model and performing network analysis calculations, so that sometimes with a significant reduction in pressure or water cut off we did not encounter in some parts of the network.

It is clear that the smaller the range of the separate area, the more accurate and reliable the results will be. But this causes an increase in the number of contours, faucets and connections required and increases the operating costs of the work significantly. For this reason, according to the scope of the separate area, a method should be considered that provides technical and economic considerations. As a primary criterion, the area should be selected in such a way that it contains about 2000–5000 branches.

After determining the scope of the separate area, different ways of isolation are considered. For this reason, after specifying the boundaries of the separate area on the maps of the urban water distribution network, the pipelines that cross the border of the area are identified, and then different ways of isolation are studied in such a way that the implementation operations, including the installation of pipes, connections, valves, and counters should be reduced to the minimum possible so as to reduce the implementation costs of the project. The use of shut-off valves instead of installing contours for some pipes is also done in the same direction.

Considering the following points in planning different options for the separation of the region can be helpful:

- Since the selection of the installation location of the faucets is of special importance, in the selection of their location, it should be done in such a way that the water pressure in the isolated area or other parts does not fall below the permissible limit.
- Even if it is possible, instead of installing the meter in some parts of the network, the shut-off valve should be used.
- Use the maximum facilities available in the network.[50–52]

3.1.58 CONSIDERATIONS REGARDING THE SELECTION OF REGIONAL CONTOURS

Basically, the regional contours should be designed to be able to measure the fluid flow from the main pipeline with a large size to the main pipeline of the area with a smaller size. Choosing the installation location of regional contours in order to determine regional consumption requires a precise analysis of the network to determine the range of fluid flow intensity at different points and to identify the optimal points for the installation of the contour.

In this context, using the following items can be beneficial:

– In order to implement intelligent pressure management, regional contours should be selected from the remote reading ultrasonic type, because for these contours only a basin should be built to install the contour (Figures 3.20 and 3.21).

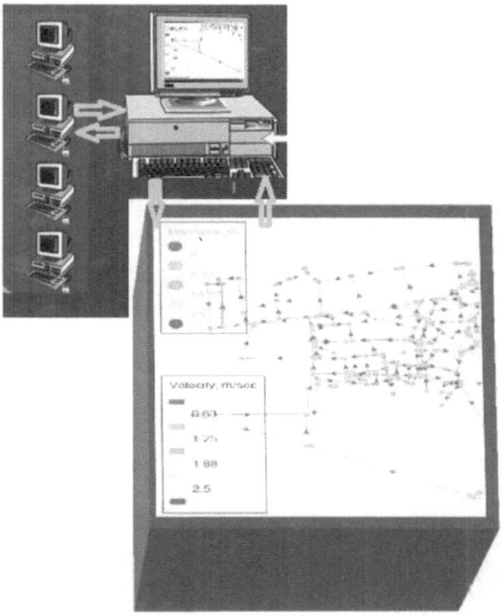

FIGURE 3.20 Intelligent pressure management.

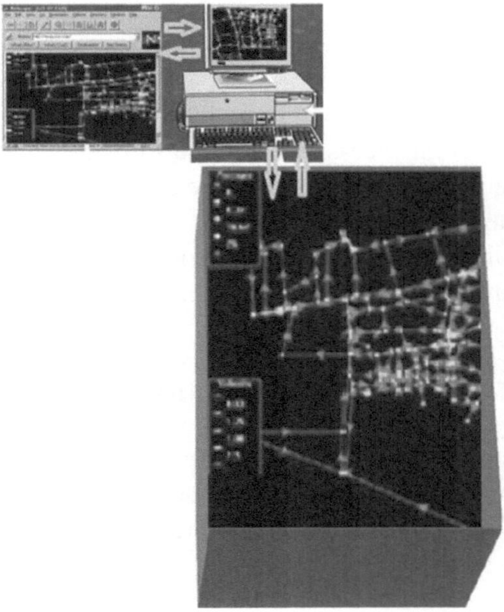

FIGURE 3.21 Implement of intelligent pressure management for pilot.

Capabilities of intelligent pressure management software:

- Measurement and monitoring
- Measuring and displaying pressure at the level of the distribution network
- Flow measurement and display at the entrance of distribution network zones
- Displaying the position and status of control valves and actuators

Warnings

- Alarm related to the entry of unauthorized people when the door is opened
- Alarms related to power outages
- Alarms related to over and under pressure in the measured points

Commands

Automatic control ability based on defined criteria

Issuing the necessary commands to control, open, or close the control valves

Ability to define control loops

- The possibility of software definition and calibration of signals by the operator
- The possibility of activating and deactivating alarms by the operator
- Having different access levels in the software
- The possibility of defining the type of input signals in the software and by the operator
- The possibility of dialing up and connecting to the system through a telephone line and

3.1.59 MODERN TECHNOLOGY FOR MONITORING

- The possibility of running the software as a client in other places connected to the network, for the use of administrators (Table 3.15)
- It is possible to define the signal measurement interval in the software and by the operator
- The possibility of defining the unit of measurement of signals in the software and by the operator

- Ability to define Set point for analog signals
- Software under Windows XP
- The ability to store information
- Ability to develop
- The possibility of communication through the Internet using ADSL with static IP
- The possibility of reporting and trending from the information stored in the form of a table
- The possibility of reporting and trending of the information stored in the form of a graph
- The possibility of reporting and trending the working hours of the pumps at different times
- Having a historical trend

TABLE 3.15 Pressure Intelligent Management System Equipment.

No.	Equipment name	Number
1	Information transfer board	11
2	Wi-Max modem under LAN	22
3	RS232 to LAN converter	11
6	Flow and pressure control valve size 250	4
7	Flow and pressure control valve size 300	1
8	Flow and pressure control valve size 500	1
9	GRP manhole	11
10	software	unlimited

3.1.60 INTELLIGENT PRESSURE MANAGEMENT PROJECT

- Scientific management by analyzing data received from sensors related to various hydraulic parameters such as pressure, discharge, and water quality polluting factors by remote reading method in the water distribution network.
- Identifying any qualitative and quantitative changes in the set of facilities and users' expenses in the least possible time and analyzing the related data in the system output.
- Creating the ability to deal with all kinds of pollutants, hydraulic instability, water loss, and unauthorized withdrawal from the network.

- Online management of water facilities while using the technique of remote reading and fast circulation of information according to GIS geographic information system.
- Scientific management of demand management and consumption management and improvement of technical and health safety factor (Figures 3.15–3.18).

In line with the implementation of intelligent pressure management and regional flow measurement of DMA, serious attention should be paid to the destructive effects of hydraulic instability in facilities and planning in order to deal with it as a complex and destructive phenomenon in the set of water supply-storage-transmission and distribution facilities. It should be the priority of the personals. In this direction and in the form of pilot studies of intelligent pressure management and DMA, intelligent pressure sensor devices and hydraulic instability data analyzer sensors with ultra-advanced technology to receive the signal according to the GIS will be installed for the facility.

Therefore, in order to manage water consumption and reduce NRW, it is necessary to analyze the information of the minimum night discharge and parallel pressure lines of distribution networks, item zoning and implementation of DMA, while using the geophone device for extensive leak detection operations and also installing remote reading logger pressure gauges in line with intelligent management, pressure should be considered. In the meantime, this operation is carried out to determine the scope of the operational pilot in order to carry out a cluster test of the subscribers' meters in accordance with the GIS in the WATERGEMS software environment based on the ArcGIS-ArcMap software and in the form of waste analysis. Therefore, according to the forecasts made in order to reduce the error of measuring equipment and correct these errors in the apparent waste section, the pilot area map of the cluster test of the subscribers' meters should be coordinated with the list of the subscribers' readings of the pilot area.

The operation of intelligent pressure management, while using the remote reading technique, makes it possible to apply online management. Therefore, the recognition of all types of complications, classes, and sub-classes in accordance with the GIS for water supply, storage, transmission, and distribution facilities is on the agenda of this system. Therefore, any qualitative and quantitative changes in the set of facilities are investigated in the minimum time, and the management of related data in the output of the system creates the ability to deal with various factors of water quality and quantity instability in the minimum possible time. The GIS in two dimensions, descriptive information and spatial information, has identified all types of complications—classes and subclasses, and with the help of the WaterGems network hydraulic analysis software and the Hammer impact analysis software under the management of ArcGIS & ArcMap software, the values and location of each of the factors causing changes. It determines the quantity and quality of water. It is worth mentioning that in line with the implementation of intelligent pressure management and DMA regional flow measurement, serious attention should be paid to the destructive effects of hydraulic instability in water facilities and planning to deal with it as a complex and destructive phenomenon in the set of water supply-storage-transmission and distribution facilities. Therefore, the following should be implemented:[53–55]

3.1.61 THE PILOT PROJECT OF INTELLIGENT PRESSURE MANAGEMENT

- The features of the pilot range should be considered in choosing the type and size of the contour. For example, in areas that do not have a storage source, the installed meters must have sufficient sensitivity to measure the intensity of low currents that occur at night.
- In the place of installing the fluid flowmeter, it should always move in one direction.
- Regional contours should be installed in the places where the maximum intensity of the current is likely. This will make the measurements have good accuracy.
- In the place of installing the fluid flowmeter, it should always move in one direction.
- Regional contours should be installed in the places where the maximum intensity of the current is likely. This will make the measurements have good accuracy.
- If the minimum flow through the pipes is so low that we have to use a contour with a small diameter due to the high loss caused by the application of the contour of the relevant connections, it is recommended to use combined contours or ultrasonic and induction contours. Combined contours include a rotary vane contour

and a one-way valve and a bypass line with a one-way valve and a contour with a smaller diameter. In the intensity of low currents, the valve of the main contour remains closed and all the flow passes through the small contour, and therefore the flow rate is measured accurately. As soon as the current intensity reaches the predetermined range, the main valve of the contour is opened due to the pressure difference and thus the current intensity is measured from both contours.

- Calibration and periodic testing of contours should be easily possible. For this reason, regional contours can be tested by installing a portable ultrasonic contour or by using a tank in the workshop.
- Usually, the reading of regional contours is done at a certain time of the day. But the real value of these meters is shown when the resulting information is presented to the management of water supply and distribution on a daily and even hourly basis. Therefore, if the regional contours have the ability to provide information in the desired forms and in desired periods, they will help a lot to control water without income.

Simulating the water distribution network by applying the necessary changes to separate areas

In order to investigate the effect of the separation of regions on the hydraulic status of the network and compare it with the current situation, the water distribution network is simulated again by a computer model. In this case, the changes presented to the model should reflect the real state of the system after the separation of regions.

Examining the information output from the computer model before and after applying the changes related to each option in different conditions (maximum and minimum consumption flow) shows the disadvantages and advantages of the proposed options and makes it possible to choose the appropriate design and option.

Since the maximum flow rate of the pipes should not exceed the maximum flow rate that can be measured by the contour, it is necessary to simulate the distribution network in the peak flow rate of consumption in two separate and current states. This will ultimately lead to the selection of the maximum flow through the pipes. It is clear that if the speed of the passing fluid is less than a certain limit, the contour lacks the necessary accuracy or does not measure any flow at all. do not the dividing line (TRANSIENT FLOWRATE) or (PARTING LINE) is a point that increases from 2 to 5% when the contour error current decreases. Therefore, in the second order, it is necessary to simulate the network based on the minimum consumption discharge in two separate and current states, and the obtained results are compared and investigated to determine the minimum discharge through the pipes.

3.1.62 SEPARATE AREAS OF THE PILOT AREA

Usually, in the pilot area, two separate areas are determined, one of which is considered to be in the points with less age and the other in the network with a long age.

According to the pressure measurement and their readings and the calibration of the hydraulic model using the results of the pressure gauges, the final design of the separate areas is given below.

The following considerations are taken into account in the final design of these areas.

- The separation of the pilot should be determined in such a way that after carrying out the mentioned operations, the distribution of urban water will not be affected in any area.
- Instead of installing a meter in some points of the network, a shut-off valve is used and these points are determined in such a way that the water pressure in isolated areas or other parts of the network does not drop. In case of disconnection of lines using valves, these cases are prepared and included in the hydraulic model and its effects are taken into consideration.
- The points considered for the installation of the meter and related equipment are determined in such a way as to cause less disturbances in the city traffic.

3.1.63 SEPARATE NETWORK AREA

The range of the proposed separate area in the numbering map and includes a certain number of common and includes entrance or exit, which should be controlled by installing appropriate connections.

Examining the computer model resulting from the changes in two modes of maximum and minimum consumption will give the desired number and type of contours.

In order to separate the separate area, a number of contours should be installed in the desired places and also a disconnect valve should be used to disconnect some lines.

If the separate lines are cut off by using the shut-off valve, there will be changes in the pressures of some nodes during peak consumption, which is not significant. The pressure changes created using the results of the hydraulic model are given in Tables 3.16–3.18.[56–59]

TABLE 3.16 Installation Location of Separate Area Contours.

No.	No pipe	Start node	End node	Diameter pipe	The address of the place where the meter is installed	Map number in maps	Executive details of the required equipment

TABLE 3.17 Taps and Connections Needed to Create a Separate Area.

No.	Pipe no	Node (start)	Node (end)	Pipe diameter	Meter location	Map number

TABLE 3.18 Pressure Changes Created in the Network if Some Lines are Cut for Isolation.

Node Number	Pressure at peak consumption in normal state	Pressure at the consumption peak in isolation mode

3.1.64 EQUIPMENT POOL FOR SEPARATE AREAS

The basin required for installing regional contours is calculated and designed according to the following considerations, and the required execution plans are also prepared. The pool of equipment for the contours of separate areas includes the following:

- Pond door
- The method of sealing the junction of the wall and the floor

- The method of sealing the place where the pipe passes through the wall of the pond
- The height of the pipe from the bottom of the pan
- To place reinforcement and their calculations
- The above items are listed separately as follows:

Pond door

The roof of the pond is structurally a two-sided slab resting on the side walls.

The slab is divided into one middle strip and two column (side) strips in each extension. The width of the middle strip is equal to one half of the slab width and the width of each of the column strips is defined as one quarter of the slab width.

The installation of the pond door in the middle of the roof slab is structurally preferable to other options and does not require any structural calculations and additional measures to control the strength of the slab, and it should only be reinforced on both sides of the hole by the number of rebar's cut in the place of the rebar hole for a sufficient length of restraint.

3.1.65 THE SEALING METHOD OF WALL AND FLOOR CONNECTION

In order to seal the connection between the walls of the pond and its floor, PVC strip seals are used, which are used for sealing in the expansion joints of the ponds and sealing between the floor and concrete walls. The location of the water stop has been specified in the executive maps.

3.1.66 THE METHOD OF SEALING THE PLACE WHERE THE PIPE PASSES THROUGH THE WALL OF THE POND

Before molding and concreting operations, clean all around the existing pipe from dust and extra materials, then two layers of sanding with a coil are done in the location inside the concrete until after concreting is finished due to concrete shrinkage and it should not be separated from the water infiltration pipe into the tap basin, which is specified in the relevant map.

In order to prevent water infiltration through the walls of the pond, due to the high level of underground water, all the outer surfaces of the walls of the pond are coated with two layers of isolation sheets.

3.1.67 THE HEIGHT OF THE PIPE FROM THE FLOOR

In the faucet basin, the required height from the bottom of the pipe to the bottom of the basin is 20 cm according to the considerations of operation, which is specified in the connection notes. Also, for the foundation under the valve, a prefabricated unreinforced concrete platform with a width of 20 cm and a height of 15 cm is used, and after placing the concrete platform under the valve, the remaining space is filled with cement mortar.

3.1.68 REINFORCEMENT AND THEIR CALCULATIONS

In walls where one surface is in contact with the ground, two layers of reinforcement are required, one of which contains at least half and at most two-thirds of the total reinforcement required for each extension at a distance of more than 5 cm and less than one-third of the thickness of the wall from the surface. should be placed outside and the other layer including the rest of the necessary reinforcement should be placed at a distance of more than 2 cm and less than one third of the thickness of the wall from the inner surface. Considering the thickness of the wall of 30 cm, the necessary distance for pouring concrete for two layers of reinforcement It is observed in the wall.[60–62]

3.1.69 CORROSION OF WATER FACILITIES AND INCIDENTS AND EVENTS

Water supply networks in contact with the soil are subject to corrosion in terms of chemistry, physics, and biology, as well as the chemical properties of the water inside the pipe and the composition of various types of pipes. This phenomenon causes leakage in water supply networks. Therefore, it is necessary to pay special attention to the corrosion of the water supply network.

Soil tests, corrosion tests, and deposition in pipes are as follows:

- Soil moisture level of subsurface water
- PM
- Soil chloride and sulfate ions
- All dissolved solids in soil extract and its oxygen
- Granulation and hydrometric test

- Soil porosity
- Determining electrical resistance or electrical conductivity of EC soil
- Determination of soil sulfate reducing bacteria in the sample
- Total count of soil bacteria of IBC sample

3.1.70 RECORDING INFORMATION ABOUT INCIDENTS AND EVENTS

Recording information about incidents and events in water supply networks is of particular importance. One of the solutions to reach the amount of wasted water is by tracking recorded incidents. This makes it possible to provide appropriate solutions in order to reduce the physical losses of the network, according to the locations of the incidents and the complementary maps of the distribution network, as well as the identity card of the pipes and equipment.

Accidents in the network occur due to different causes such as excessive fluid pressure, extra loading on the pipe or impacts due to traffic load, improper installation of connections, network pressures, ram impact, corrosion, and clogging of pipes.

3.1.71 COMPLETION OF GIS MAPS OF THE DISTRIBUTION NETWORK AND PREPARATION OF ID CARDS FOR PIPES AND FAUCETS

In order to hydraulically analyze the distribution network and obtain the pressure in the consumption nodes as well as the amount of leakage through pressure and repair and repair the distribution network, it is necessary to have complete information about the network and the way of connections and the diameter of the pipes, as well as their type and life.

All basins of open faucets and the diameter of the pipes were recorded in the form.

Along the route, the locations of the buried faucets were identified and recorded so that the faucets can be made visible in the next stages.

It should be mentioned that during the survey, the places with visible leakage and the places suspected of leaking were identified and recorded in a form.

The faucet specifications form includes complete information such as the specifications of the pond, the dimensions of the pond, the type of valve in terms of use, the condition of the pond, the condition of the soil around the pond, the condition inside the pond, the number of valves and how they are placed, the location, the diameter of the inlet and outlet pipes, and the correctness of the valve installation; in the location, the replacement of the valve, the sketch of the place and the details of the basin, the characteristics of the valve, the type of connection, the type of valve, the types of replacement, knee and brain, and the condition of the valve.

About the valve information collection form, in addition to general information, this form also contains information about the survey equipment, controller, date, and specific information page for each valve, including valve code, diameter, type of valve, usage, status (unburied, buried, leaky) diameter, he chamber, the depth of the valve, and the specific descriptions of each valve.

Regarding the form for collecting information related to the pipelines of the distribution network, general information as well as specific information of each pipe, including the first node code, the end node code, the first tap code, the end tap code, diameter, gender, pressure class, thickness, depth, and age or age are included. The occurrence of the accident (including the number and type of the accident), the number of branches, length, and address and considerations were entered.

3.1.72 MEASUREMENT OF WATER PRODUCTION

One of the most important parameters of unaccounted water measurement is the amount of production or water injected into the pilot areas, so that the amount of unaccounted for water (UFW) in the pilot areas will be determined from the difference of the water entering the pilot areas with the number recorded according to the contour reading of the area.

A pressure gauge has been installed next to the mentioned contours so that simultaneously with the measurement of water production, the pressure gauge of the inlet is also recorded.

The method of bimonthly reading was done as 2 h in the first two weeks, and after 6 weeks it was recorded as daily reading. It should be mentioned that in the 2-h reading, the night discharge, which represents a major percentage of the network outage, and also the peak hour coefficient of the network will be determined.

3.1.73 HOME METER (BRANCHES METER) TESTING

Inaccuracy in water measurement includes a significant percentage of unaccounted for water. Therefore, it is necessary to test a number of home meters to control the measurement accuracy. The most common cause of inaccuracy in contours is the incorrect selection of the correct type and diameter of the contour for installation. Also, the life of the contours and the conditions of the installation place have a great effect on the accuracy of the contours. It is also necessary to mention that the accuracy of the contours decreases over time. In the following chapters, detailed descriptions of the contours are provided.

3.1.74 TEST EQUIPMENT RELATED TO MEASURING THE ERROR OF CONTOURS

There is a solution for measuring the error of contours. One in the place and another in the laboratory. The first solution which is schematically shown below, basically two methods are used, in the first method, a scale is used to determine the passing water through the water meter to the branches, and in the second method, a reference meter is used. The second method is generally more effective. Of course, it is possible to combine two methods or use a graduated bucket instead of a scale, but the weakness of these methods is the error of measuring the weight or volume of water in the bucket or the error of the meter itself.

It should be mentioned that there is suitable equipment in the market that can be used to test the performance of measurement accuracy and hydrostatic pressure testing of household water meters. In this method, home contours should be selected, and after opening, they should be transferred from the place to the laboratory. The technical specifications and capabilities of these devices are as follows:

Capabilities

- Ability to test semi-automatically (electro pneumatic).
- Ability to test performance in flow rates (Qn, Qt, Qmin).
- Ability to perform hydrostatic test up to 30 times manually.
- Testing capacity of 10 water meters simultaneously in one line.
- It has a clamping mechanism for contours using a pneumatic system.

- Water supply unit for low flows by fixed head tank
- Flow measurement and regulation unit with %2FS accuracy.
- Accurate volume measurement tanks with 0.2% accuracy.
- Water supply unit for high flows by pump.
- It has a closed circuit water supply system and automatic filling.
- It has an automatic water flow cut-off system.

To test the error of the contours, the average measurement method of the above device was used in the laboratory. The form was also used to record the information and specifications of the contours.

– Pressure measurement and drawing lines of equal pressure in the pilot area:

The amount of leakage from the network is directly related to the amount of pressure in the network. For this reason, measuring and controlling the hydraulic pressure in the urban water distribution network can be an effective factor in measuring the performance of the network. By carrying out pressure measurement operations and its regular control and the fact that pressure has a direct relationship with some parameters such as consumption in the network or the amount of water injected into the network, it is possible to investigate them.

In order to obtain the pressure in different points of the distribution network, two methods have been used—the first method is direct measurement in some points of the network in a scattered manner, which includes main and subsidiary lines, and the second method is the hydraulic analysis of the distribution network; and finally, the two methods have been compared with each other.

3.1.75 NECESSARY SUPPLIES TO INSTALL THE PRESSURE GAUGE

In order to install the pressure gauge in different parts of the region, you must first prepare equipment so that the work can be done at a higher speed during the installation, considering that the pressure gauge is usually installed on the taps used by the subscribers, there is a need for special interfaces to make it possible. The connection of the pressure gauge to the faucets in the houses of the subscribers should be realized so that the mentioned pressure gauge can be easily placed in place of the common faucet.

In general, the equipment used in the pressure measurement operation pilot area are as follows:

- Monomeric pressure gauge
- Conversion of half an inch
- Teflon tape for waterproofing
- Faucet
- Pipe wrench and French wrench
- Washers in various others
- Kneeling, changing for emergencies

Due to the pressure of less than 6 atm. in the network, hand pressure gauges with a pressure range of 6 atmospheres were used for installation on the distribution network lines and domestic appliance valves.

3.1.76 PLACE AND NUMBER OF PRESSURE GAUGE INSTALLATION

It is necessary to observe the following points in choosing the installation locations of the pressure gauge:

- The installation site should not be exposed to atmospheric and damaging factors.
- It should be installed in a place where it is possible to tamper and control the reading easily.
- It is necessary to have a number of pressure gauges in high points and low points in the area and be placed in topographical conditions to control the static pressures of the network.
- Pressure gauges should be installed in the vicinity of industrial centers as well as high water consumption centers in order to influence the above consumptions.
- The pressure gauge should be installed next to the main network lines that feed the desired area.
- In any case, it is necessary to install the pressure gauges in such a way that homogeneous conditions are created in its distribution.

In many cases, water loss reduction is not taken into account in the studies. It is better to install pressure gauges on the houses of the subscribers or branches. In this situation, the following points should be observed:

- The pressure gauge is installed as much as possible on the paths where its use is minimized.

- To be installed on high water consumption taps.
- It should only be installed on subscribers who can be entrusted with their reading day and night by the person who owns the property.
- During the day, there should be houses belonging to the subscribers that the agents can control and defy in order to ensure the correctness of the operation.

Of course, in water studies, pressure measurement is done in a short period of time. But it is better to install the pressure gauges in places where it is stable that it is possible to measure the pressure on the network weekly, monthly and seasonally and record it. Awareness of the above information, in addition to the accuracy of the studies, shows the issues and problems of the network in a more effective way.

What is the problem being that the more the number of pressure gauges, the more accurate the results will be?

In the pilot area, this number was chosen according to the study time to obtain appropriate results and according to the topography of the project area.

There were many limitations in the choice of houses for subscribers. However, due to the courtesy given by the respected employer, we tried to measure the pressure of houses in the vicinity of the selected places.[63,64]

It should be mentioned that the location mentioned is suggested and the exact location of the installed pressure gauges.

The form related to recording the pressure measurement results:

In order to record the results of pressure measurement at different places and at different times, the form is prepared, which includes the form number, common name, subscription number, address, map number, date (including the date of installation of the pressure gauge along with the pressure of the connection time, the cut-off date of the pressure gauge along with the cut-off time (pressure)), and the maximum air temperature.

The aim of current work was the definition of modeling of water system for corrosion by networked sensors and internet of things (IoT) in compliance with GIS. The other works in this field, generally investigated the corrosion phenomena by classic methods. This work used the combination of advanced technology due to investigation of water loss for water systems.

3.2 MATERIALS AND METHOD

The current work was conducted in order to reduce NRW and to investigate the percentage of corrosion and determine the leakage points in the distribution network and water transmission lines through the definition of the model for the distribution network and water transmission lines. The basic definition of this procedure is described as follows.

3.2.1 SYSTEM

A network of interconnected components capable of responding to one or more stimuli is called a system, such as a water transmission and distribution system, a sewage collection system, etc.

3.2.2 MODEL

A model is something that describes a system with something other than the system itself, or in other words, any organized method or process that is used to analyze problems and predict the system's behavior is called a model. The models are categorized in different types, as shown below.

3.2.3 MATHEMATICAL MODEL

If a model defines and explains the system using mathematical relationships and algorithms, it is referred to as a mathematical model.

3.2.4 DETERMINISTIC MATHEMATICAL MODEL

A model that is based on a series of accepted mathematical equations and methods in order to produce output or system responses to stimuli or spatial and temporal inputs is called a deterministic mathematical model. Most commercial models are of this type.

3.2.5 STOCHASTIC MATHEMATICAL MODEL

Stochastic mathematical model is a model in which statistical relationships are used to predict the current state of the system in different situations in order to generate a series of random responses based on the characteristics and statistical records of system responses. In the application of this model, the following points are relevant.

- Usually, there is a need to monitor the system in a wide range of imagined states.
- Usually, they are prepared for a specific application and a specific system. Therefore, the stochastic model prepared for one system cannot be used in other systems.
- The statistical methods used to prepare the stochastic model of a system can be used for other systems.

3.2.6 COMPUTER MODEL

A model in which mathematical operations are performed by a computer is called a computer model. In these models, using numerical methods, a large number of equations are solved iteratively (Hardy Cross) or simultaneously (Newton-Raphson and linear theory methods) to obtain results.

3.2.7 SIMULATION

The application of the model is called simulation and it is done in the following two ways:

3.2.7.1 SINGLE-EVENT SIMULATION

Using the model to simulate the response of the distribution system to a specific state of flow is called single-event or static simulation. In this case, for example, the consumption of different nodes in the network and the water level inside the reservoirs are assumed to be constant, and the response of the network is investigated in a special case. Considering that in this case, the simulation is related to a moment or event such as the maximum hourly current or the minimum hourly current, so it cannot be used in other moments and events.

3.2.7.2 CONTINUOUS SIMULATION

Continuous simulation is also called dynamic or extended period simulation. The model is used to simulate the responses of the system to a series of flow events or hydraulic processes. In other words, in this method, different flows are simulated during multiple time intervals with the assumption that the flow rate and pressure are constant in each time interval during the day and night, so they can be used in order to obtain the flow rate and pressure values between the maximum and minimum flow hours.

Due to the fact that the responses in continuous simulation can be studied and analyzed as time series, therefore it provides more information and results than other methods, especially statistical methods, which can be used in the following cases:

- Determining the size of facilities (reservoirs, pumping stations, etc.)
- Hourly inlet and outlet flow rates from the reservoirs
- Hourly rate of pumping
- System exploitation studies

3.2.8 VARIABLE

A quantity that can be changed at each step of working with the model.

3.2.9 PARAMETER

A quantity that has a fixed value at each step of working with the model and is unchangeable.

3.2.10 ALGORITHM

The calculation methods or steps are called algorithm.

3.2.11 OBJECTIVES AND APPLICATION OF MODELING

The water distribution network model is used to achieve different goals, in other words, the model can be used for different purposes and applications, including:

3.2.11.1 SYSTEM DESIGN RELATED TO IMPROVEMENT AND DEVELOPMENT

The model provides a suitable tool for evaluating the existing water distribution network and its deficiencies, therefore, the designer will be able to review and evaluate the options for improvement and development of the system.

- Training of users: The model provides a very suitable educational tool for training and helping users in understanding the effects of applying policies and operating methods on system efficiency
- Designing the operation method: By the model, the experts are able to analyze the existing system and according to it, they can compile the optimal operation strategies and methods that ultimately lead to the establishment of useful rules and guidelines
- Evaluating the efficiency of system operation: A model provides useful information to the operators in order to evaluate and control the system.
- Review of new development plans
- Optimizing the system in emergency and critical situations: For example, when a part of the network goes out of circuit and the load of part of it is transferred to another part
- Solving issues and problems raised in hot seasons and times of water shortage
- Study, evaluation, and analysis of the system simultaneously with operation
- Study related to energy management, which ultimately leads to the prevention of its water loss
- Study and economic evaluations of the distribution system
- Qualitative studies: Determining the retention time of solutes and chemicals soluble in water and analyzing its effects

- Determining the size and type of various facilities: Such as reservoirs and pumping stations
- Determining the place of construction of various facilities: Such as reservoirs and pumping stations
- Determining the size, type, and location of faucets
- Determining the appropriate size of pipes
- Studies related to determining the amount of water required for fire fighting
- Determining different pressure areas and analyzing the specific issues of each area
- Determining water needs and network capacity during different times
- Determining the amount of load entering the network at different points
- Evaluation and determination of network capacity in different places
- Studies related to comprehensive water plans
- Communication between adjacent networks and load adjustment
- Pursuing resources
- Availability of water

3.2.12 SIMULATION LIMITATIONS

Due to the existence of some inherent limitations in each model, its use in different situations has limitations. Among the limitations in using the model to simulate water transmission and distribution systems, the following can be mentioned:

- The computer model does not improve the quantity and quality of basic information.
- The model can only extract the required information of the subject from the basic information and is never able to overcome the lack of information.
- No model is able to produce completely accurate results, because every model has shortcomings and deficiencies in showing the whole system. (Usually all pipelines and other network components cannot be included in the model.) In other words, because every system is affected by complex and diverse processes, it is not possible to put all of them in one model.

- The numbers and results obtained by the computer model are not more accurate than manual calculations, and the only difference is in the speed of operations and the accuracy of calculations.
- Generally, the results of a model are in the form of numbers, graphs, and figures, and the final interpretation of them is the responsibility of the user according to the local conditions and other existing conditions. Therefore, the model is not just an application software that, after being installed on a computer, can only be used by an operator. It is necessary to enter and exit information, but the user must be familiar with modeling.
- Using the model to predict states and responses in the distribution network to a series of defined conditions requires the analysis and precise interpretation of the different components of the model.
- The simulation of a network with respect to different states requires the use of assumptions, the choice of which has an important effect on the results.
- Despite the above facts, a model is a valuable tool in showing the behavior of the system under different conditions, and it can answer questions that are very difficult to answer without a model.[17]

3.2.13 MODEL SELECTION

Considering the variety and abundance of existing models, the first step in choosing a model is to determine the problem and the subject in question, and knowing its response and behavior in different states of the system is considered. Since every model inherently has shortcomings, defects, limitations, as well as advantages and capabilities, therefore, in choosing it, it is necessary to define the subject in question precisely.

In addition to the above, several factors influence the choice of model, the most important of which are as follows:

- The amount of available and accessible information
- Type of required outputs or in other words software capabilities
- Required data and information
- Technical and economic considerations
- Software requirements
- Hardware requirements
- Trained personnel available

Due to the fact that in recent years, modeling software has expanded quantitatively and qualitatively, and due to their great variety, choosing a software that matches the physical conditions of the system, modeling goals, available and obtainable data, and the financial resources of the person responsible for preparing the model is challenging, and it is a relatively difficult thing to have adequate software and hardware support facilities, expert staff, etc. Therefore, all the above aspects should be carefully considered in choosing a model.

3.2.14 TYPES OF MODELS AND AVAILABLE SOFTWARE

Since the analysis and simulation of a water distribution network, even if it is small, requires solving many equations and it generally takes a lot of time to solve it, therefore since the late 1960s, many software and models have been used in the management of water distribution systems. Water distribution has been arranged.

The initial software was only able to simulate the main components of water distribution networks and their execution was only possible on mother computers and supercomputers. With the passage of time and the growth and development of information technology, from the early 1970s to the mid-1980s, various software packages have reached the world markets, which can be implemented on personal computers and some of them are currently being used.

Today, the available software is able to model and analyze the distribution networks, which consist of a large number of pipes, tanks, pumps, different control and pressure valves, flowmeters, etc.

3.2.15 MODELING STEPS

Before considering and choosing a model, some information is necessary to show and describe the idea and problem in the existing system. For example, the results of pressure measurement in different points of the water system shows a lack or high pressure, local measurements related to the amount of water consumption and entering different areas indicate a lack of water or the existence of losses. Such observations provide valuable information in recognizing the problem in the estimation system. In modeling, the following matters are considered:

- Basic input information
- Sensitivity analysis
- Information related to the verification and approval of the model

The modeling process starts with determining the subject and after collecting the basic information according to the type of selection software, it is calibrated and finally approved. The desired model can be used according to the set goal. According to the above, the different stages of modeling are explained below.

3.2.16 BASIC INPUT INFORMATION

According to its structure, each model needs its own set of input information. The basic input information has been classified into two categories: quantitative and qualitative.

3.2.17 QUANTITATIVE INFORMATION

Quantitative information depends on the physical conditions of the system and the studied area and is studied under four branches as follows:

- Network information
- Information related to exploitation
- Information about the amount of consumption
- Information about applications and covered areas
- In the following, each of the above cases is explained.

In general, the hydraulic model of water facilities is prepared in order to achieve the following goals:

- Water quality studies
- Estimation and evaluation of water distribution system
- Energy waste prevention studies
- Improvement and development of water distribution system
- Connecting with nearby networks
- Updating existing facilities

The process of choosing the model and the type of software to prepare the hydraulic model is as follows:

- Hydraulic analysis
- Water quality analysis
- Simulation features and specifications
- Graphic features
- Hardware requirements
- Software requirements
- Software capabilities
- Required data
- Cost

The information needed in the preparation of the hydraulic model of the water distribution system includes information about pipes, faucets, pumps, nodes, operating conditions, water consumption and needs, consumption fluctuations and water consumption patterns in the pilot, and loss coefficient. Therefore, quantitative information required for hydraulic modeling and analysis is prepared and studied under three branches as follows:

- Network information
- Information related to exploitation
- Information about consumption

Information about the network including pipes is as follows:

- Pipe code
- Flow direction in the pipe
- Node code of the beginning and end of the pipe
- Diameter
- Length
- Roughness coefficient
- Coefficient C
- Partial loss coefficients
- Working pressure and design pressure

The information about the pumps is as follows:

- Start and end node code
- The diameter of the vane suction pipe
- Pump head
- Pump capacity
- Altitude number of the place
- Pump installation height figure

3.2.18 CORROSION MODELING FOR DISTRIBUTION NETWORK

For the distribution network, a model based on fast return of information and compatible with the GIS was considered. In several independent DMAs, based on the maps in CAD and SCAN formats, the range of selected pilots for DMA implementation was determined. Then, for the CAD and SCAN maps related to the selected pilots, the GPS coordinates of two points were taken from the pilot. After preparing the Shape-file, all the maps were geo-referenced. The spatial and descriptive information fields were created. Hydraulic analysis in the WATERGEMS software and ArcGIS9-ArcMap software were provided. The data received by flowmeters and pressure gauges with remote reading. The actual existing model of water distribution networks was calibrated with the network analysis model. In this work, the modeling of the actual water loss of the water distribution network includes the process of computer statistics in the simulation of the network.

3.2.19 CALIBRATED MODEL OF WATER DISTRIBUTION NETWORKS

Modeling of water distribution networks is considered a very useful and valuable tool in estimating and evaluating different states of distribution systems. Calibration of water distribution networks model is the process of using a computer to simulate a real water distribution network to perform network analysis calculations. Today, the modeling of water distribution networks is common in the world. As an independent branch of fluid mechanics and a combination of different sciences, it is used in many ways. Some of the common terms in hydraulic modeling are summarized as follows:

3.2.20 PARAMETERS OF THE DISTRIBUTION NETWORK

Pipe information includes:

- Pipe code
- Code of the initial and final node of the pipe
- Diameter

- Length
- Friction coefficient and roughness
- Index of parallel pipes
- Area or facility identification number
- Partial loss coefficients
- Installation date
- Type of pipe
- Work pressure and design pressure
- Piping depth

Faucet information includes:

- Code of the start and end node of the pipe on which the valve is located
- Diameter
- Length
- Friction coefficient and roughness
- Type of milk
- Low pressure valve
- High valve pressure
- One-way valve index
- Changes in the head
- Installation date
- Type of milk
- Altitude number

Pump information includes:

- Node code of the beginning and end of the pipe on which the pump is located
- Suction and thrust pipe diameter
- Pump head
- Pump capacity
- Installation date
- Altitude number of the place
- Pump installation height figure
- Pump impeller diameter

Node information includes:

- Node type (source nodes or constant slope and consumption nodes or constant flow)
- Node code

- Pipe diameter changes at the knot location
- Changes in the type of pipe
- Altitude number
- Location of the knot
- Areas covered by each node
- Additional information includes:
- Open or closed state of pipes
- Pumps on and off
- Status of flow control valves (open, closed, semi-open)
- Information related to operation

Tanks

- The height of the water level inside the tank
- The amount of changes in the water level inside the tank
- Tank capacity
- The diameter of the inlet and outlet pipes of the tank
- Maximum and minimum water level inside the tank

Pumps

- Suction and thrust pressure
- Flow rate

Pressure regulating valves

- Upstream pressure
- Lower hand pressure
- Status and position of milk

Flow control valves

- Lower hand pressure
- Status and position of milk
- Flow rate
- Tubes
- The amount of flow in the pipes

Information related to consumption

- Consumption changes, including:
  - Changes caused by weather
  - Seasonal changes

- Current coefficients
- Daily (minimum, average, maximum)
- hourly (maximum and minimum)

Water needs

- Per capita consumption
- Type of use
- Residential (home and irrigation purposes)
- Industrial
- Commercial
- Various departments, organizations, and bodies
- Entertainment

Experimental coefficients

- Daily maximum to daily average
- Maximum hourly to daily average
- Maximum hourly to maximum daily
- Minimum one hour to maximum one hour
- Water loss coefficient
- Maximum, minimum, and average current coefficients

Unaccounted for water caused by

- Pipes (pipelines, tanks, pumps, faucets, etc.)
- Fractures
- Firefighting expenses
- Washing tanks
- Overhead tanks
- Unauthorized branches
- Inaccurate readings
- Expenses not read
- Unauthorized branches

Information about applications and covered areas

- Identifying and defining applications
- Knowing the covered areas
- Knowing the communication valves between two pressure areas and their status

3.2.21 QUALITATIVE INFORMATION

If the preparation of the quality model of the water distribution network is considered, it is necessary to have information in this connection, such as unit load, mean, and standard deviation of the quality components of relationships, gradient, parameters of creation and deterioration of materials, statistics, and information related to water quality analysis, colorimetric information, etc., to be collected.

3.2.22 SENSITIVITY ANALYSIS

- Sensitivity analysis is done in order to measure the sensitivity of the model to the change of parameters. Generally, before calibrating and confirming and approving the model, the user should familiarize himself with the capabilities and differences that exist in the models.
- The sensitivity analysis determines the parameters that have the most impact on the responses and outputs of the model in different aspects. In this way, by changing certain key parameters during the analysis, the answers of the model are examined.
- The effective factors in calibration are identified and therefore the parameters that should be changed during the calibration process are determined. For example, in different urban areas, most of the models have a high sensitivity to the consumption pattern according to the usage of the region rather than per capita consumption.
- After calibrating the model and using it, in order to predict estimates, it is very important to conduct a test that enables the effect of changing parameters. Because some of the parameters and variables are doubtful to the user, and therefore it is necessary to specify the range of expected responses to the change of parameters in an error range and its effect on the model outputs.
- Through the sensitivity analysis, the ambiguities in the parameters are clarified.
- According to the above-mentioned cases, the model must first be designed and launched with a specific and very simple structure for a specific purpose and gradually expanded and the range of input information added to it, until it gradually matches the existing reality.

3.2.23 VALIDITY AND CONFIRMATION OF THE MODEL CREDIBILITY MODEL

Generally, due to insufficient and incomplete knowledge of the natural system, the models give an incomplete picture of the physical system. Therefore, it is necessary to carry out tests on it so that the parameters and variables on which the model is based are closer to the existing reality. In general, it is only after this stage that the model can be used well for the purposes mentioned at the beginning. In the final approval and validation of the prepared model, the following should be done:

3.2.24 MODEL CALIBRATION

- In order to analyze and make decisions, after building the model, it should be calibrated to match the reality of the system as much as possible. Model calibration is one of the most important
- Activities in model preparation.
- The purpose of calibration is generally to increase the validity of the model, the possibility of predicting the future, collecting information about the current state of the system, and discovering problems and ambiguities in the system. In other words, after the construction and design of the model, the parameters of the model need to be adjusted to more accurately express the behavior of the system in a wide range of existing and possible conditions. In general, a model is not complete until it is calibrated.

Among the goals of calibrating a model, the following items can be mentioned:

- Validating the model
- Obtaining a tool for comparative uses in different model executions, or in other words, providing the possibility of predicting the future situation.
- Obtaining accurate results regarding the current state of the system
- Discovery and diagnosis of system problems and ambiguities

During the calibration process, it should be noted that the physical parameters are not set outside their acceptable range. For example, the roughness coefficient for asbestos cement pipes is in a certain range, so if during the calibration process, the results showed other than this, an error has occurred in entering the information of some variables or parameters.

Among the problems that cause confusion for the user is the preparation and creation of primary information in contrast to what the model is calibrated against. In this connection, questions such as how much and how much information is needed in relation to the amount of flow and where it should be used, or what quantity or quality, should be investigated and accurately defined the issues of the model and available information sources should be answered. In general, the reviews and results of the model in different departments should be complementary and a guide for other departments, according to another plan.

The initial for each part should be modified and completed with the progress of the project during the calibration stage. At this stage, the relevant information and situations should be precisely specified and calibrated for different flow conditions and different points in the system.

Due to the fact that a model works incompletely in understanding the physical system, therefore, a model that is calibrated for one flow state (for example, the maximum daily flow) may be correct and accurate for another flow state (for example, the minimum daily flow). According to this, the model is calibrated according to the intended need. During the calibration of the model, it is necessary to consider the following points.

3.2.24.1 KNOWING THE ALGORITHMS AND IMPORTANT PARAMETERS OF THE MODEL

In each special application, in a specific model, the user is in contact with a series of important and unimportant parameters and algorithms. Therefore, algorithms and parameters should be considered that do not waste time and money.

3.2.24.2 CLASSIFICATION OF PARAMETERS

Due to the wide range of parameters included in the model, their classification is necessary and important. For example, it is very difficult to determine and estimate some parameters (such as determining the amount of losses) and some others are easily possible (such as determining the area covered by each pipeline). and can be defined and classified.

3.2.24.3 VOLUMETRIC PARAMETERS

The parameters involved in generating the flow rate must be accurately identified and arranged, otherwise many problems will arise when working with the model.

3.2.24.4 DETERMINATION OF EXPERIMENTAL COEFFICIENTS

After simulating the flow according to the basic information, the maximum coefficients and the shape of the flow, it should be adjusted and adapted according to the existing facts.

3.2.24.5 CALIBRATION FREQUENCY

Models should be calibrated once in a while, especially when they are used in operation and design. In general, the model should be re-calibrated in the following cases:

- With any important change in the distribution network
- Changing the operating methods in new and old networks
- Basic changes in the flow rate and pattern
- Annually regardless of the changes made in the system

3.2.25 APPROVAL OR CONFIRMATION OF THE MODEL VERIFICATION MODEL

Sometimes the approval and confirmation of the model is ignored. Validation of the model includes all the activities in which the components of the model are examined for their expected performance and natural characteristics. For example, the graph of current intensity with respect to time in a distribution network for a certain amount of consumption has a specific shape, so if we do not achieve this specific shape by using the model, the structure of the model is inappropriately designed and therefore the components need to be re-inspected.

In general, in the approval or verification phase, the model is tested and controlled for specific situations, and in each case, its shortcomings are identified and finally, action is taken to fix it.

After completing this step, the prepared model is ready to be used for the determined subject and will somehow predict different states according to the subject and purpose.

3.2.26 THE PREPARATION OF THE WATER DISTRIBUTION NETWORK MODEL

The preparation process of the practical model of the water distribution network are:

Calibration

- Calibration of the model
- Complete information for model calibration
- Current coefficients

Information about the distribution network

- Incorrectly entering pipe information
- Wrong definition of the network
- Incorrect way of connecting different points of the network
- Inaccuracy of the boundaries of the pressure areas

Current information

- Estimation of the total flow rate of the system
- Amount of incoming and outgoing current to each node
- The information obtained from the control stations
- Spatial distribution of current
- Definition of the pattern of flow into or out of the areas according to the population pattern (day population compared to night) for example in administrative and commercial areas
- Definition of the connection point of important industries to the system
- Daily flow changes throughout the year

Information related to exploitation

- Control and data acquisition system (SCADA)
- Monitoring the flow
- The devices used to calibrate the model, pressure gauges, local measurement contours, etc.

Output information in a water distribution network model

- Hydraulic information, including:
  - Flow in pipelines and nodes
  - Pressure in pipelines and nodes and drawing pressure curves

Optimizing the size of pipes

- Determining the flow of maximum consumption
- Slope of hydraulic line
- Speed
- Head drop
- The condition of the pumps

Quality, including

- Water life in pipelines, transfer time
- The amount of concentration or decomposition of substances soluble in water
- Presentation of water quality curve

Consumption management, including

- Placement of load nodes
- Recalculation of expenses based on the type of consumers, geographical areas
- Operation cost
- Estimation of network price

Modeling of water distribution networks is not considered in the water reduction plan

- In the unaccounted water loss reduction plan, network analysis calculations are done to achieve the following goals:
- The minimum number of incoming and outgoing flows from each area is determined
- Study, evaluation, and analysis of the system simultaneously with operation
- Study and economic evaluations of the distribution system
- Determining the size, type, and location of faucets, regional contours, pressure booster pumps, etc.
- Examining different network isolation options and choosing the best option in terms of operational cost and optimal efficiency.

- Determining the amount of load entering the network and the capacity of the network at different points
- Optimizing the system to adjust pressure and flow in order to reduce physical and non-physical leaks
- Determining the amount of flow passing through different sections and the maximum pressure in the pipe, speed and flow direction
- Providing the necessary information to estimate water loss caused by broken pipes and other incidents
- Comparison of field measurements of water flow and pressure in different points of the network or the results obtained from the computer model and determining the areas with the most leakage.

In general, the purpose of the network analysis in the water loss reduction plan is to estimate and evaluate the distribution system and match the theoretical parameters with the existing reality.

The choice of model and the way of modeling the distribution network is considered in the water loss reduction plan.

There are many issues in choosing a model and the ability of models to meet these needs will depend on them. In choosing a model, there are various questions that can be answered easily by reviewing the capabilities of the available software, and the desired software can be selected and used.

The hydraulic model of the distribution system has been prepared by WATERGEMS software, which is one of the software for hydraulic analysis of water distribution networks.

3.2.27 WATERGEMS SOFTWARE

Hydraulic model in compliance with GIS in WATERGEMS software in the form of water loss analysis in compliance with ArcGIS-ArcMap software.

Minimum Night Flow (MNF)

Unavoidable water loss (total) =

Allowable unavoidable water loss without income + Actual unavoidable water loss + Apparent unavoidable water loss

Unavoidable water loss (total)

$$= \left(\left(CD/1000 \right) 25 + C\left(0.8\right) + 18\left(A+B\right) \right) 4E \tag{3.1}$$

A—The length of the transmission line (km)
B—The length of the distribution network (km)
C—Number of subscribers (No.)
D—Average branching length (m)
E—Average pressure (meter-water)

$$Q_L = Q_{MNF} - (Q_{NC} + Q_{BL}) \tag{3.2}$$

$$Q_{BL} = (0.25)\,(Q_{MAXC}) \tag{3.3}$$

$$Q_{NC} = (0.6)\,N_P \tag{3.4}$$

Q_{MNF} – Minimum Night Flow (MNF) of water distribution network
Q_{NC} – Night Allowable Consumption
Q_{BL} – Night Field Leakage
Q_L – Unreported leakage of water distribution network, water transmission line, consumers water branches

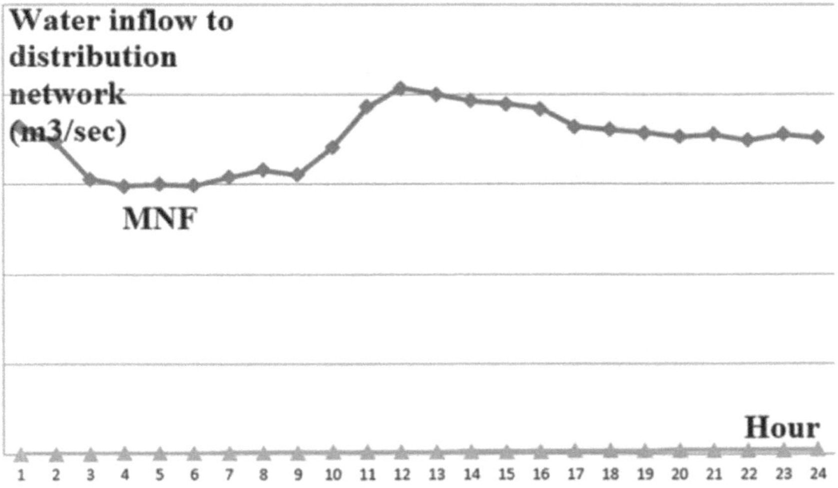

FIGURE 3.22 Analysis of the consumption curve and the minimum night flow (MNF).

3.2.28 CORROSION MODELING FOR WATER TRANSMISSION LINE

The hydraulic model of the present research was defined in compliance with the GIS in the HAMMER & WATERGEMS software environment in the form of waste analysis and under the management of ArcGIS9-ArcMap software (5–28). This hydraulic model was designed to achieve the following goals:

- Estimation and evaluation of water distribution system
- Determining the amount and position of the leak
- Improvement and development of water distribution system
- Updating existing facilities

3.2.29 AMOUNT OF WATER LOSS

In order to determine the amount of water loss due to pipe breakage in transmission lines, by solving the continuity and Bernoulli relations, the water velocity equation when exiting the broken section and the flow rate of water loss from the broken section are calculated:

$$V_2 = \left(V_1^2 + \frac{2\Delta P}{\rho} \right)^{\frac{1}{2}} \tag{3.5}$$

V_2 — Water velocity when exiting the fracture section (m/s)
About 50% of the decrease in the pressure inside the pipe after the break compared to the $-\Delta b$
V_1 — Velocity in the normal state (m/s)
ρ — Water density (Kg/m^3)
P_1 — Pressure inside the pipeline (pa)

$$Q = A * V_2 \tag{3.6}$$

Q — Flow rate of water loss from the fracture section (m^3 /s)
A — Fracture cross section (m^2)
(The cross-sectional area of broken or created pores is 5% of the total cross-sectional area of each pipe)

3.2.30 CORROSION AND WATER HAMMER

3.2.30.1 CORROSION MODELING FOR WATER TRANSMISSION LINE

Water hammer causes the corrosion in water transmission line. In this work, corrosion investigated for water transmission line approaches to transient flow. Model was defined by the method of characteristics "MOC" based on the finite difference (FD) technique. The pressures were computed along with the pipe for each time step. The combined elasticity of both the water and the pipe walls was characterized by the pressure wave speed.

The method of characteristics "MOC" approach transforms the water hammer partial differential equations into the ordinary differential equations along the characteristic lines.

Balancing the energy across two points in the system applied the energy or Bernoulli equation for steady-state flow. The components of the energy equation can be combined to express two useful quantities (the hydraulic grade line and the energy grade line):

$$(P_1/\gamma)+Z_1\left(V_1^2/2g\right)+h_p=(P_2/\gamma)+Z_2+\left(V_2^2/2g\right)+h_L \tag{3.7}$$

$$(g/a)(dH/dt)+dv/dt+\left(f\,v|v|2d\right)=0\Rightarrow(ds/dt)=c^+ \tag{3.8}$$

$$-(g/a)(dH/dt)+dv/dt+\left(f\,v|v|2d\right)=0\Rightarrow(ds/dt)=c^- \tag{3.9}$$

The method of characteristics as a finite difference technique where pressures were computed along the pipe for each time step. Calculation automatically sub-divided the pipe into sections (intervals) and selected a time interval for computations.

$$(dp/dt)=(\partial p/\partial t)+(\partial p/\partial s)(ds/dt) \tag{3.10}$$

$$(dv/dt)=(\partial v/\partial t)+(\partial v/\partial s)(ds/dt) \tag{3.11}$$

P and *V* changes due to time are high and due to coordination are low then it can be neglected for coordination differentiation:

$$(\partial v/\partial t)+(1/\rho)(\partial p/\partial s)+g(dz/ds)+(f/2D)v|v|=0, \quad \text{(Euler equation)} \tag{3.12}$$

$$C^2(\partial v/\partial s)+(1/P)(\partial P/\partial t)=0, \quad \text{(Continuity equation)} \tag{3.13}$$

By linear combination of Euler and continuity equations in characteristic solution method:

$$\lambda\left[(\partial v/\partial t)+(1/\rho)(\partial p/\partial s)+g(dz/ds)+(f/2D)v|v|\right]+C^2(\partial v/\partial s)+(1/p)(\partial p/\partial t)=0\ \lambda={}^+c\ \&\ \lambda={}^-c \tag{3.14}$$

$$(dv/dt)+(1/cp)(dp/ds)+g(dz/ds)+(f/2D)v|v|=0 \tag{3.15}$$

$$(dv/dt)-(1/cp)(\partial p/\partial s)+g(dz/ds)+(f/2D)v|v|=0 \quad (3.16)$$

Method of characteristics drawing in (s-t) coordination:

$$(dv/dt)-(g/c)(dH/dt)=0 \quad (3.17)$$

$$dH=(c/g)dv, \text{ (Joukowski formula)} \quad (3.18)$$

By Finite Difference method:

$$\begin{aligned} c+ &: ((vp-v_{Le})(Tp-0))+((g/c)(Hp-H_{Le})/ \\ &(Tp-0))+((fv_{Le}|v_{Le}|)_e|/2D)=0| \end{aligned} \quad (3.19)$$

$$\begin{aligned} c- &: ((vp-vRi)(Tp-0))+((g/c)(Hp-HRi)/(Tp-0)) \\ &+((fvRi|vRi|)/2D)=0| \end{aligned} \quad (3.20)$$

$$c+:(vp-v_{Le})+(g/c)(Hp-H_{Le})+(f\Delta t)(fv_{Le|}|v_{Le}|)/2D)=0 \quad (3.21)$$

$$c-:(vp-vRi)+(g/c)(Hp-HRi)+(f\Delta t)(v_{Ri}|v_{Ri}|)/2D)=0 \quad (3.22)$$

$$V_P=1/2\begin{pmatrix}(V_{Le}+V_{ri})+(g/c)(H_{Le}-H_{ri}) \\ -(f\,\Delta t/2D)(V_{Le}|V_{Le}|+V_{ri}|V_{ri}|)\end{pmatrix} \quad (3.23)$$

$$H_P=1/2\begin{pmatrix}C/g(V_{Le}-V_{ri})+(H_{Le}+H_{ri}) \\ -C/g(f\,\Delta t/2D)(V_{Le}|V_{Le}|-V_{ri}|V_{ri}|)\end{pmatrix} \quad (3.24)$$

$$\Delta p=\left(\frac{C.\Delta V}{g}\right), \text{ (Joukowski Formula)} \quad (3.25)$$

$$C=\sqrt{\frac{g.\frac{E_W}{\rho}}{1+\frac{d}{t_W}.\frac{E_W}{E}}}, \text{ (pressure wave velocity)} \quad (3.26)$$

The schematic curves of water transmission line modeling; hydrulic grade line; maximum transeint pressure; minimum transeint pressure; flow variation; maximum transeint pressure relief; pressure relief at leakage location; flow relief at leakage location; pressure variation at leakage location as shown in Figures 3.23–3.30 emphasized the water hammer causes the corrosion in water transmission line in.[17,18]

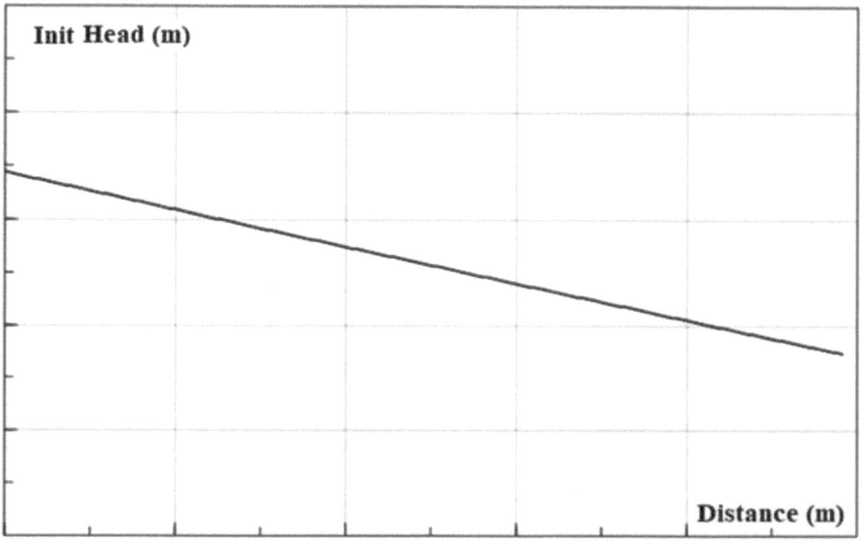

FIGURE 3.23 Schematic curve of water transmission line modeling: haydrulic grade line.

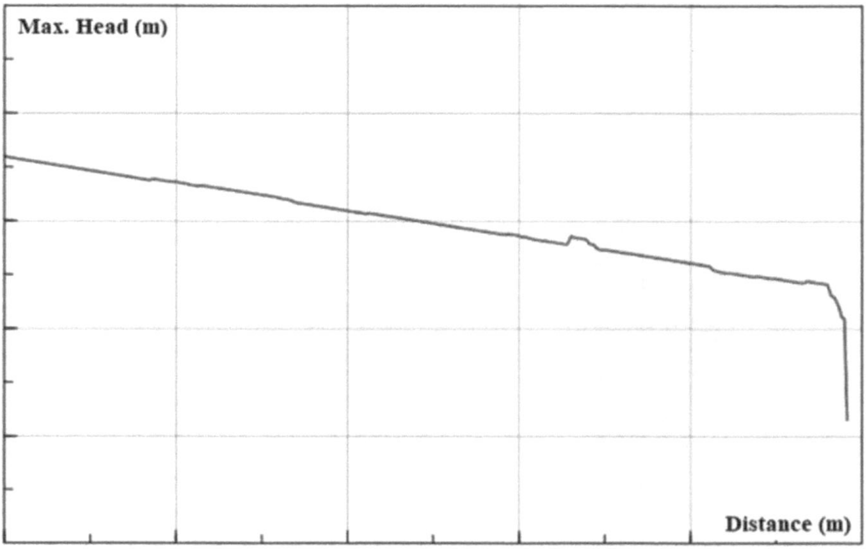

FIGURE 3.24 Schematic curve of water transmission line modeling: maximum transeint pressure.

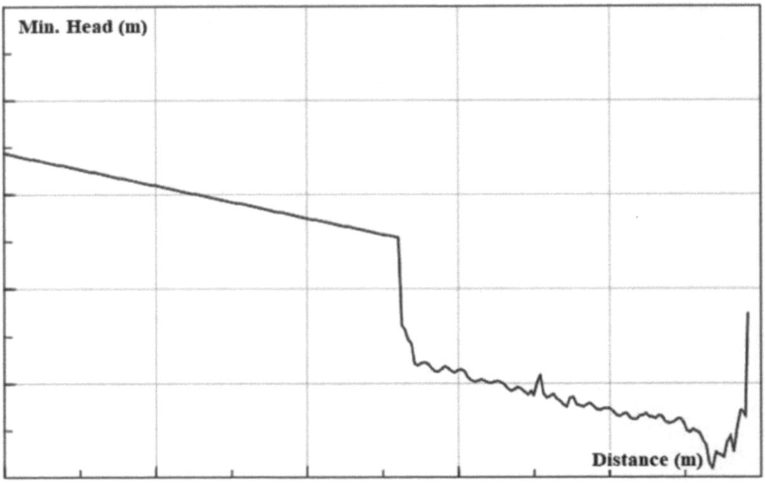

FIGURE 3.25 Schematic curve of water transmission line modeling: minumum transeint pressure.

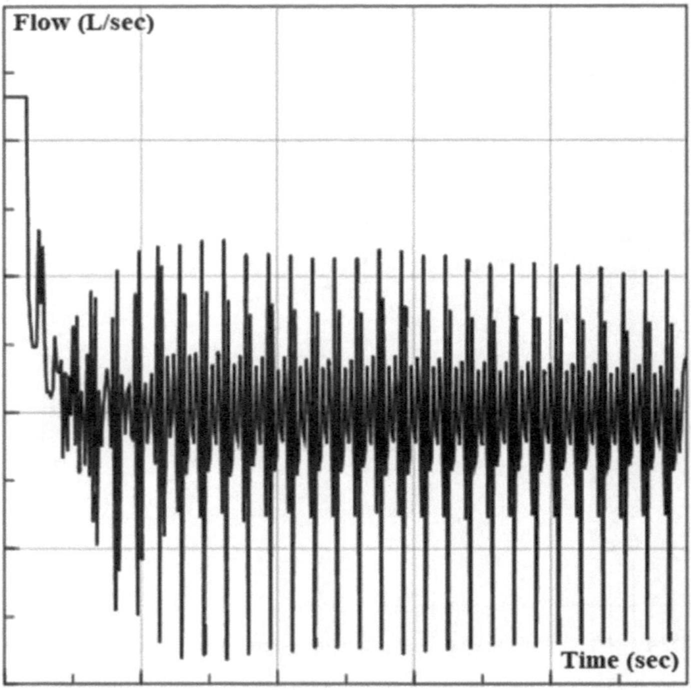

FIGURE 3.26 Schematic curve of water transmission line modeling: flow variation.

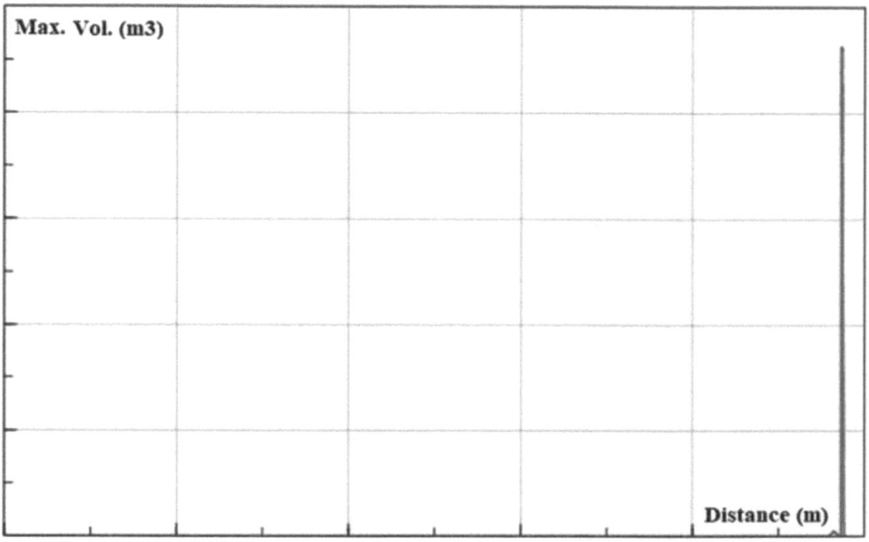

FIGURE 3.27 Schematic curve of water transmission line modeling: maximum transeint pressure relief.

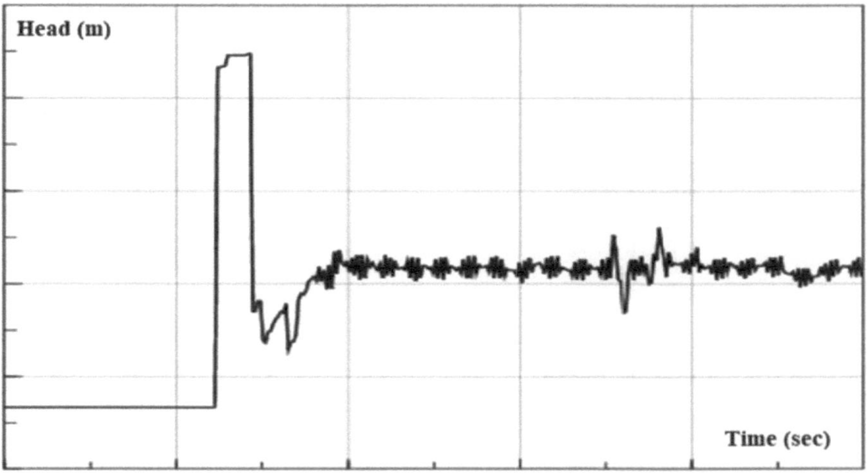

FIGURE 3.28 Schematic curve of water transmission line: pressure relief at leakage location.

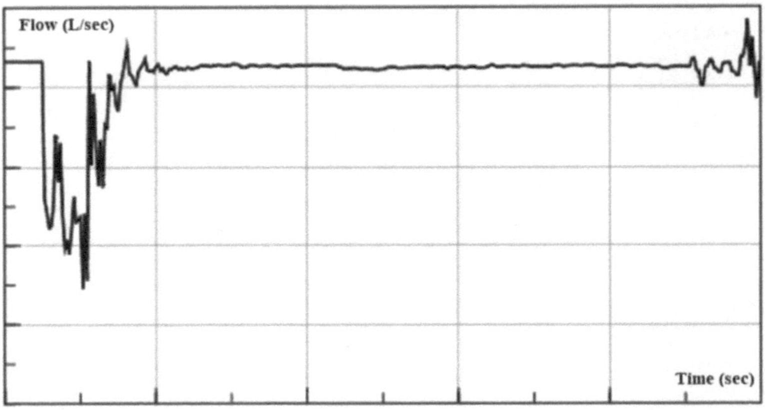

FIGURE 3.29 Schematic curve of water transmission line: flow relief at leakage location.

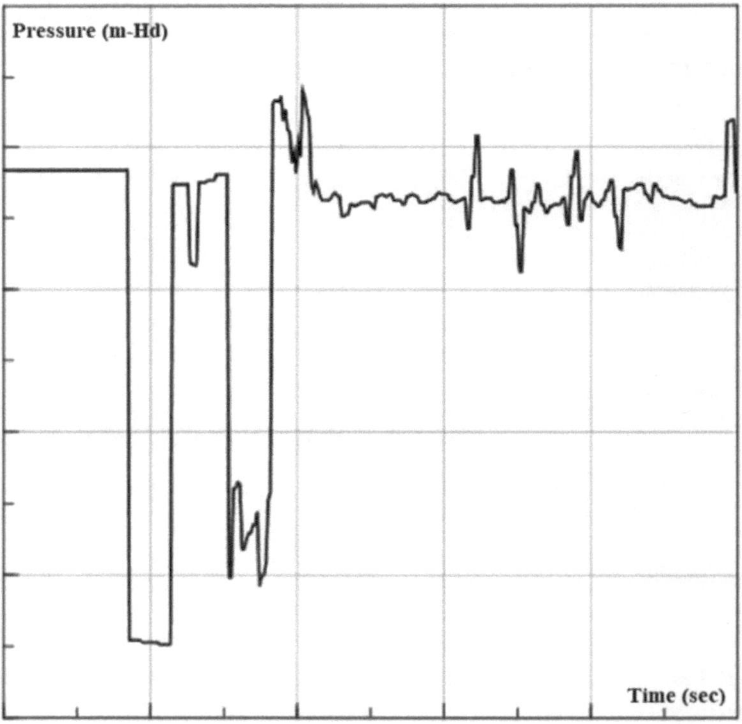

FIGURE 3.30 Schematic curve of water transmission line: pressure variation at leakage location.

In this work, the Hazen-Williams formula was used in the design and analysis, and to determine the amount of water loss due to pipe breakage, by solving the continuity and Bernoulli relations, the water velocity equation when exiting the broken section and the flow rate of water loss from the broken section were calculated:

$$V_2 = \left(V_1^2 + \frac{2\Delta P}{\rho} \right)^{\frac{1}{2}} \tag{3.27}$$

V_2 — Water speed when leaving the fracture section (m/s) about 50% of the decrease in the pressure inside the pipe after the break compared to the

ΔP — Pressure in the normal state (Bar)

V_1 — Water velocity inside the pipe (m/s)

ρ — Water density (Kg/m^3)

P_1 — Pressure inside the pipeline (pa)

$$Q = A * V_2 \tag{3.28}$$

Q — Flow rate of water loss from the fracture section (m^3/s)

A — Fracture cross section (m^2)

The cross-sectional area of the fracture or created pores is 5% of the total cross-sectional area of each pipe. The usual range of roughness coefficients for different types of pipes (Table 3.19) shows that these coefficients can only be used for new pipes.

TABLE 3.19 Coefficient of Roughness of New Pipes.

No.	Manning	Darcy-Weisbach	Hazen-Williams	Material
	N	E (mm)	C	
1	0.012–0.015	0.26	130–140	Ductile iron
2	0.012–0.017	0.3–3.05	120–140	Concrete or pre-stressed
3	0.015–0.017	0.152	120	Galvanized iron
4	0.011–0.015	0.00152	140–150	Plastic
5	0.015–0.017	0.046	140–150	Steel

3.3 RESULTS

The results of intelligent pressure management model for water facilities are as follows:

The process of analyzing the hydraulic model in this work, were considered in order to achieve the following results:

- Creating a smart network and water transmission line control system by introducing tolls.
- Creating traceability and performing hydraulic analysis of the network and water transmission line.
- Network and water transmission line modeling in ARC-GIS system based on RS for flow rate and pressure.
- Hydraulic model of the WATERGEMS network by ArcGIS9-ArcMap software.

Relationship classes between spatial and non-spatial data is a tool for creating a spatial relationship between the problem and the data related to water loss. Therefore, the relationship class for the desired problem was determined according to the type of relationship (simple or complex). With this method, it was possible to access the leakage points in this research. Therefore, the results of this activity were:

Determining leakage locations using the definition of relationship classes for complications.[65–81]

3.4 CONCLUSION

This work investigated the modeling of water system for corrosion by networked sensors and IoT in compliance with GIS. This work investigated of leakage phenomena in the water system due to the corrosion. With the aim of determining the location of the leakage, the results of calibration for network and transmission line led to the leakage positions. The real water loss has been determined. During this work, the consumption curve and the minimum night flow (MNF) were considered. The isobar of water distribution network provided along the hydraulic modeling and intelligent pressure management in the context of GIS. The hydraulic models of distribution networks as a useful tool in this work were considered.

The analysis of the MNF in accordance with GIS by receiving the signals through the remote reading technology enabled accurate control of water loss. The significant result was the effectiveness of water pressure entering the water distribution network in reducing the minimum night discharge. This result showed the recuing of water loss and led to savings in water consumption up to 5 o'clock in the morning.

ACKNOWLEDGMENT

The author thanks to all specialists for their valuable observations and advice, and the referees for recommendations that improved the quality of this paper.

NOMENCLATURE

f	Friction
C	Slope (deg.)
V	Velocity (m/s)
t	Time (s),
H	Head (m)
V	Aeration Tank volume (m^3)
LO	Total influent pollution (kg - *BOD 5*)
B_V	Tank volume rate (kg - BOD 5/m^3.day)
B_V	Volume rate (kg - BOD 5/m^3.day)
MLSS	Mixed liquor suspended solids (kg)
MLVSS	Mixed liquor volatile suspended solids (kg - BOD 5/day)
R_t	Detention time (h)
H	Aeration Tank depth (m)
Q	Influent flow rate (m^3/h)
V_S	Sedimentation velocity (m/sec)
F/M	Coefficient is the food to Micro - Organism ratio
SMLSS	Sludge Mixed liquor suspended solids
MLSS	Mixed liquor suspended solids (kg/m^3)
V	Aeration tank volume (m^3)
F	Atmospheric pressure coefficient = 1 (at sea level) and for every 1000 (m) height the "*F*" value decrease .1
SSE	Effluent suspended solids density (kg/m^3)
Q	Influent (m^3/day)
MLSS	Mixed liquor suspended solids
RS	Return sludge
Q_{Max}	Max influent in without rain case.
SLS	Effluent sludge due to suspended solids (kg)
A	Constant (.35–.55) average ~ .5
B	Constant (.07–.15) average ~ .1

y	Aeration tank efficiency
ON	Need of oxygen for nitrogen zings
B_V	Tank volume rate (kg-BOD5/m^3.day)
$MLSS$	Mixed liquor suspended solids (kg /m^3 day)
Q_V	Amount of oxygen per 24 hours for one cubic meter of aeration lagoon (kg/m^3.day).
C_X	Dissolved oxygen density (mg/L)
C_S	Amount of water dissolved oxygen in saturation condition (mg/L)
Q_C	Oxygen required per L (m^3) of aeration tank per 24 hours (kgO_2 / m^3 day)
λ0	Unit of length
p	Pressure (N/m^2)
V	Velocity (m/s)
S	Length (m)
D	Diameter of each pipe (m)
R	Pipe radius (m)
γ	Specific weight (N/m^3)
ν	Fluid dynamic viscosity (kg/m.s)
hp	Head gain from a pump (m)
hL	Combined head loss (m)
P	Surge pressure (m)
ρ	Density (kg/m^3)
C	Velocity of surge wave (m/s)
P/γ	Pressure head (m)
z	Elevation head (m)
$V2/2g$	Velocity head (m)
P	Pressure (N/m^2),
γ	Specific weight, (N/m^3)
Z	Elevation (m)
Hp	Surge wave head at pipeline points—intersection points of characteristic lines (m)
Vp	Surge wave velocity at pipeline points—intersection points of characteristic lines (m/s)
g	Gravitational acceleration constant (m/s^2)
Vri	Surge wave velocity at right hand side of intersection points of characteristic lines (m/s)
Hri	Surge wave head at right hand side of intersection points of characteristic lines (m)

Vle	Surge wave velocity at left hand side of intersection points of characteristic lines (m/s)
Hle	Surge wave head at left hand side of intersection points of characteristic lines (m)
C^-	Characteristic lines with negative slope
C+	Characteristic lines with positive slope
Min.	Minimum
Max.	Maximum
Lab.	Laboratory

KEYWORDS

- **corrosion**
- **networked sensors**
- **non-revenue water**
- **geography information system**
- **water loss**
- **Internet of Things**

REFERENCES

1. Kevin, A. Internet of Things. *RFID J.* 2009.
2. Parker, D. S.; Merrill, M. S. Oxygen and Air Activated Sludge: Another View. *J. Water Pollut. Control Fed.* **1976,** *48* (11), 2511–2528.
3. Hariri Asli, K. Nonlinear Dynamics Modeling by Geospatial Information System (GIS). *Map. Geo. Inform. J. Guilan* **2019,** *3* (1).
4. https://en.wikipedia.org/wiki/Artificial_intelligence
5. Kelly, K. *New Rules for the New Economy*. First published in 1998 by Viking Penguin, 1998.
6. https://fa.wikipedia.org/wiki/ Geographic Information System
7. Wylie, E. B.; Streeter, V. L. *Fluid Transients*. Feb Press: Ann Arbor, MI, 1983, corrected copy: 1982; pp 166–171.
8. Lee, T. S.; Pejovic, S. Air Influence on Similarity of Hydraulic Transients and Vibrations. *ASME J. Fluid Eng.* **1996,** *118* (4), 706–709.
9. Yoon, S. H.; Park, C. S. Objective Building Energy Performance Benchmarking Using Data Envelopment Analysis and Monte Carlo Sampling. *Sustainability 9,* 780. DOI: 10.3390/su9050780, 2017

10. Hariri Asli, H.; Nazari, S. Water Age and Leakage in Reservoirs; Some Computational Aspects and Practical Hints. *Larhyss J.* **2021,** *48,* 151–167.
11. Hariri, Asli, H.; Hozori, A. Non-Revenue Water (NRW) and 3d Hierarchical Model for Landslide. *Larhyss J.* **2021,** *48,* 189–210.
12. Kensek, K. BIM Guidelines Inform Facilities Management Databases: A Case Study over Time. *Buildings* **2015,** *5,* 899–916.
13. Wang, S. K. *Handbook of Air-Conditioning and Refrigeration.* McGraw-Hill, Inc.: 1994.
14. https://fa.wikipedia.org/wiki/ Internet of Things _IoT.
15. Ferras, D.; Manso, A. P. A.; Schleiss, A. J.; Covas, D. I. C. One-Dimensional Fluid-Structure Interaction Models in Pressurized Fluid-Filled Pipes: A Review. *Appl. Sci.* **2018,** *8,* 1844–1856.
16. Vardy, A. E.; Brown, J. M. B.; He, S.; Ariyaratne, C.; Gorji, S. Applicability of Frozen-Viscosity Models of Unsteady Wall Shear Stress. *J. Hydraul. Eng.* **2015,** *141,* 1–13.
17. Asli, K. H.; Aliyev, S. A. O.; Thomas, S.; Deepu, A. *Handbook of Research for Fluid and Solid Mechanics: Theory, Simulation, and Experiment*, 1st Ed.; Apple Academic Press, 2017.
18. Asli, K. H.; Naghiyev, F. B. O.; Haghi, R. K.; Asli, H. H. *Advances in Control and Automation of Water Systems*, 1st Ed.; Apple Academic Press, 2012.
19. Lema, M.; Peña, F. L.; Buchlin, J. M.; Rambaud, P.; Steelant, J. Analysis of Fluid Hammer Occurrence with Phase Change and Column Separation due to Fast Valve Opening by Means of Flow Visualization. *Exp. Therm. Fluid Sci.* **2016,** *79,* 143–153.
20. Zecchin, A. C.; Maier, H. R.; Simpson, A. R.; Leonard, M.; Nixon, J. B. Parametric Study for an Ant Algorithm Applied to Water Distribution System Optimization. *IEEE Trans. Evol. Comput.* **2005,** *9,* 175–191.
21. Andrade, D. M., de Freitas Rachid, F. B. A Versatile Friction Model for Newtonian Liquids Flowing Under Unsteady Regimes in Pipes. *Meccanica* **2022,** *57,* 43–72.
22. Sarkardeh, H.; Zarrati, A. R.; Jabbari, E.; Marosi, M. Numerical Simulation and Analysis of Flow in a Reservoir in the Presence of Vortex. *J. Eng. Appl. Comput. Fluid Mech.* **2014,** *8* (4), 598–608.
23. Zhao, L.; Yang, Y.; Wang, T.; Han, W.; Wu, R.; Wang, P.; Wang, Q.; Zhou, L. An Experimental Study on the Water Hammer with Cavity Collapse under Multiple Interruptions. *Water* **2020,** *12,* 2566.
24. Duan, H. F.; Pan, B.; Wang, M.; Chen, L.; Zheng, F.; Zhang, Y. State-of-the-Art Review on the Transient Flow Modeling and Utilization for Urban Water Supply System (UWSS) Management. *J. Water Supply Res. Trans.* **2020,** *69,* 858–893.
25. Zhou, L.; Liu, D.; Karney, B. Investigation on Hydraulic Transients of Two Entrapped Air Pockets in a Water Pipeline. *J. Hydraul. Eng.* **2013.** DOI: 10.1061/(ASCE)HY.1943-7900.0000750
26. Möller, G.; Detert, M.; Boes, R. M. Vortex-Induced Air Entrainment Rates at Intakes. *J. Hydraul. Eng.* **2015.** DOI: 10.1061/(ASCE)HY.1943-7900.0001036
27. Hariri Asli, K.; Nagiyev, F. B.; Haghi, A. K. Interpenetration of Two Fluids at Parallel Between Plates and Turbulent Moving in Pipe. In *Computational Methods in Applied Science and Engineering.* Nova Science Publisher: USA, 2009; Chapter 7, pp 115–128. https://www.novapublishers.com/catalog/product_info.php?products_id=10681
28. Hariri Asli, K.; Nagiyev, F. B.; Haghi. A. K. Some Aspects of Physical and Numerical Modeling of Water Hammer in Pipelines. In *International Journal of Nonlinear*

Dynamics and Chaos in Engineering Systems, ISSN: 0924-090X (print version) ISSN: 1573-269X (electronic version) Journal no. 11071 Springer: USA, 2009. http://nody.edmgr.com/

29. Hariri Asli, H.; Arabani, M.; Golpour, Y. Reclaimed Asphalt Pavement (RAP) Based on a Geospatial Information System (GIS). *Slovak J. Civil Eng.* **2020,** *28* (2), 36–42.
30. Hariri Asli, H. Investigation of the Factors Affecting Pedestrian Accidents in Urban Roundabouts. *Transac. Civil Environ. Eng.* **2022,** *8,* 2226, 1–4.
31. Hariri Asli, H.; Arabani, M. Analysis of Strain and Failure of Asphalt Pavement, Computational Research Progress in Applied Science & Engineering. *Transac. Civil Environ. Eng.* **2022,** *8,* 1–11, Article ID: 2250.
32. Monzavi, M. T. *Wastewater Treatment*. Tehran, Iran, 1991; pp 7–162.
33. Filion, Y.; Karney, B. W. In *A Numerical Exploration of Transient Decay Mechanisms in Water Distribution Systems*, Proceedings of the ASCE Environmental Water Resources Institute Conference, American Society of Civil Engineers, Roanoke, Virginia, 2002.
34. Shirzad, S.; Hassan, M. M.; Aguirre, M. A.; Mohammad, L. N.; Daly, W. H. Evaluation of Sunflower Oil as a Rejuvenator and Its Microencapsulation as a Healing Agent. *J. Mater. Civil Eng.* **2016,** *28* (11), 04016116.
35. Chen, M.; Leng, B.; Wu, S.; Sang, Y. Physical, Chemical, and Rheological Properties of Waste Edible Vegetable Oil Rejuvenated Asphalt Binders. *J. Constr. Build. Mater.* **2014,** *66* (15 Sep), 286–298.
36. Fini, E. H.; Kalberer, E. W.; Shahbazi, A.; Basti, M.; You, Z.; Ozer, H. Chemical Characterization of Bio-Binder from Swine Manure: Sustainable Modifier for Asphalt Binder. *J. Mater. Civil Eng.* **2011,** *23* (11), 1506–1513.
37. Hassan, M.; Essam, T.; Yassin, A. S.; Salama, V. Optimization of Rhamnolipid Production by Biodegrading Bacterial Isolates using Plackett–Burman Design. *Int. J. Biol. Macromol.* **2016,** *82* (4), 573–579.
38. https://en.wikipedia.org/wiki/Computational_physics
39. NZS 3106. *Design of Concrete Structures for the Storage of Liquids*, Cement and Concrete Association of New Zealand, Standards Association of New Zealand, 2009.
40. Recommended Standards for Water Works, Great Lakes-Upper Mississippi River Board of State and Provincial Public Health and Environmental Managers, Health Research Inc., Health Education Services Division, PO Box 7126, Albany, NY 12224, 102-SA Water TG, 2012.
41. Planning and Infrastructure. Above Ground Circular Concrete Tank, Manager Engineering, 2008. 103-Selected Earthquake Engineering Papers of George W. Housner, American Society of Civil Engineers, 1990.
42. Specification for Construction of Concrete Reservoirs. *Standard Specifications and Drawings Specification for Construction of Concrete Reservoirs*. Gold Coast: Australia, SS 11, 2004.
43. TM 5-809-10. *Seismic Design Guidelines for Essential Buildings*, US Department of the ARMY: Washington, D.C., 1986.
44. USEPA Finished Water Storage Facilities. U.S. Environmental Protection Agency, 2002.
45. Zecchin, A. C.; Maier, H. R.; Simpson, A. R.; Leonard, M.; Nixon, J. B. Parametric Study for an Ant Algorithm Applied to Water Distribution System Optimization. *IEEE Trans. Evol. Comput.* **2005,** *9* (2), 175–191.

46. Walski, T. M. Hydraulic Design of Water Distribution Storage Tanks. In *Water Distribution Systems Handbook*; Mays, L. W., Ed.; McGraw-Hill: New York, 2000; Chapter 10.
47. Farmani, R.; Walters, G.; Savic, D. Evolutionary Multi-Objective Optimization of the Design and Operation of Water Distribution Network: Total Cost vs. Reliability vs. Water Quality. *J. Hydroinform.* **2006,** *8* (3), 165–179.
48. Farmani, R.; Savic, D. A.; Walters, G. A. The Simultaneous Multi-Objective Optimization of Anytown Pipe Rehabilitation, Tank Sizing, Tank Siting and Pump Operation Schedules. In *Critical Transitions in Water and Environmental Resources Management*. ASCE, American Society of Civil Engineers: Salt Lake City, 2004; pp 4663–4672.
49. Haarhoff, J.; Van Zyl, J. E.; Nel, D.; Van der Walt, J. J. Water Supply and Water Quality in Rural Areas: Stochastic Analysis of Rural Water Supply Systems. *Water Supply* **2000,** *18* (1), 467–470.
50. Maier, H. R.; Simpson, A. R.; Zecchin, A. C.; Foong, W. K.; Phang, K. Y.; Seah, H. Y.; Tan, C. L. In *Ant Colony Optimization for the Design of Water Distribution Systems*, Proceedings of World Water and Environmental Resources Congress 2001, ASCE, The Rosen Plaza Hotel: Orlando, Florida, United States, 2001.
51. Maier, H. R.; Simpson, A. R.; Zecchin, A. C.; Foong, W. K.; Phang, K. Y.; Seah, H. Y.; Tan, C. L. Ant Colony Optimization for Design of Water Distribution Systems. *J. Water Resour. Plann. Manag.* **2003,** *129* (3), 200–209.
52. Walski, T. M.; Chase, D. V.; Savic, D. A.; Grayman, W.; Beckwith, S.; Koelle, E. *Advanced Water Distribution Modeling and Management*. Haestad Press: Watertown, Conn, 2003.
53. Hashemi, S. S. Optimization of Water Networks with Minimization of Pumping Energy. Master Thesis, University of Tehran: Iran, 2010.
54. Mau, R. E.; Boulos, P. F.; Clark, R. M.; Grayman, W. M.; Tekippe, R. J.; Trusell, R. R. Explicit Mathematical Models of Distribution Storage Water Quality. *ASCE J. Hydraul. Eng.* **1995,** *121* (10), 669–709.
55. Rossmann, L. A. EPANET 2 User's Manual. Water Supply and Water Resources Divison. U.S. 2000. Environmental Protection Agency. Cincinnati, EE.UU. http://www.epa.gov/ORD/NRMRL/wswrd/epanet.html
56. Rossmann, L. A.; Grayman, W. M. Scale-Model Studies of Mixing in Drinking Water Storage Tanks. *J. Environ. Eng.* **2000,** *125* (8), 755–761.
57. Yeung, H. Modeling of Service Reservoirs. *J. Hydroinform.* **2001,** *3* (3), 165–172.
58. Martínez-Solano, F. J.; Lopez-Jimenez. P. A.; Iglesias-Rey, P.; Perez, R. Utilization of Fluid Dynamics Co. 2006.
59. Mahmood, F.; Pimblett, J. G.; Grace, N. O.; Grayman, W. M. Evaluation of Water Mixing Characteristics in Distribution System Storage Tanks. *J. Am. Water Works Assoc.* **2005,** *97* (3), 74–88.
60. Marek, M.; Stoesser, T.; Roberts, P. J. W.; Weitbrecht, V.; Jirka, G. H. CFD Modeling of Turbulent Jet Mixing in a Water Storage Tank. XXXII IAHR Congress, Venice (Italy), July 1/6, 2007, 2007.
61. Van Zyl, J. E.; Haarhoff, J. Reliability Analysis of Municipal Storage Reservoirs Using Stochastic Analysis. *J. S. Afr. Inst. Civ. Eng.* **2007,** *49* (3), 27–32.

62. Arturo Leon, S. Improved Modeling of Unsteady Free Surface, Pressurized and Mixed Flows in Storm-Sewer Systems. *Submitted in Partial Fulfillment of the Requirements for the Degree of Doctor of Philosophy in Civil Engineering in the Graduate College of the University of Illinois at Urbana-Champaign*, 2007; pp 57–58.
63. Hariri Asli, K.; Nagiyev, F. B.; Haghi. A. K. Physical Modeling of Fluid Movement in Pipelines. In *Nanomaterials Yearbook*. USA, 2009. https://www.novapublishers.com/catalog/product_info.php?products_id=11587
64. Leon, A. S.; Ghidaoui, M. S.; Schmidt, A. R.; Garcia, M. H. Godunov-Type Solutions for Transient Flows in Sewers. *J. Hydraul. Eng.* **2006,** *132,* 800–813.
65. Feizbahr, M.; Mirhosseini, S. M.; Joshaghani, A. H. Improving the Performance of Conventional Concrete Using Multi-Walled Carbon Nanotubes Express. *Nano Lett.* **2020,** *1* (1), 1–9.
66. Asli, K. H.; Asli, K. H. Isolated Pressure Zones Based on GIS as a Solution for Water Network Problems. *Water Prac. Technol.* **2022,** *2022,* wpt2022119..
67. Asli, K. H.; Asli, K. H. Minimum Night Flow (MNF) and Corrosion Control in Compliance with Internet of Things (IoT) for Water Systems. *Water Prac. Technol.* **2023,** *18* (3), 608–625.
68. Nazeryan, M.; Feizbahr, M. Seismic Evaluation of the Cheng and Chen Modified Model Using Shear Keys in Steel Beam-to-Concrete Column Connections. *Transac. Civil Env. Eng.* **2022,** *8* (2), 2787, 1–6.
69. Asli, K. H.; Asli, K. H. Smart Water System and Internet of Things. *J. Mod. Ind. Manuf.* **2023,** *2,* 5.
70. Shukla, A.; Asli, K. H.; Rawat, N. K.; Abraham, A. R.; Haghi, A. K. *Technological Advancement in Clean Energy Production*. Apple Academic Press: Hard, 2024.
71. Pourhashemi, A.; Asli, K. H.; Abraham, A. R.; Haghi, A. K. Sustainable Water Engineering. In *Smart and Emerging Technologies*. Apple Academic Press: Hard, 2024.
72. Abraham, A. R.; Susanto, H.; Haghi, A. K.; Asli, K. H. *Sustainability in Energy and Environment Engineered Materials and Smart Computational Techniques*. Apple Academic Press: Hard, 2024.
73. Khanna, S.; Abraham, A. R.; Asli, K. H.; Haghi, A. K. Sustainable Environmental Engineering Water Security. In *Energy Conservation, and Green Processes*. Apple Academic Press: Hard, 2024.
74. Asli, K. H. Technological Advancement in HVAC&R System Energy Efficiency. In *Technological Advancement in Clean Energy Production*. Apple Academic Press: Hard, 2024.
75. Feizbahr, M.; Jamshidi, M.; Jayaprakash. Review on Various Types and Failures of Fibre Reinforcement Polymer. *Middle East J. Sci. Res. 13* (10), 1312–1318.
76. Kian Hariri, A. IoT-Based Smart Water Leak Detection System for a Sustainable Future: A Case Study. In *Sustainable Water Engineering; Smart and Emerging Technologies*. Apple Academic Press: Hard, 2024.
77. Feizbahr, M.; Bhattarai, S. R. Some Aspects of Research for Reclamation of Asphalt Based on Remote Sensing (RS). *CRPASE: Transac. Civil Environ. Eng.* **2022,** *8* (3), 2810, 1–6.
78. Kian Hariri, A. Measurement, Modeling, and Sustainable Control of Environmental Contaminants Using Smart Computational Technologies: A Case Study, Sustainability in Energy and Environment; Engineered Materials and Smart Computational

Techniques. Apple Academic Press: Hard, 2024. https://www.appleacademicpress.com/sustainability-in-energy-and-environment-engineered-materials-and-smart-computational-techniques/9781774916209#bios

79. Mohammad, N.; Brake, N. A.; Adesina, M.; Rahman, A.; Selvaratnam, T.; Snyder, H. A.; Günaydın-Sen, O. Reutilization of Oil and Gas Produced Water in Cement Composite Manufacturing. *J. Clean. Prod. 381,* Part 1,2022,135113. DOI: 10.1016/j.jclepro.2022.135113
80. Kian Hariri, A. Fuzzy Logic for Smart Water and Wastewater Systems. In *Sustainable Environmental Engineering Water Security, Energy Conservation, and Green Processes*. Apple Academic Press: Hard, 2024. https://www.appleacademicpress.com/sustainable-environmental-engineering-water-security-energy-conservation-and-green-processes/9781774916902
81. Oruji, S.; Brake, N. A.; Hosseini, S.; Adesina, M.; Nikookar, M. Enhancing Recycled Aggregate Concrete Using a Three-Stage Mixed Coal Bottom Ash Slurry Coating. *J. Mater. Civil Eng. 35* (5). DOI: 10.1061/(ASCE)MT.1943-5533.0004723

PART II
Pollution Control and Treatment

CHAPTER 4

Sustainable Approach for the Evacuation of Heavy Metal Pollutants by Bioremediation

SUBBULAKSHMI GANESAN, PRIYANSHI ROSHAN, and RITIKA MATHUR

Department of Chemistry, Jain University, Bangalore, Karnataka, India

ABSTRACT

Heavy metals pose a considerable threat to the environment and ecosystems due to their hazardous and non-biodegradable identity. The plants and animals are adversely affected as they can cause growth inhibition, accumulation of molecules in leaves and roots, and multiple organ damage even at low concentrations. The effluents from the industries are an alarming cause for the heavy metals to be introduced into water bodies. Current research interests involve the removal of heavy metals using bioremediation. It emphasizes removing the environmental pollutants through biological systems. The initial categories involved under bioremediation are in situ and ex situ bioremediation. We have primarily focused on algae-based remediation, mycoremediation, and phytoremediation. This report also includes bioremediation through symbiotic interactions of algae with microbes (e.g., *Micrococcus* sp.) and fungi (lichens, e.g., *Diploschistes muscorum*). The operational activities of these biological systems are broadly stratified, such as biosorption, bioaccumulation, Rhizofiltration, phytotoming, extracellular precipitation, and other processes. Microalgae also find their application in composite material for making nanoparticles employed in heavy metal elimination.

Sustainable Water Treatment and Ecosystem Protection Strategies. Hossein Hariri Asli, Tamara Tatrishvili, Ann Rose Abraham, & A. K. Haghi (Eds.)

Owing to the treatment of water, the algae-laden water passes under high filtration parameters. The filtration is done through a high-gradient magnetic separation, ultrafiltration, membranes, etc. These tools are productive as the performance is more than 95%. The evolving technology tested in the bioremediation of arsenic in the groundwater of Gangetic plains in India and the hyperaccumulation of lead and arsenic present in drinking water in Mexico. Here, we estimated the BOD and removal percentage of elements done between the conventional method and bioremediation of treating wastewater. It was observed that BOD was successively very low when the water is remediated through the biological system. This shows that it is a very effective sustainable technology. Despite the advantages of this cost-effective system, some challenges remain. These need to be monitored in the prospects for full-scale use.

4.1 INTRODUCTION

"Heavy metals" is an expression used for metals having a density higher than 5 g/cm^3. They possess high atomic weight and specific gravity. Some of them are required in trace amounts by organisms as micronutrients but become toxic if consumed in excess.[1] From the view of environmental pollution, heavy metals are eminently hazardous contaminants to biological systems. Their concentration in natural habitats and ecosystems has increased over time. It is due to the ever-rising anthropogenic activities. Wastewater from oil refineries, smelting industry, coal-fired power stations, petrochemical plants, and other such chemical industries contain high concentrations of heavy metals. These industrial domains discharge lead, arsenic zinc, copper mercury, nickel, etc. The heavy metal elements have the characteristic of bioaccumulation. When they get concentrated beyond the permissible limit, it may lead to neurotoxicity, carcinogenesis, loss of cellular function, and damage to DNA and its repair mechanisms. They also harm cell membrane lipids and cause aggregation and misfolding of proteins. The bioavailability of heavy metals in the food chain causes biomagnifications. It means that as the heavy metal passes from one organism to the other via predation or consumption of dead biomass, its concentration keeps on increasing. Various conventional treatment methods like membrane filtration, electrochemical treatment, chemical precipitation, etc., have been used. But they consume high energy, have less removal efficiency, and are not cost-effective.[2] Therefore, environmental remediation is shifting to bioremediation methods as they are nature-friendly, inexpensive, and show better removal efficiency. Biosorbents such as algae and fungi are used extensively in treating wastewater. They also form associations with suitable microorganisms for enhanced remediation ability. Among phytoremediation methods, mostly phytofiltration is involved in purifying heavy metal contaminated wastewater. Other methods, such as phytovolatilization, phytostabilization, and phytoextraction can be used to detoxify both soil and water. The chief detoxification process that holds paramount importance is biosorption. It is employed by both fungi and algae. Bioaccumulation and biotransformation follow afterward (in algae). Fungi have several other mechanisms, namely, mycoprecipitation, bioreduction, compartmentalization, etc. These biological detoxifying agents have high heavy metal tolerance due to several defense mechanisms that keep them shielded. After the completion of treatment, the biological agent is separated from purified water.

4.2 SOURCES AND TOXICITY OF HEAVY METALS

The emissions of heavy metals, such as arsenic, cadmium, chromium, copper, mercury, lead, nickel, and zinc get discharged into the water bodies from an unending list of sources. These sources level up from daily wastes to industrial wastes. The heavy metals from daily wastes include ashes, plastics, household effluents, vehicle exhaustion, medical wastes, debris, and electronic wastes called e-waste, and other such examples. Industries from which toxic metals are released include mining and smelting, agricultural industries, textile industries, plastics and paint industries, extraction of ores, and fuels from fossil fuel industries, automobile industries, electronic industries, pharmaceutical industries, refineries, and it continues. Although after the treatment of water, some heavy metals remain in the water. With the concern related to public health, the permissible amount of heavy metals in water has been determined. It is represented in the following graph. Any values of the toxic metals varying from the limiting values are considered unfit for human usage and the survival of the aquatic ecosystem. If the tolerance of these heavy metals is above the threshold limits, the toxicity can lead to adverse effects.

4.3 TOXICITY IN HUMANS

Toxicity in humans can have definite acute and chronic effects. Exposure to arsenic in humans can cause diarrhea, diabetes, hyperkeratosis, and cancer of the skin, lungs, and bladder. Lead can cause copper-deficient anemia and osteoporosis. Cadmium and chromium generally affect the lungs and cause pulmonary fibrosis and cancer. Mercury affects voluntary muscle functions, neuron systems, and hypersensitivity. Likewise, the acute effects remain the same but in the long term, genes are also affected and later on, the exposure can cause genetic disorders in the coming generations.

4.3.1 TOXICITY IN AQUATIC LIFE

The dissolved mercury is methylated in water to form methylmercury. It is considered to be a fatal entity in aquatic life. It decreased the hatching rate of the fish eggs. Lead has affected the respiratory system of the fishes and inhibited the growth of algae in water bodies. Salmon is found to be the most sensitive to lead. Copper and mercury have affected the pawning conditions. There is a reduction in the growth of the fishes and the availability of food nutrients. The metals can also accumulate in the bodies of fishes and later can lead to biological magnification. Besides fishes, the population of other organisms in the water bodies has substantially decreased.

4.3.2 TOXICITY IN PLANTS AND ANIMALS

Water absorbed by the plants through their roots affects the conduction system, xylem, and phloem of the plant. Due to chromium, photosynthesis does not function properly in terms of CO_2 fixation, electron transport, and photophosphorylation.[3] It can also induce changes in enzymatic activity, metabolic modification, and plant pigmentation. Lead accumulation can increase oxidative stress. Cobalt is seen to decrease the transpiration in plants causing the closing of stomatal pores and inflammation of the stems. All these toxic effects will lead to the death of the plants. Similarly in animals, the consumption of heavy metals affects their gastrointestinal system, respiratory, and nervous systems. Tumors and inflammation of the cells are common among cows and dogs due to arsenic and lead. The population of the animals has been reduced. The quality of milk products acquired from the cows has decreased and the accumulation has caused biological magnification in humans. The following graph displays the permissible limits for significant heavy metals in water.

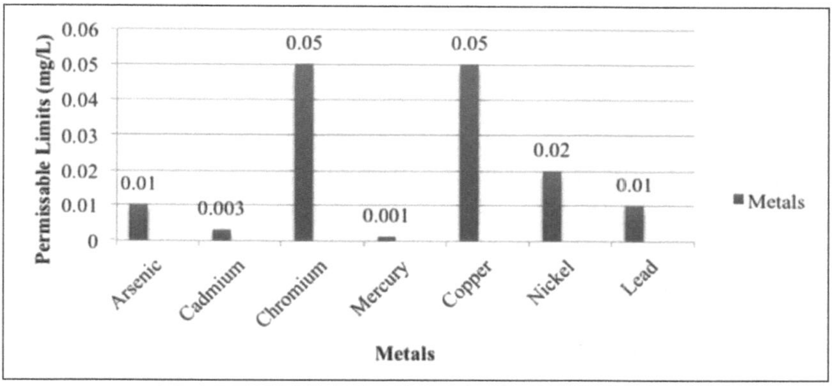

FIGURE 4.1 Permissible limits for heavy metals.

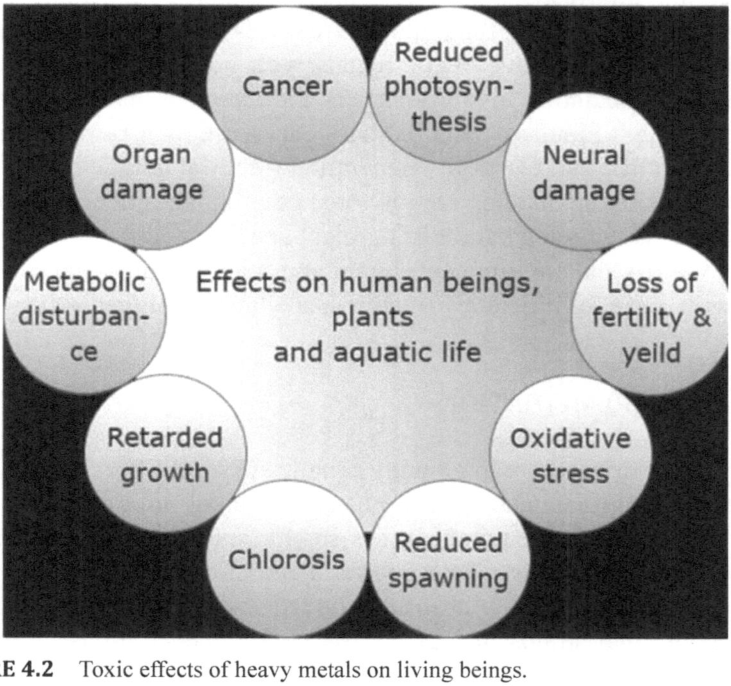

FIGURE 4.2 Toxic effects of heavy metals on living beings.

4.4 BIOREMEDIATION

Bioremediation is a broad term used for treatment techniques to protect the environment from damage. They essentially use only naturally occurring organisms for waste management. These biological entities convert hazardous substances into nontoxic or less toxic materials. Bioremediation can be in situ or ex situ depending on the site of treatment.[4] Compared with conventional methods, bioremediation has cost-effective and environmentally friendly nature. There is not any single method that can be used for treating all kinds of solid waste and effluents. We decide the suitable bioremediation method according to various parameters, such as the extent of cleanup needed, heavy metal concentration, and effluent diversity. Different species of organisms, such as microbes, algae, plants, and fungi find application in wastewater bioremediation. This article mainly focuses on algal, plant, and fungal bioremediation. Microorganisms utilized for treating the contaminated site can be indigenous or exogenous (based on origin). Indigenous (autochthonous) microbes are naturally present at the pollution site. Exogenous (allochthonous) microorganisms are those injected into the waste management site for treatment. When algal diversity is engaged to purify the polluted area, it is called phytoremediation. This method can remove heavy metals with no secondary contamination. Phytoremediation involves tertiary biotreatment that also results in beneficial algal biomass. This biomass can serve as a raw material for biofuel, biochar, and electricity generation (renewable bioenergy). Other patterns of bioremediations are phytoremediation and mycoremediation. The former is the remediation using plant species, and the latter is using fungi to purify the environment. Fungi and microbes also make associations with algae for efficiently treating heavy metal polluted water.

4.5 PHYCOREMEDIATION

Algae characterize as predominantly aquatic eukaryotic organisms that are autotrophic. Nitrogen and phosphorus are the chief limiting nutrients for algal growth as their requirement are 10–40 times more than normal for proper protein synthesis. The scope for algae in bioremediation stretches vastly due to the ability to absorb noxious substances and the long-term survival of algal strains in high heavy metal concentrations.[5] They can also use several wastewater pollutants like organic chemicals, heavy metals, and pesticides as nutrient sources. Algae act as an ally against heavy metals using two processes–absorbing heavy metals through their cell surface (physical adsorption) and degrading the metals into nutrients using enzymes (chemisorption). Algae have two informal divisions drawn on cellularity–microalgae (unicellular species) and microalgae (multicellular species). Both forms of the algae find their way into heavy metal bioremediation.

4.6 BIOREMEDIATION BY MACROALGAE

Macroalgae commonly refer to the macroscopic marine algae, also called seaweeds. They are found in the littoral zone and are known to have around 20% higher carbohydrate content than the terrestrial flora. Seaweed biomass (viable and non-viable) is highly considerable for detoxification as its macroscopic size provides a larger adsorption area. It enhances their biosorption ability. It is also low-cost biomass or biosorbent that has high availability. If the construction of the biosorption systems is according to the influencing experimental factors, they can show high bioremediation potential against heavy metals. Additionally, macroalgae are classified according to the color of the pigment they contain, namely, green, red, and brown algae. The presence of different pigment mixtures in their chloroplast provides them with their exclusive color. Different biomass compositions essentially mean various cell wall structures. The cell wall is the main structural layer taking part in biosorption, hence, its various forms influence biosorption ability. The biomass has several organic molecules, such as carbohydrates, proteins, and lipids that provide binding sites for the biosorption process.

4.7 BIOREMEDIATION BY MICROALGAE

Microalgae are microscopic and photosynthetic algae invisible to unaided eyes. They are also called microphytes and consist of phytoplankton, diatoms, dinoflagellates, etc. Microalgae can survive independently as a unicellular entity or dwell as colonies. They are the main species that are involved in phycoremediation. This biological material has been studied for green application in the past four decades. They are predominantly abundant and are found in both freshwater and saltwater. Microalgae's high retention capacity for a vast range of toxic heavy metals makes them a suitable fit for further research in remediation strategies.[6,7] Dead microalgae biomass can also be used for remediation and is preferred over viable biomass. Though living microalgae will apply passive adsorption as well as enzymatic degradation, their extent of action is somehow limited by various environmental factors. Dead algal biomass has almost equal or sometimes more biosorption ability when compared with viable microalgae. Non-living microalgae have several other benefits, such as easy storage under standard room conditions, unaffected by heavy metal toxicity, and retaining biosorption potential for a long duration. The high protein content of microalgae may also be responsible for their strong binding with heavy metals. Transgenic strategies for microalgae have also been adopted to improve the degradation of heavy metals.

4.8 PHYTOREMEDIATION

Phytoremediation refers to the techniques where plants and associated microbes help to refine the biosphere contaminated by either natural or anthropogenic activities. It divides into particular subcategories, such as phytovolatilization, phytostabilization, phytoextraction, and rhizofiltration. All the methods are focused on detoxifying soil and water. Phytovolatilization is a process where there is an uptake of water pollutants by the roots, and then they get transpired as volatiles (gaseous state). This process is practiced to eliminate both organic compounds and heavy metals. Phytostabilization is the remediation of heavy metals where plants immobilize them in the soil. It decreases their bioavailability that terminates any possibilities of toxifying water or entering any food chain. Phytoextraction is serviceable in detoxifying soil and water. Plants assimilate the toxins through their roots and transfer them to the shoot region. The toxins are usually heavy or toxic metals, other inorganics, etc.[8] The method of rhizofiltration exposes plant roots to contaminated water for adsorption and absorption of pollutants. It is originally for groundwater but also helps in decontaminating wastewater. Root exudation helps in absorbing heavy metals. Different plant species use distinct mechanisms, and the selection of the suitable plant depends on various parameters, such as heavy metal tolerance, easy cultivation, uninviting for herbivores, high absorption ability, etc. In wastewater bioremediation, we use aquatic plants, such as water hyacinth (*Eichhornia crassipes L.*) and ornamental plants like cattails (*Typhadomingensis*). Employment of enhancement techniques using natural and synthetic agents results in better environmental reparation.

4.9 MYCOREMEDIATION

The multicellular eukaryotic natural decomposer commonly referred to as fungi are involved in the bioremediation of wastewater. Fungi grow fast in their suitable conditions and can withstand unfavorable conditions by forming spores. The cell wall of fungi plays an important role in wastewater treatment. The rigid cell wall comprises chitin (a polysaccharide known as N-acetyl glucosamine), lipids, phosphates, glycans (such as glucans, galactans, and mannans), and proteins. Among these, polysaccharide residues comprise the bulk of the composition. It behaves as a metal-binding ligand ($-NH_2$, -OH, -COOH, SO_4^{2-}, and $RCOO^-$). The functional group amine is an active ligand that uptakes the heavy metals via interaction and eliminates them. Fungi can perform in both dry or dead and live conditions. The interaction of fungi as a living mass follows bioaccumulation involving membrane transport and intercellular accumulation as the path for bioremediation, whereas the dead and live biomass follow biosorption which involves physical adsorption, precipitation, ion exchange, and metal complexation.[9] Live cells are metabolic-dependent and hence are inefficient when compared with dead cells as it is metabolic independent. In parallel to algal cells, fungi have greater adaptability, serve a larger surface area, and withstand high flow pressure and agitation. Other fungi–metal interaction involves mobilization, uptake, and transformations as well. Here, the combination of processes happens simultaneously, the bioaccumulation and biosorption vary with cell age, and the maximum uptake is at the lag phase.

The following table is an account of different types of microalgae, fungi, and plants used in bioremediation and their target heavy metals.

4.10 OPERATION

4.10.1 BY PHYCOREMEDIATION

The removal of heavy metals by microalgae involves a complex physicochemical process. It generally takes place using three mechanisms–biosorption, bioaccumulation, and biodegradation.

TABLE 4.1 Bioremediation Species (Microalgae, Fungi, and Plants) and Their Targeted Heavy Metal.

Microalgae	
Chlorella sp. (*Chlorella vulgaris, Chlorella sorokiniana*)	Au, Ni, Pb, Hg, Cd, Cr, Zn, Cu
Cladophora sp.	Zn, Cu
Spirogyra sp.	Cd, As, Hg, Pb, Co, Fe, Cr
Oscillatoria sp.	Cd, Co, Pb, Ni, Zn
Fucus sp.	Cd, Cu, Ni, Pb, Zn
Ulothrix sp.	As, Cu
Fungi	
Aspergillus sp.	Cr, Cu, Ni, Pb, Zn
Candida parapsilosis	Hg
Paecilomyces sp.	Cr, Cd
Pycnoporous sp.	Pb, Cu, Cd
Penicillium sp.	As, Hg, Pb, Cd
Rhizopus sp.	As, Ni, Zn, Cd, Pb
Plants	
Pteris vittata	Zn, As
Miscanthus	Pb, Cd, Cu, Zn
Sedum sp.	Pb, Cd, Zn
Thalaspi sp.	Cd, Ni, Zn
Microcystis sp. (Algal Bloom)	As, Pb, Cu, Cr, Zn

4.10.2 BIOSORPTION

Biosorption is a passive, reversible, and rapid step in algal bioremediation that involves adsorption. Both living and dead algal biomass can perform biosorption. It happens promptly when microalgae or macroalgae biomass encounters heavy metals. It generally lasts for 5–10 min in the case of microalgae. Here, the algal cells adsorb the heavy metals through their surface. The algal cell wall is composed of several constituents, such as cellulose, hemicellulose, starch, and glycogen. These molecules possess functional groups, such as carboxyl, thiol, amino, phosphoryl, and many others.[10] The carboxyl group is regarded as one of the most important components for adsorbing heavy metals. It has two distinct methods (type

I and type II) for binding with heavy metal ions. Metal binding can be either covalent or ionic.

For macroalgae, the biosorption phase is further defined into two stages. The first 5–30 min of biosorption is regarded as stage I. During this stage, initially, all the functional groups are geometrically free for binding with the metal ions, hence it is rapid. The removal percentage reaches almost 90% due to ease in conducting electrostatic interactions and ion exchange processes. Stage II is the later duration of biosorption that may take up to 3–4 h. It is a slow stage as most of the functional groups get occupied with the heavy metal ions. The rest of the ions require binding to those functional groups that lie within the pores of the macroalgae biomass. The amount of heavy metal deposited on the biomass surface does not vary much (around 5–10%). The pathway of biosorption may vary depending on the nature of the heavy metals, the complexion of the functional groups on the algal cell surface, and the properties of the aqueous solution (pH, ionic strength, etc). Different methods of biosorption are as follows:

4.10.3 ION EXCHANGE METHOD

It is the predominant method of biosorption that involves a simple exchange of ions between the algae and its environment. It consists of both physical and chemical interactions in the form of Van der Waals or electrostatic force as well as ionic or covalent bonding. Algae contain several metal ions in their anatomy (Ca^{+2}, Na^{+}, Mg^{+2}, K^{+}). They are mobile and associated with the acid functional groups of the algal cell wall. During biosorption, the heavy metal ions [like Cd (II), Zn (II), Pb (II), Cu (II)] replace them and bind with the acid functional group. It can be represented as follows:

$$R\text{-}X^{+} + M^{+} \leftrightarrows R\text{-}M^{+} + X^{+}$$

Here, R– = acid functional group, X^{+} = mobile metal ion, M^{+} = heavy metal ion

4.10.4 COMPLEXATION

It involves the formation of superficial complexes on the algal surface. The complex consists of algal functional groups and heavy metal ions.

Complexation may occur as a single-step reaction initially when there is a high heavy metal ion concentration. Interaction between the reactants can be ionic, covalent, and/or coordinative. The resulting complex has greater stability in comparison with those formed during ion exchange. This property may be a hindrance in the regeneration process for which strong complexing agents like EDTA become indispensable.

4.10.5 MICROPRECIPITATION

This operation mode of biosorption takes place in solutions where the concentration of heavy metals has reached the saturation point. Another situation in which microprecipitation occurs is the sudden rise of solution pH during biosorption. In either case, the heavy metals get precipitated from the aqueous solution and their deposition on the biomass surface follows. It may or may not determine the nature of algae used. It tends to disrupt the actual results of biosorption as we cannot determine the original heavy metal ion uptake by the algae.

4.10.6 BIOACCUMULATION

Bioaccumulation is the process of active contaminant uptake by living cells and involves metabolism. It happens as the concentration of the adsorbate exceeds in the algal tissue compared with their concentration in the environment. Organic and inorganic impurities, such as heavy metals, phosphates, and nitrates are withdrawn from the aqueous solution and get transported into viable cells that constitute the algal biomass.[11,12] Live microalgae biomass helps in reducing heavy metal toxicity by the production of metal-detoxifying chelators. They include metallothionein (MT) and phytochelatin (PC). The efficiency of bioaccumulation can be understood by studying the organism's bioconcentration factor (BCF) specific to the target pollutant. BCF is a peculiar case of BAF or bioaccumulation factor. BAF is the ratio of the concentration of the pollutant in the living organism to its concentration in the environment. BAF takes into account all the potential uptake mechanisms. BCF is identical to BAF, the only difference is that BCF considers the uptake of pollutants only from the surrounding medium.

4.10.7 BIODEGRADATION OR BIOTRANSFORMATION

It is a significant method of detoxification wherein algae take up the pollutants and break them down into smaller molecules. These intermediate molecules are then utilized by the algal biomass as a nutrient source. Algae also promote biodegradation by enhancing the degradation capability of the surrounding microbial population. It is also involved in biotransforming the pollutant into significant metabolites. Algae achieve this through bioaccumulation of the target pollutant followed by degradation through transformation and mineralization. The following graph shows the uptake capacity for three distinct microalgae species out of which, *Fucus* sp. demonstrated better efficiency for all heavy metals taken into account.

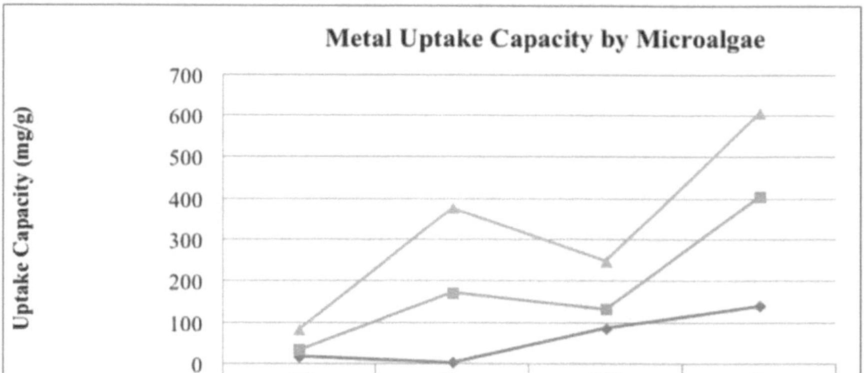

	Ni	Pb	Cd	Cr
Fucus sp.	50	204.1	114.9	200
Spirogyra sp.	17.5	170	47	265
Chlorella sp.	15.6	2	85.3	140

FIGURE 4.3 Metal uptake capacity for various microalgae.

4.10.8 BY PHYTOREMEDIATION

There are several mechanisms for the sequestration of toxins under phytoremediation. Plants use one or a combination of the methods based on the contaminant type, form, and media. The mechanisms are as follows:

4.10.9 DISSIPATION (PHYTOVOLATILIZATION)

This method is employed to eliminate highly volatile contaminants through vegetation and release them into the atmosphere. It is a peculiar type of phytotransformation. Toxins, such as mercury, selenium, and arsenic are removed from soil and water using different biophysical mechanisms. There is a mass transport of the pollutants using the root system. Then, they get converted into a lesser toxic and volatile form. The transpiration from roots, shoots, and leaves follows. The phytovolatilization of several organic and inorganic forms of selenium to the volatile form of methyl selenate has been highly observed in the Indian mustard (*Brassica juncea*).

4.10.10 PHYTOSTABILIZATION

It is also called Phytoimmobilization or in place inactivation of the pollutants. It is more focused on soil remediation and removes toxins from within the rhizosphere. Through this mechanism, the harmful impurities are either immobilized or prevented from migrating. It is majorly used for heavy metals, such as lead, zinc, cadmium, chromium, copper, and arsenic.[13] Plants achieve phytostabilization of contaminants by adsorbing and accumulating them through roots. They can also carry out their precipitation within the root zone. Physically stabilizing the soil also reduces the bioavailability of the contaminants.

4.10.11 PHYTOEXTRACTION

Also known as phytoabsorption, this technique is the most applied cost-effective phytoremediation strategy for the removal of heavy metals. It occurs in the shallow region of the root zone. The plants absorb and store the toxic metals through translocation in their root and shoot tissues. It is not effective for some metals like lead but can be used for the elimination of arsenic, nickel, and cadmium. Exudates from the plant root and rhizosphere microorganisms also affect the efficiency of phytoextraction. The plants involved require routine harvesting and maintenance.

4.10.12 RHIZOFILTRATION

It is a form of Phytofiltration which is mostly used when a large volume of water is contaminated with a low concentration of toxins, for example, groundwater. It can operate in both in situ and ex situ mode. Hyperaccumulator plant species are used that retain the heavy metals in their roots and have a rapidly growing root system.[14] Distinct physical and biochemical processes take part in detoxification through rhizofiltration. Examples of major processes are bioaccumulation, adsorption, precipitation, and rhizodegradation. Other lesser significant mechanisms happening simultaneously are intracellular uptake and deposition in vacuoles (biological processes).

The following graph is a representation of the uptake capacity for three different fungal species.

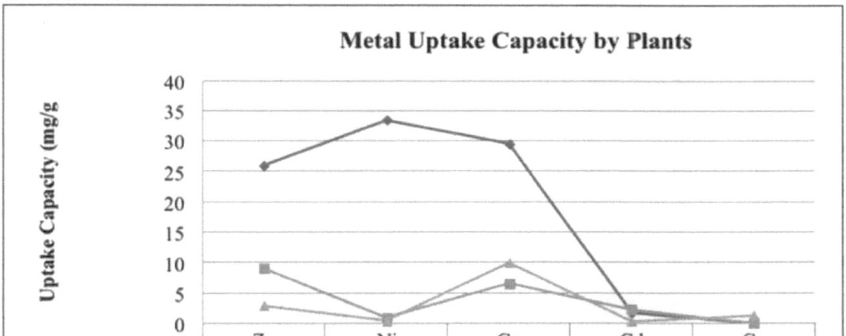

	Zn	Ni	Co	Cd	Cu
Allysum sp	26	33.45	29.5	1.68	0.03
Silene sp.	9	0.885	6.49	2.24	0.019
Cyperus sp.	2.88	0.45	10	0.34	1.31

FIGURE 4.4 Metal uptake capacity for various plant species.

Here, *Alyssum* sp. showed maximum removal efficiency for most of the heavy metals taken into consideration.

4.10.13 BY MYCOREMEDIATION

The mechanism of action varies for viable fungi and dead fungal biomass. While characteristics like heavy metal tolerance are required only for living biomass, the adsorption ability of the fungal species for the target pollutant is significant for both forms.[15] The main mechanisms involved in mycoremediation are as follows:

4.10.14 METAL BIOSORPTION

The fungal cell wall is distinct from other organisms and consists of chitin, polyphosphates, carbohydrates, lipids, glucans, chitosan, and glycoproteins. It is the cause of the surface phenomenon of biosorption. The polysaccharides present on the cellular surface comprise several phosphodiester bridges.[16] They are responsible for the abundance of negative charges on the fungal cell surface. It aids the cell in adsorbing the positive metal ions using the electrostatic force of attraction. Many proteins in the cell wall are associated with each other by disulfide bonding. They make metal–thiol complexes with heavy metals, another form of metal sorption. The presence of several functional groups, such as hydroxyl, thiol, carboxyl, amino, and methoxy on the cell wall is the chief means of heavy metal adsorption.

4.10.15 METAL TRANSPORTATION

When fungi and other organisms come in contact with metals in their environment, the metals get bound to their surface. Then, the bioaccumulation process takes place either through an active or passive mechanism. The passive operation is the accretion of the metal through processes that do not require metabolism, such as ion exchange and osmotic mechanism. The dynamic operations require metabolic processes involving several agents for the process to occur. They are called transporters and are assigned to transport the metal from the cytoplasm to either the periplasm or vacuoles of the fungal cell. They can be present as transmembrane proteins and derive the energy for the transportation cycle through ATP hydrolysis. Some examples are ATP-binding cassette (ABC) transporter, copper transporter (CTR).

4.10.16 METAL COMPARTMENTALIZATION

It is an intracellular detoxification process that plays a significant role during metal accumulation in fungal biomass and differs with various species. Compartmentalizing a metal is subject to factors, such as its ionic radius and electronegative character. Vacuoles of fungi contain organoligands. They reduce the heavy metal toxicity by implementing complexation. Some examples of organoligands found in fungi are peptides, organic alkali rich in sulfur, organic acids, etc.

4.10.17 BIOREDUCTION

Through bioreduction, some fungal species are capable to reduce the oxidation state of heavy metals. This leads to a decrease in their toxicity as in the case of chromium which exists in different oxidation states in the environment. Cr (VI) is more toxic than Cr (III). *Aspergillus tubingensis* is involved in producing glucose oxidase, a glycoprotein belonging to the class oxidoreductase. It can convert Cr (VI) to Cr (III) by utilizing glucose as this process requires a high amount of energy. Existence of organic acids in the environment like citrate favors this transformation reaction. This indicates that chromium bioreduction gets enhanced at a lower pH.

4.10.18 MYCOPRECIPITATION

It is a form of bioprecipitation where the positively charged heavy metal ion reacts with an anion. The source of anion may be the fungi or the surrounding medium. After reacting with the anionic species, it gets precipitated as the resulting mineral of the particular anion. The anions may be phosphates, chlorides, carbonates, hydroxides, etc. Mycoprecipitation can be of two types: intracellular and extracellular. In intracellular precipitation, the heavy metal ion enters the cell through the fungal cell wall and then reacts with the anions. Then precipitation follows. The process is called extracellular precipitation when the heavy metal ion reacts with the anionic species outside the cell. The anion may be donated by the fungus, or the fungi may provide the surface to the metal ion for binding with the anion provided by the surrounding medium.

The following graph depicts metal uptake capacities for three fungal species and *Penicillium* sp. proves to be effective against most of the heavy metals.

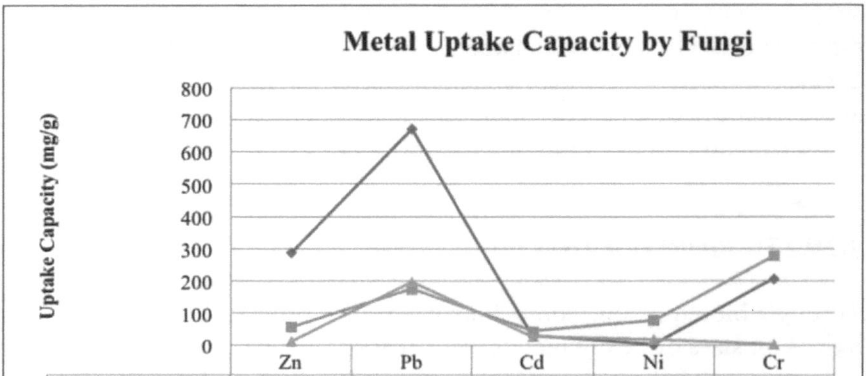

	Zn	Pb	Cd	Ni	Cr
Aspergillus sp.	287.8	670	27.3	0.53	208
Penicillum sp.	57	176.3	43	77.96	279.9
Rhizopus sp.	13.5	200	26.8	18.7	2.55

FIGURE 4.5 Metal uptake capacity for various fungal species.

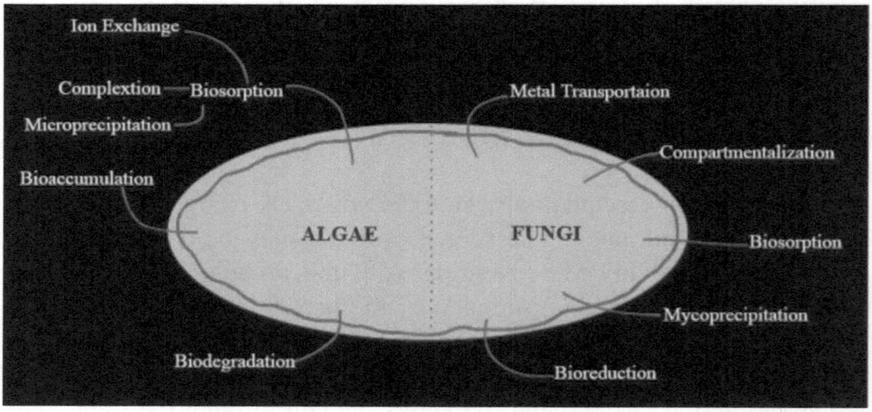

FIGURE 4.6 Mechanism of action for bioremediation by algae and fungi.

4.10.19 ASSOCIATIONS

Apart from the individual aspects of bioremediation, these also work in association to perform the same function. These associations were found to be interacting in four consortiums. This relationship between the organisms had mutualism, commensalism, and parasitism. Such association was handy and efficient in bioremediation as metabolites and nutrients were passed onto the other organism. These associations include:

4.10.20 LICHEN

The relationship found between algae and fungi. Fungi protect the algal cells, gather nutrients and provide moisture to grow. On the other hand, algae utilize the nutrients to produce food and supply some to the fungi. The process of bioremediation in a lichen is dynamic because the consortium is sensitive to heavy metals. Heavy metals interact with lichens at plasma lemma. The photobiont part is majorly involved in the process and the mycobiont part is less likely to be affected in its operation.

4.10.21 MYCORRHIZAE

It is the consortium between the plant's root and fungi. Here the mutualistic function remains the same as seen in lichen. Based on their attachment, it can be ecto-based or endo-based. In ectomycorrhiza, the fungal cells surround the roots and in endomycorrhiza, the fungal cells invade the roots. This association is an enhanced form of phytoremediation as the support of fungi enhances its activity.

4.10.21.1 ASSOCIATION BETWEEN PLANTS AND BACTERIA

In the relationship between plants and bacteria, bacteria gather nutrition for the plants that are not in reach of the plant due to their non-locomotory property. When microbes form a symbiotic relationship, it produces a metal-binding protein which boosts the accumulation of toxic metals. For their cooperation, plants communicate via signaling molecules and secondary metabolites. Due to the hyperaccumulation of metals in plants microbes play a crucial role to strengthen the defense mechanism of the plant.

4.10.21.2 ASSOCIATION BETWEEN ALGAE AND BACTERIA

The relationship between these two varies drastically from mutualism, commensalism, and parasitism. Generally, bacteria use up the oxygen produced by the algae, and these two exchange nutrients for their function. Compared with single-cell bioremediation, these two have shown advanced uptake of heavy metals without affecting their morphological conditions. It helped to achieve a high amount of harvested biomass.

TABLE 4.2 Uptake of Suitable Heavy Metal by Mycorrhizal Association.

Mycorrhizal Association		
Plant partner	**Mycorrhizal partner**	**Heavy metal**
Triticum aestivum	*Glomus etunicatum*	As
Pteris vittata L	*Glomus mosseae*	As
Phragmites australis	*Rhizopus irregularis*	Cd
Trigonella foenumgracum	*Gigaspora intraradices*	Zn
Oenothera picensis	*Claroideoglomus claroideum*	Cu
Lichen		
Parmelina tiliaceae		Pb, Cr
Plant and microbe association		
Plant partner	Microbe partner	
Brassica napus	*Pseudomonas fluorescens*	Pb
Cicer arietinum	*Pseudomonas* sp.	Ni
Eichhornia crassipes	*Pseudomonas diminuta*	Cr
Helianthus annuus	*Pseudomonas fluorescens*	As

4.10.22 ANALYSIS OF BOD

When the water is polluted by heavy metals, it interferes with the aerobic microorganism in water to break down the organic compounds. The microorganism utilizes these compounds as their source of food. They use the dissolved oxygen present in the water to break down their compounds so the depletion of the dissolved oxygen in water increases their total biological oxygen demand. Interference of the regular system in the water by heavy metals increases the biological oxygen demand and decreases the dissolved oxygen. The analysis of Chemical oxygen demand goes parallel with the biological oxygen demand. They define the quality of water present.

With the concern of managing the wastewater, conventional methods were devised. The disposal of these chemicals after treatment also posed some problems related to pollution. Now new bioremediation methods are being tested to replace the old conventional method that uses chemicals to treat wastewater. The comparison of the BOD using conventional and bioremediation method was studied, and it was observed that the BOD of water purified by bioremediation was low. The removal efficiency was also at a high rate. The comparison graph along with the removal efficiency is given. The removal efficiency of BOD and COD can be calculated by

$$\text{Removal efficiency } \% = \frac{\text{influent} - \text{effluent}}{\text{influent}} \text{ X } 100$$

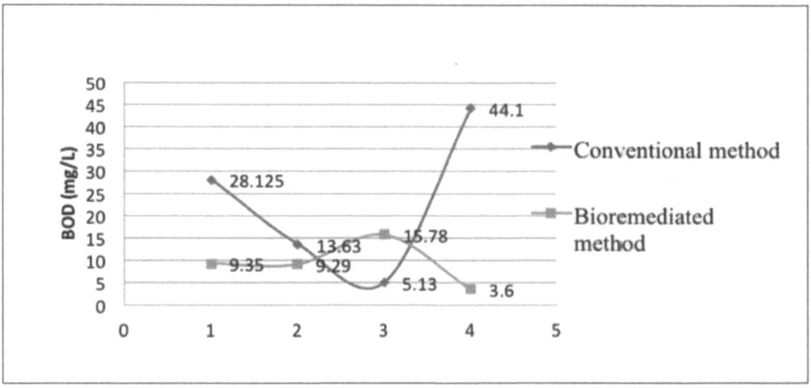

FIGURE 4.7 Comparison of BOD values obtained via bioremediation and conventional methods.

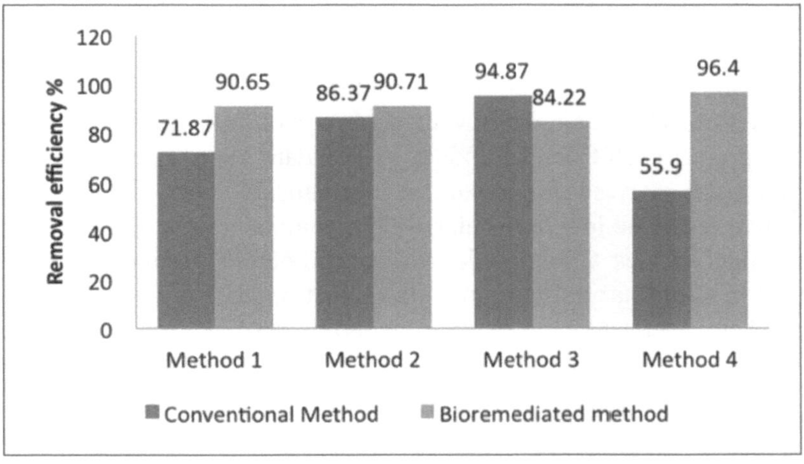

FIGURE 4.8 Comparison of removal efficiencies of conventional and bioremediation methods.

Here, the influent (initial concentration) values are all standardized to 100 mg/L before calculating the BOD and removal efficiency. We conclude that bioremediation is more efficient than other conventional methods. Another factor of bioremediation follows the toxicity index that represents the limit to which the algal cells, fungal cells, and plants can intake the heavy metal above which the toxic effects can be observed. They carry out their series of defenses to prevent themselves from the toxicity effect and function constantly with their regular functions. The equation for tolerance index can be calculated by the given equation.

$$\text{Tolerance Index} = \frac{\text{Growth in the presence of metal}}{\text{Growth without metal exposure within same time limit}}$$

4.10.23 RESPONSE UNDER STRESS

4.10.23.1 ALGAE

The ability of algae as a bioremediation agent for heavy metals relies on its degree of tolerance to heavy metals. Though low metal concentrations can promote the growth of algae (hormesis phenomenon), heavy metals pose toxicity to the algal species as well. This includes the generation of oxidative stress and a decrease in the cellular antioxidant level. Oxidative stress in an organism is the increase in the level of reactive oxygen species (ROS) and its accumulation at the cellular level that promotes oxidative damage. ROS are oxygen radicals or unstable molecules that contain oxygen and can cause damage to other molecular structures in a cell. Some examples of ROS are singlet oxygen, hydroxyl radical, superoxide anions, etc. To protect itself from the toxicity and oxidative stress caused by heavy metals, algae exercise certain defense mechanisms.

There are some low molecular weight composites naturally occurring in the algal species. Carotenoids are a natural pigment found in algae that protect the light-harvesting pigments against photochemical damage and ROS. They target O_2 and RO_2•. Flavonoids restrain lipid peroxidation and aim toward eliminating OH• and HOCl. Other important compounds include ascorbate and α-tocopherol. Ascorbate acts as an electron donor for OH• and protects against O2• and HO2•. Vitamin E or α-tocopherol targets peroxy radicals. Several high molecular weight compounds (antioxidant enzymes) also contribute to safeguarding the algae. The chief enzymes involved are peroxiredoxin, superoxide dismutase (SOD), glutathione reductase, ascorbate peroxidase, catalase, glutathione peroxidase, and thioredoxin. SOD serves as the first line of defense against oxidative stress by catalyzing the transformation of superoxide anions into molecular oxygen and hydrogen peroxide. Alkyl hydroperoxides are converted into water and its analogous alcohol by the enzyme peroxiredoxin. It is a thiol-specific antioxidant. Glutathione peroxidase speeds up the breakdown of ROOH. Other mechanisms to reduce toxicity are gene regulation, chelation, and heavy metal immobilization.

4.10.23.2 PLANTS

Plants have developed a significant tolerance level against heavy metals due to various mechanisms that protect the underlying biological machinery from stress. Usually, each mechanism protects the plant from particular metal stress that affects it in a specific region. The first line of defense the plant ensures to have when the heavy metal concentration exceeds the permissible limit is to stop metal ion uptake through the root region. They are usually absorbed by the root system through pumping mechanisms or ion channels that may be energy-dependent. The plants restrict these processes by bringing a change in the rhizosphere pH by secreting root exudates. It causes the precipitation of heavy metals. Due to the exudates, the metals bind to the cell wall, accumulate in the apoplast region, and become unable to enter the cell. Chelating effects have been observed in plants through malate and oxalate exudates. Nitric oxide, a strong antioxidant helps in reducing oxidative stress in plants due to heavy metals. It is involved in inducing transcriptional changes to eliminate ROS and can detoxify many oxygen radicals. It also stimulates the production of another antioxidant, glutathione. Numerous highly tolerant plant varieties locally distribute the heavy metal to different plant parts and compartmentalize them so that they get separated from the cytosol. Vacuoles are the principal metal storage sites in the plant cell and act as a compartment for the metal ions. Various proton pumps help in transporting these metal ions into the cell vacuole. Some heat shock proteins are known to reduce oxidative stress. Plants that show a greater level of tolerance usually contain hyperaccumulators. They help in migrating heavy metals from roots to shoots through the xylem.

4.10.23.3 FUNGI

While fungi absorb heavy metals, they undergo stress conditions. So to prevent themselves from the action of toxic effects, they have a series of defenses working inside the cell continuously. Morphologically, some changes in the structure of the hyphae and mycelium were observed. The feature includes swollen hyphae, smooth and irregular surfaces of mycelium, bulbous hyphae tips and conidia, and the formation of different textures on the cell wall. These demonstrate that fungi have successfully undergone their mechanism to eliminate heavy metals. At a molecular level, it has three lines of defense for detoxification.

- Extracellular chelation of the metal ions and binding it to the cell wall by the metabolites and proteins is followed for the detoxification process. To maintain homeostasis, it secretes GSH underexposure. Hence, structural modification is observed.
- Compartmentalization of the conduction system by separating the regular pathway for ions uptake and heavy metals uptake is another process for detoxification. This keeps the toxic elements away from the center of the cell to its exterior region. Sometimes oxalates are produced and form crystals to block the pathway up taking the toxic metals.
- Under stress conditions, fungi produce antioxidants to prevent themselves from any stress injury. The antioxidants may work as GSH independent or GSH dependent by neutralizing the toxins in the cells.

In association with fungi and plants, all the defense lines go hand in hand by distributing the workload. Since the toxicity can cause molecular dysfunction among them, so the mycorrhizal fungi stop the dislocation of biomolecules, binding the metals and inhibiting the overproduction of enzymes.

4.10.24 SEPARATION OF ALGAE AND FUNGI

Now that the wastewater has got treated successfully, bioremediating algal cells need to be separated before releasing the water into the water bodies. Generally, the cells settle at the bottom of the chamber and get harvested through the outlet. If the algal or fungal cells do not sediment, they are immobilized using a very low concentration of nutrients such as Ca^{2+}. These ions do not affect the quality of water. Flocculants are also introduced in the reactors to create large agglomerations of the cells making them easier to separate. Some chemical-based flocculants such as aluminum sulfate and natural flocculants such as chitosan are used for flocculation. Using natural flocculants has not affected agglomeration in any manner. Instead, it showed the same result as the chemical flocculant. It shows the accumulation of a variety of algal and fungal cells. Electro-separation is another way to separate the cells from the water. It was devised because after the bioremediation, the cells carry heavy metal ions that are conducted and separated from the water. Magnetic separation follows this principle as well. The cells that bear heavy metals form several layers known as biofilms on the electrodes. Hence, the water is released effortlessly after the separation. Apart from this, the harvested cells undergo a defined process for extraction of lipids as these cells have a higher content of lipids. The lipids find their application as fuel.

4.11 CASE STUDY

4.11.1 NATURAL BIOREMEDIATION WITH THE HELP OF ALGAE

In the Gangetic plains of North India, the groundwater which was used for irrigation via tube wells had a heavy metal concentration of more than 300 μg/L. Even the drinking water after had more than 50 μg/L. The area which reported such a concentration of heavy metals especially Arsenic was considered a contaminated zone. A range of algal and plant species growing in that area were identified as potent bio-indicator of arsenic in the aquatic system. Soon, the use of algae and plants in bioremediation was recognized and researched worldwide. It was understood that the growth of such species in that zone has some accumulation capacity, high tolerance potential, and better response under stress conditions for bioremediation. It was noted that the crops growing in close contact with these species had less accumulation and fewer adverse effects. Hence, to evaluate the potential of the natural growing species for remediation purposes in the future, some soil samples from the irrigated land along with a few algal and plant species were isolated. The level of contamination in the soil, in algal and plant species, was thoroughly studied for its evaluation. The data of microalgal spp. *Hydrodictyon reticulatum* and plant spp. *Eichhornia crassipes* in bioremediation of arsenic in four different irrigated soil samples are represented.

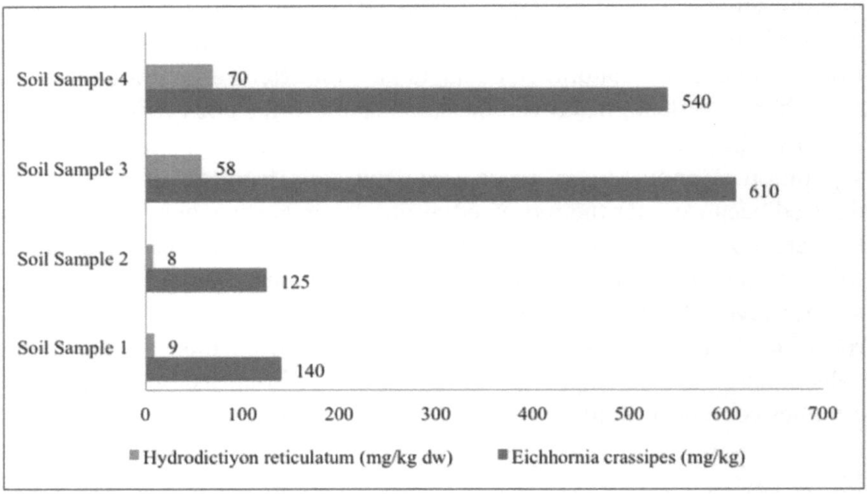

FIGURE 4.9 Study of bioaccumulation of arsenic by *Hydrodictyon reticulatum* and *Eichhornia crassipes* in different soil samples.

*For plant species, only accumulation in the shoot part is considered.

Thus, the level of arsenic is consistent due to continuous irrigation of the contaminated land. The toxicity in the Gangetic plains is mainly because of natural erosion. The impact of the toxicity is well known to counter this, the natural occurrence of these algae and plants acts as a potential accumulator.[17,18] They may also work in a consortium with an interdisciplinary approach for bioremediation. With the development of this approach, it can provide safe agricultural practices and for drinking purposes.

4.11.2 BIOREMEDIATION OF HYPERACCUMULATED ARSENIC AND LEAD IN DRINKING WATER

Arsenic and lead occur naturally in Mexico but the presence of the elements in the water is due to mining and volcanic activity. These elements are dissolved in water and soil and seep into the interiors. All this water was used for all human activities and posed more threats than consumption.

The metals seeped through the river source water, surface water, and tap water as well with the help of mobilization and desorption. Although the permissible amount, according to the public health department was set at less than 25 μg/L, the concentration of arsenic in drinking water was 50 μg/L. Other consumable items had 20% more concentration than the drinking water.[19] Common filtration techniques such as boiling did not have any effect on its concentration. Some zones were surveyed and 58 wells in that zone reported a concentration of around 700 μg/L. This hyperaccumulator was considered highly toxic for usage. Phytoremediation was examined and it resulted that the efficiency was more than phycoremediation and mycoremediation. Plants, such as *Pteris vittata, Nicotiana glauca*, and *Sorghum* were considered hyperaccumulators in the bioremediation process. They can translocate toxic metals at a rapid rate, increasing their efficiency to accumulate arsenic and lead.[20] The concentration of arsenic in leaves, stems, and roots was approximately around (4.3–12 mg/kg), (0.26–4.61 mg/kg), and (0.26–4.61 mg/kg), respectively. The uptake of arsenic in plants in the form of As(V) was preferred over As(III), so as As(V) reached the roots, it was reduced into As(III) in the leaves. However, Sorghum accumulated As(V) more than As(III). Some test was also run for genetically modified crops and came out as a potential mode of bioremediation. These are yet to be implemented after the considerations.

4.12 CONCLUSIONS

Heavy metals have pernicious effects on humans, plants, and aquatic life. Therefore, to make sure that the water is safe for usage, the permissible limit of each metal is determined. Conventional methods that are in use right now do not offer permanent removal of toxins and are immoderately priced. By 2050, annual debris formation is expected to rise by 73%. It is an alarming concern and points toward the need for green technology. Bioremediation is a sustainable method that uses natural living organisms to treat wastewater and has potential for future innovations. These methods have not yet been commercialized and are under study for improving their efficiency. This makes it an important domain of discussion for better handling of waste. Various processes are followed under these methods to manage the wastewater out of which biosorption is the most effective one. Along with this, information about the species along with their targeted metals was provided. It is significant to select suitable species to treat different types of wastewater. The analysis of BOD when the water is treated with this method and conventional method helped us to understand that bioremediation is an efficient way of water management. The accumulation of heavy metals can be toxic to the organism itself but they have a line of defense to shield themselves. Also, algae and bacteria have a small generation time, therefore, the formation of continuous batch culture or self-replenishing culture for bioremediation is possible. After the successful bioremediation, the cells are separated from the water and the harvested cells are processed further. The cells that are harvested have high lipid content which can be used as fuel. Bioremediation was tested out in India and Mexico to evaluate the remediation of hyperaccumulation of lead and arsenic present in drinking water. It was a successful attempt.

KEYWORDS

- **contaminants**
- **heavy metals**
- **water**
- **treatment**
- **remediation techniques**

REFERENCES

1. Bhati, T.; Gupta, R.; Yadav, N.; Singh, R.; Fuloria, A.; Waziri, A. et al. Assessment of Bioremediation Potential of Cellulosimicrobium sp. for Treatment of Multiple Heavy Metals. *Microbiol. Biotechnol. Lett.* **2019,** *47*, 269–277. DOI: 10.4014/mbl.1808.08006.
2. Bissonnette, L.; St-Arnaud, M.; Labrecque, M. Phytoextraction of Heavy Metals by Two Salicaceae Clones in Symbiosis with Arbuscular Mycorrhizal Fungi During the Second Year of a Field Trial. *Plant Soil* **2010,** *332*, 55–67. DOI: 10.1007/s11104-009-0273-x.
3. Chellaiah, E. R. Cadmium (Heavy Metals) Bioremediation by Pseudomonas Aeruginosa: A Minireview. *Appl. Water Sci.* **2018,** *8*, 1–10. DOI: 10.9734/ajsspn/2017/36868.
4. Ciarkowska, K. Assessment of Heavy Metal Pollution Risks and Enzyme Activity of Meadow Soils in Urban Area Under Tourism Load: A Case Study from Zakopane (Poland). *Environ. Sci. Pollut. Res.* **2018,** *25* (14), 13709–13718. DOI: 10.1007/s11356-018-1589-y.

5. Cumberland, S. A.; Douglas, G.; Grice, K.; Moreau, J. W. Uranium Mobility in Organic Matter-Rich Sediments: A Review of Geological and Geochemical Processes. *Earth Sci. Rev.* **2016,** *159*, 160–185. DOI: 10.1016/j.earscirev.2016.05.010.
6. Dhokpande, S. R.; Kaware, J. P. Biological Methods for Heavy Metal Removal—A Review. *Int. J. Eng. Sci. Innov. Technol.* **2013,** *2* (5), 304–309.
7. Diep, P.; Mahadevan, R.; Yakunin, A. F. Heavy Metal Removal by Bioaccumulation Using Genetically Engineered Microorganisms. *Front. Bioeng. Biotechnol.* **2018,** *6*, 157. DOI: 10.3389/fbioe.2018.00157.
8. Galiulin, R. V.; Galiulina, R. A. Removing Heavy Metals from Soil with Plants. *Herald Russ. Acad. Sci.* **2008,** *78*, 141–143. DOI: 10.1134/s1019331608020044.
9. Geva, P.; Kahta, R.; Nakonechny, F.; Aronov, S.; Nisnevitch, M. Increased Copper Bioremediation Ability of New Transgenic and Adapted *Saccharomyces cerevisiae* Strains. *Environ. Sci. Pollut. Res.* **2016,** *23*, 19613–19625. DOI: 10.1007/s11356-016-7157-4.
10. Malaviya, P.; Singh, A. Physicochemical Technologies for Remediation of Chromium-Containing Waters and Wastewaters. *Crit. Rev. Environ. Sci. Technol.* **2011,** *41* (12), 1111–1172. DOI: 10.1080/10643380903392817.
11. Mani, D.; Kumar, C. Biotechnological Advances in Bioremediation of Heavy Metals Contaminated Ecosystems: An Overview with Special Reference to Phytoremediation. *Int. J. Environ. Sci. Technol.* **2014,** *11* (3), 843–872. doi: 10.1007/s13762-013-0299-8.
12. Mohamed, R. M.; Abo-Amer, A. E. Isolation and Characterization of Heavy-Metal Resistant Microbes from Roadside Soil and Phylloplane. *J. Basic Microbiol.* **2012,** *52* (1), 53–65.
13. Ojuederie, O. B.; Babalola, O. O. Microbial and Plant-Assisted Bioremediation of Heavy Metal Polluted Environments: A Review. *Int. J. Environ. Res. Public Health* **2017,** *14* (12), 1504. DOI: 10.3390/ijerph14121504.
14. Olaniran, A. O.; Balgobind, A.; Pillay, B. Bioavailability of Heavy Metals in Soil: Impact on Microbial Biodegradation of Organic Compounds and Possible Improvement Strategies. *Int. J. Mol. Sci.* **2013,** *14* (5), 10197–10228. DOI: 10.3390/ijms140510197.
15. Rani, M. J.; Hemambika, B.; Hemapriya, J.; Kannan, V. R. Comparative Assessment of Heavy Metal Removal by Immobilized and Dead Bacterial Cells: A Biosorption Approach. *Glob. J. Environ. Res.* **2010,** *4* (2), 23–30.
16. Sivakumar, D.; Kandaswamy, A.; Gomathi, V.; Rajeshwaran, R.; Murugan, N. Bioremediation Studies on Reduction of Heavy Metals Toxicity. *Pollut. Res.* **2014,** *33* (3), 553–558.
17. Tchounwou, P. B.; Yedjou, C. G.; Patlolla, A. K.; Sutton, D. J. Heavy Metal Toxicity and the Environment. *Exp. Suppl.* **2012,** *101*, 133–64.
18. Thapa, B.; Kc, A. K.; Ghimire, A. A Review on Bioremediation of Petroleum Hydrocarbon Contaminants in Soil. *Kathmandu Univ. J. Sci. Eng. Technol.* **2012,** *8* (1), 164–170. DOI: 10.3126/kuset.v8i1.6056.
19. Tyagi, S.; Sarma, K. Assessment of Groundwater Quality in Different Land Uses in Ghaziabad District of Uttar Pradesh, India. *Environ. We Int. J. Sci. Tech.* **2018,** *13* (2), 99–117.
20. Wuana, R. A.; Okieimen, F. E. Heavy Metals in Contaminated Soils: A Review of Sources, Chemistry, Risks and Best Available Strategies for Remediation. *Int. Sch. Res. Notices* **2011,** *2011*, 20. DOI: 10.5402/2011/402647.

CHAPTER 5

Adoption of Green Technology for Environmental Sustainability

SUBBULAKSHMI GANESAN

Department of Chemistry, Jain University, Bangalore, Karnataka, India

ABSTRACT

It is possible to employ biological processes to remove toxic contaminants from the environment using phytoremediation. A combination of causes, including industrialization, agricultural operations, the excessive use of fertilizers, untreated wastes, and unregulated lab operation wastewaters, has resulted in soil and environmental degradation. A substantial role is played by toxic metals, which play an essential part in plant growth. As a direct participant and vital component in several critical enzymes, they often participate in a variety of reductions, and nucleic acid metabolism relies heavily on oxidation processes. Toxic metals must be present in a normal developing medium at a certain level, but too much may have multiple deadly consequences. As a result, eliminating poisonous and hazardous metal ions is now a top priority for achieving sustainable development objectives. In this study, the goal of soil reclamation techniques is to discover an environmentally sustainable alternative using phytoremediation. Toxic metals may be removed from polluted soil using a variety of physical and chemical methods; however, because of specific restrictions, plant-based technology is favored. Metal ions can be eliminated via plants' metabolic pathways with this method, which has been employed for the last two decades. In general, plants employed as phytoremediators are hyperaccumulators, able to collect metal ions at

Sustainable Water Treatment and Ecosystem Protection Strategies. Hossein Hariri Asli, Tamara Tatrishvili, Ann Rose Abraham, & A. K. Haghi (Eds.)

concentrations more than 1000 ppm, and they must have particular qualities, such as a branching root system, low biomass, and ease of harvesting, in order to be effective. Hyperaccumulating plants are the center of this review article, which examines their origins, adverse consequences, and alternative solutions for eradicating toxic metals.

5.1 INTRODUCTION

Phytoremediation is the combination of phyto (plant) and remedium (removal). Phytoremediation is a well-developed technology for removing toxins from our environment and restoring their natural balance utilizing plants as remediants in a variety of biological ways. Soil, sewage, ground water, surface water, or any other pollutant source may be treated using plants or plant components to remove several organic and inorganic contaminants. Land degradation as a result of urbanization and industrialization is making it unsuitable for agriculture and even unsuitable for creatures, such as plants and animals.[1] In the natural environment, toxic contamination from sources such as untreated industrial wastewater poured into close by waterways and household garbage deposited in landfills is slowly causing the extinction of plants and animals.[2] Phytostabilization, rhizodegradation, rhizofiltration, phytodegradation, and phytovolatilization are some of the techniques used to remove organic pollutants. In order to ease, discharge, and mix the pollutants in the soil matrix, root plants exude. Phytostabilization is the collective name for these steps. There are crops that have been bred to absorb and adsorb contaminants from the soil and groundwater and store them in their roots or the root zone of the plant. For organic and metal contaminants found in the soil, sediment, or sewage environment, which are used in this procedure. Phytoextraction mechanisms may cause some crops to collate excessive amounts of metal pollutants or additional nutrients in their harvestable vegetative growth tissue.

5.2 STUDY ON TOXIC METAL POLLUTION IN SOIL

Toxic metals may infiltrate the soil from both anthropogenic and geological activity. Metals from industrial and residential sewage, metallurgy, mining, fuel manufacture, and the use of agro- chemicals are all types of toxic metals in soil.[3] Environmental impacts from the transportation and production of building materials, as well as energy, contributes to the release of harmful metals into the atmosphere. For example, aerosols of toxic metals are present in the air and are subsequently washed down to the ground by rain. Figure 5.1 depicts the soil sources of toxic metals. An excess of agricultural inputs, such as activated sludge, fertilizer, insecticide, and weed killer produces chemical accumulation in the soil and absorption by plants.

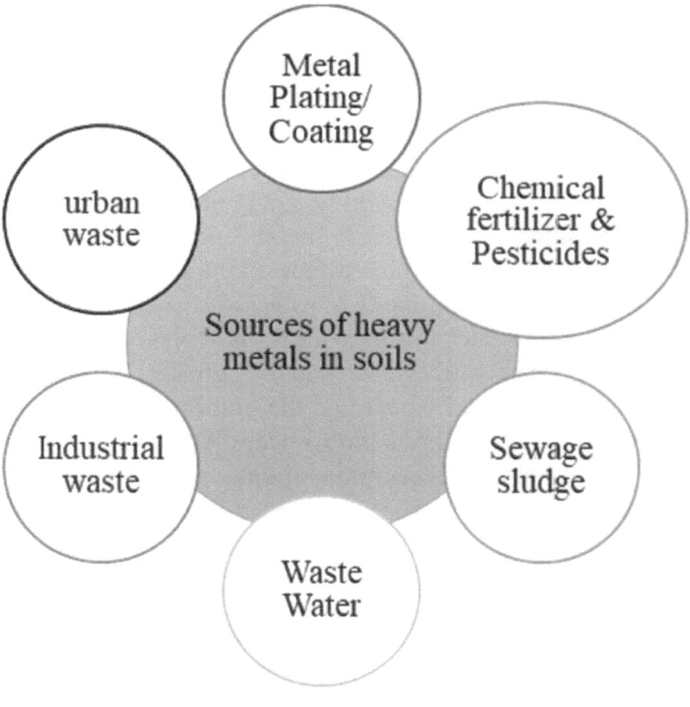

FIGURE 5.1 Soil toxic metal formation.

In agricultural areas, toxic metal levels grow when biosolid manures, such as decomposition, sewage sludge, and animal dung are used. The organic substance is known as activated sludge containing various toxic metals, such as lead, cadmium, mercury, and chromium. As a result, groundwater might be contaminated by toxic metals that are deposited in the soil. These toxic metal concentrations have been determined in wastewater and fertilizers.

5.3 TOXIC METALS THAT ARE ENVIRONMENTALLY HARMFUL AND THEIR CATEGORIZATION

Toxic metals may be divided into vital and non-vital categories. Copper, cobalt, chromium, zinc, iron, nickel, and manganese are among the key toxic metals, while lead, cadmium, and mercury are also included.[4] The degree of toxicity of toxic metals may be divided into three categories: substantially high toxic, substantially fairly toxic, and substantially less hazardous. Cadmium is the most bioaccessible and bioaccumulative among the non-vital toxic metals. As a divalent ion, cadmium has no biological role and is found in compounds. Most cadmium is found in combination with zinc, a vital micronutrient in soil that has been polluted by toxic metals. Cadmium, on the other hand, often produces phytotoxicity in plants at high concentrations of zinc. Because of its ability to absorb and retain high concentrations of cadmium, vetiver grass may be used to cleanup soil that has been polluted with the heavy metal. Cadmium is sequestered by microbes in the rhizosphere, which affects plant absorption of cadmium. High glutathione and glutathione-S-transferase concentrations have been linked to detoxification mechanisms in the roots of plants, used to remediate cadmium-contaminated soil. Maize had the greatest concentrations of cadmium and zinc, whereas wheat had the highest concentration of lead among edible crops.[5] Wheat has a higher cadmium absorption than maize; hence using soil polluted with cadmium might lead to food contamination. Antioxidative enzymes, namely, glutathione reductase, may be suppressed by cadmium.

5.4 ENVIRONMENT-HARMING IMPACTS OF CONTAMINATION WITH TOXIC METALS

Leaching and no degradability are two of the key causes of toxic metal contamination. The bioaccumulation and biomagnifications of toxic metals in soil have a negative impact on life forms. Soil contains a wide spectrum of toxic metals, ranging from 1 to 100,000 mg/kg, all of which have a direct effect on micro and macro species.[6,7] When it comes to toxic metals, the surface soil tends to have the highest amounts of these pollutants, owing to the greater binding capacity of organic layers. As a result, plants watered with polluted water occupy high amounts of lead and cadmium in the surface of the soil, where root exudates and microbes in the rhizosphere boost the movement of toxic metals. This significantly impacts the uptake of vital nutrients. It is possible that lead and cadmium compete for the same sites or activities as the cations cupric ion, ferrous iron, zinc (II) ion, and manganous ion. Because of their extensive root systems in the soils surface layer, metal-contaminated crops, such as vegetables, cannot be eaten by humans.

For freshwater, soil, and plants, the maximum permitted levels of harmful toxic metals. Plants contaminated with toxic metals may cause cancer and other long-term health problems in people. Respiratory problems, as well as harm to the heart, brain, and kidneys, are all consequences of lead, cadmium, and zinc exposure in humans.[8] When replacing base metal ions while generating reactive oxygen species (ROS) while blocking functional groups, toxic metals are harmful to biomolecules in three separate ways. This results in stunted development, reduced yield, and abnormal metabolic functions in plants that are contaminated with toxic metals.

5.5 REMEDIATION METHODS FOR SOIL CONTAMINATED WITH TOXIC METALS

Environmental sustainability legislation focuses on removing degraded soils' toxic metals. There are two methods for cleaning up contaminated soils: ex situ and in situ. Using standard physicochemical procedures, dirty soils must be decontaminated ex situ. However, the bioaccessibility of toxic metals determines the economics and environmental benefits of in situ remediations of soils impacted by toxic metals.[9] Soils contaminated with toxic metals may be remedied using a variety of methods, including chemical, biological, and physical. Several methods are suggested for cleaning up soil that has been contaminated with toxic metals. Two key ideas underlie remediation techniques. The first step is to remove all toxins (green technology) from the environment, and the second step is to change contaminants into less hazardous forms employing engineering technology. Conventional physical and chemical soil remediation treatments can only be used to limited areas of substantially heavy metal impacted soil. High energy consumption and negative impacts on soil structure and production restrict their use. Soil that has been poisoned by toxic metals may be restored using microorganisms.[10] To remove toxic metals and reestablish healthy levels to the soil, plants may be used for reclamation.

5.6 CHEMICAL AND PHYSICAL METHODS FOR ELIMINATING TOXIC METALS FROM CONTAMINATED SOIL

Among the physical techniques for reclaiming soil are soil replenishment and thermal desorption, respectively. Using contaminant-free soil to replace contaminated soil is part of the process of soil replenishment. There are three distinct methods for the replenishment of soil:

a) Removing soil that is contaminated and replacing it with fresh soil.
b) Digging up soil that is contaminated and spreading pollutants to the deepest layers.
c) Soils that are contaminated are mixed with newly imported soil.

Pollutants may be volatilized by heating the polluted soil. Thermal desorbers are used to remove and remediate contaminated soil. Gas collecting equipment is used to catch contaminated gases before they escape into the air. High and low are two types of thermal desorption. A combination of soil immobilization methods and soil washing are used in the chemical rehabilitation of heavy metal-contaminated soil. Soil pollution may be stabilized using binding agents, either organic or inorganic.

5.7 VITRIFICATION

To reduce the mobility of toxic metals in contaminated soil, high-temperature vitrification is needed. Organic molecules and certain metals, such as mercury, may be volatilized when heated to high temperatures, which should be collected and the rest are transformed into an inert glass-like product throughout this method.[11] Electrical, thermal, and plasma technologies are the three primary types of vitrification processes. Ex situ vitrification is not recommended; instead, in situ vitrification is the ideal method, which may be powered by a variety of sources. Electrodes put into the soil are used to conduct an electric current through polluted soil. Using an array of four electrodes, a single melt can eliminate up to 1000 ppm of dirt from the soil.

5.8 SOIL WASHING

Toxic metal-contaminated soil is cleaned by using a combination of solvents and mechanical processes. By separating the fine dirt from the coarse soil, this procedure significantly minimizes the quantity of polluted soil. Clean soil may be used as backfill while the contaminated soil can be dealt with using other cleanup methods.[12] Soil washing may remediate a variety of pollutants, including toxic metals. Toxic metals are first extracted from the soil slurry using an ex situ extraction process, then flushed in situ using washing co-solvents, and surfactants and other agents are also used. The efficacy of washing surfactants or co-solvents depends on their capacity to break down toxic metals in the soil. Depending on the kind of soil contamination, different solvents are used.

5.9 A BIOLOGICAL METHOD TO REMOVING TOXIC METALS FROM SOIL: AN ECOLOGICALLY FRIENDLY SOLUTION

5.9.1 PHYTOREMEDIATION

Using fast-growing plants to remove, retain, or deliver harmless pollutants from the soil or water is a green method known as phytoremediation. It is an attractive, cost-effective, and environment-friendly method of removing toxins.[13] As long as the plant that is utilized can absorb and accumulate significant concentrations of metal pollutants in its shoot sections, phytoremediation may succeed.

5.9.2 PHYTOIMMOBILIZATION

For soil that has been polluted with metals, this method of inactivation uses plants which are also known as phytostabilization. Furthermore, reducing the metals' mobility and bioaccessibility, as well as their absorption into the water and the food system, is the goal of this technology. Plants perform a secondary function in phytostabilization due to soil amendments.[14,15] Incorporation of different soil amendments and subsequent absorption by the roots immobilize metal pollutants via physical and chemical means. Metal pollutants may be physically and chemically immobilized with this method, which includes the adsorption of metals onto roots and the incorporation of soil nutrients. Soil amendments including phosphatic fertilizers, biosolids, organic material, and clay minerals are best for preventing toxic metals from entering the soil. Minimize percolation of water, restricting contamination, controlling soil erosion, and decreasing the movement of pollutants are the primary goals of plants. Since metal pollutants remain in the soil after phytostabilization, it is a management approach rather than a long-term solution.

5.9.3 RHIZODEGRADATION

Organic contaminants in the rhizosphere are broken down by increased microbial activity, this process known as phytostimulation.[16] The rhizosphere is the area of the soil around the root that is directly impacted by the activity of the root. Microbial activity in the rhizosphere is improved in a variety of ways:

a) Innate microbes benefit from root exudates, which contain amino acids and carbohydrates;
b) In order for aerobic changes to take place, the rhizosphere must be supplied with oxygen;
c) Root biomass increased the amount of organic carbon that is available;
d) A fungus called mycorrhizae is able to degrade substances that bacteria cannot;
e) Increased microbial populations may thrive in the presence of plants.

5.9.4 PHYTODEGRADATION

No microbial community is required for phytodegradation (also known as phytotransformation), which breaks down organic substances through plant metabolic pathways or the production of enzymes by plants. As a result, plants might be thought of as the biosphere's green liver. Only organic chemicals can be broken down by phytodegradation.[17] Volatile chemicals are emitted into the atmosphere by plant transpiration, a process known as phytovolatilization. Toxins in the soil are absorbed by plants, altered to volatile chemicals, and then released back into the atmosphere. Only organic molecules and toxic metals, mercury and selenide may be analyzed using this method.

5.10 RHIZOFILTRATION, BLAST FILTRATION, AND CAULOFILTRATION

Plant roots may be used to clean water supply systems and sewage while removing low levels of pollutants. Before rehabilitation can begin, plants must be exposed to polluted water in order to acclimatize themselves. Harvesting occurs when the roots are fully wet. It is possible to use plant roots, seedlings, or shoots as a means of phytofiltration. By absorbing, dissolving, or accumulating these contaminants, the amount of pollution that ends up in the groundwater is reduced significantly. The pH of the rhizosphere may be altered by root exudates, which can lead to precipitation of metals on the roots of plants.

5.10.1 PHYTOMINING

Toxic metals are removed from the environment by using fast-growing plants. There are two ways by which phytoextraction are applied: naturally occurring and chemically generated phytoextraction. Natural phytoextraction involves removing soil toxic metals via a network of roots before directing the extraction to the plant's higher tissues. Biogas may be produced from plant biomass, and it can also be burned. Metal recovery, brick making, or land disposal are all possible uses for burned plant biomass. Phytomining or biomining is another name for phytoextraction.

5.11 ENVIRONMENTAL PHYTOREMEDIATION: BENEFITS, DRAWBACKS, AND THE TENDENCIES OF THE FUTURE

The most cost-effective method for treating soil that is polluted with toxic metals is phytoremediation. The cost of adopting this innovative technology is 60–80% cheaper than that of more traditional chemical and physical processes. Following phytoextraction, the collected plant biomass has the potential to be used in the generation of bioenergy. A potential method for cleaning up contaminated soil caused by toxic metals is called phytoremediation. However, this method also has significant drawbacks, such as a slow rate of increase of plant accumulation of toxic metals, a limited bioaccessibility of toxic metals, and the phytoremediation process taking a long time. As a relatively new topic of study, phytoremediation has only seen a few case studies conducted in the field. It is possible that the outcomes obtained in a greenhouse will vary from those obtained in a field. At the field level, a variety of variables influence the phytoremediation process.[18] These include uneven pollution distribution, pH of the soil, microorganisms, nutrients, precipitation, and temperature. This technique relies on the isolation and screening of different species of plants with maximum growth production and toxic metal accumulation potential to be successful.

5.12 CONCLUSIONS

According to this study, toxic metals pose one of the greatest hazards to the environment and human health. From a variety of sources, these metals enter the environment. Remediation using conventional methods is both expensive and harmful to the environment. As a result, cleaning up soils contaminated with toxic metal cannot be done other than by applying low-cost and environmentally favorable technology. One of the most efficient plant-based methods for removing pollutants from polluted regions without harming the soil's structure is phytoremediation. Hyperaccumulators are plants that can absorb toxic metals up to a few per cent of their dry shoot biomass. Fines should be imposed on new plant species that can remove toxic metals from polluted soils. Polluted soil may benefit from strategies that increase metal bioavailability. For example, the application of bioaugmented acidified manure lowers the soil pH and then increases the availability of toxic metals to plants.

KEYWORDS

- **hyperaccumulation**
- **phytoremediation**
- **toxic metals**
- **wastewater**
- **soil**
- **ground water**
- **sewage**

REFERENCES

1. Acar, Y. B.; Alshawabkeh, A. Principle of Electrokinetic Remediation. *J. Environ. Sci. Technol.* **1993,** *27* (13), 2638–2647.
2. Acosta, J. A.; Gabarron, M.; Faz, A.; Martinez-Martínez, S.; Zornoza, R.; Arocena, J. M. Influence of Population Density on the Concentration and Speciation of Metals in the Soil and Street Dust from Urban Areas. *Chemosphere* **2015,** *134*, 328–337.
3. Adrees, M.; Ali, S.; Rizwan, M.; Zia-ur-Rehman, M.; Ibrahim, M.; Abbas, F.; Farid, M.; Qayyum, M. F.; Irshad, M. K. Mechanisms of Silicon-Mediated Alleviation of Heavy Metal Toxicity in Plants: A Review. *Ecotoxicol. Environ. Saf.* **2015,** *119*, 186–197.
4. Banerjee, R.; Goswami, P.; Pathak, K.; Mukherjee, A. Vetiver Grass: An Environment Clean-Up Tool for Heavy Metal Contaminated Iron Ore Mine-Soil. *Ecol. Eng.* **2016,** *90*, 25–34.
5. Bani, A.; Pavlova, D.; Echevarria, G.; Mullaj, A.; Reeves, R. D.; Morel, J. L.; Sulce, S. Nickel Hyperaccumulation by the Species of Alyssum and Thlaspi (Brassicaceae) from the Ultramafic Soils of the Balkans, 2010.
6. Beckers, F.; Rinklebe, J. Cycling of Mercury in the Environment: Sources, Fate, and Human Health Implications: A Review. *Crit. Rev. Environ. Sci. Technol.* **2017,** *47* (9), 693–794.
7. Beiyuan, J.; Awad, Y. M.; Beckers, F.; Tsang, D. C.; Ok, Y. S.; Rinklebe, J. Mobility and Phytoavailability of As and Pb in a Contaminated Soil Using Pine Sawdust Biochar Under Systematic Change of Redox Conditions. *Chemosphere* **2017,** *178*, 110–118.
8. Basta, N. T.; Ryan, J. A.; Chaney, R. L. Trace Element Chemistry in Residual-Treated Soil: Key Concepts and Metal Bioavailability. *J. Environ. Qual.* **2005,** *34* (1), 49–63.
9. Clemens, S.; Palmgren, M. G.; Kramer, U. A Long Way Ahead: Understanding and Engineering Plant Metal Accumulation. *Trends Plant Sci.* **2002,** *7* (7), 309–315.
10. Coetzee, J. J.; Bansal, N.; Chirwa, E. M. Chromium in Environment, Its Toxic Effect from Chromite-Mining and Ferrochrome Industries, and Its Possible Bioremediation. *Expo. Health* **2018,** 1–12.
11. Conceicao, E. T.; Bonotio, D. M. Radionuclides, Toxic Metals and Fluorine Incidence at Tapira, Phosphate Rocks. Brazil and Their Industrial By-Products. *Environ. Pollut.* **2006,** *139*, 232–243.
12. Dada, E. O.; Njoku, K. I.; Osuntoki, A. A.; Akinola, M. O. A Review of Current Techniques of Physico-Chemical and Biological Remediation of Toxic Metals Polluted Soil. *Ethiop. J. Environ. Stud. Manag.* **2015,** *8* (5), 606–615.
13. Danh, L. T.; Truong, P.; Mammucari, R.; Tran, T.; Foster, N. Vetiver Grass, *Vetiveria zizanioides*: A Choice Plant for Phytoremediation of Toxic Metals and Organic Wastes. *Int. J. Phytoremediat.* **2009,** *11* (8), 664–691.
14. Davis, C. M.; Vincent, J. B. Chromium Oligopeptide Activates Insulin Receptor Tyrosine Kinase Activity. *Biochemistry* **1997,** *6*, 4382–4385.
15. De Souza Freitas, E. V.; Do Nascimento, C. W. A. The Use of NTA for Lead Phytoextraction from Soil from a Battery Recycling Site. *J. Hazard. Mater.* **2009,** *171* (1–3), 833–837.

16. Emenike, C. U.; Jayanthi, B.; Agamuthu, P.; Fauziah, S. H. Biotransformation and Removal of Toxic Metals: A Review of Phytoremediation and Microbial Remediation Assessment on Contaminated Soil. *Environ. Rev.* **2018,** *26* (2), 156–168.
17. Erakhrumen, A. A. Phytoremediation: An Environmentally Sound Technology for Pollution Prevention, Control and Remediation in Developing Countries. *Edu. Res. Rev.* **2007,** *2*, 151–156.
18. Etesami, H. Bacterial Mediated Alleviation of Heavy Metal Stress and Decreased Accumulation of Metals in Plant Tissues: Mechanisms and Prospects. *Ecotoxicol. Environ. Saf.* **2018,** *147*, 175–191.

CHAPTER 6

Applications of Biotechnology in Wastewater Treatment for a Green Future

SUBBULAKSHMI GANESAN, G. PADMA PRIYA, and IZEGAEGBE DANIEL OMOIKHOJE

Department of Chemistry, Jain University, Bangalore, Karnataka, India

ABSTRACT

Due to rapidly increasing industrialization all around the world, water pollution also had increased rapidly. Due to the waste water which had released into the water bodies, the contamination in the lakes, rivers, and ponds had occurred. This may lead to bad impacts for both plants as well as for animals. Eutrophication, spread of diseases is few of the environmental as well as ecological problems that occur due to the release of waste into the river. Biotechnology can be applied in an innovative way, in order to treat the waste water effectively. These application methods are eco-friendly and cost-effective, they consist of anaerobic treatment, biosorption, usage of membrane filters, application of biocatalysts, and usage of hydroponic system are the methods used to treat wastewater. Moreover, in developed countries, the use of advanced methods in genetic engineering is also improvised. Therefore, the above-mentioned methods can act as the promising solution to stop the bad impacts which are affected from wastewater.

Sustainable Water Treatment and Ecosystem Protection Strategies. Hossein Hariri Asli, Tamara Tatrishvili, Ann Rose Abraham, & A. K. Haghi (Eds.)

6.1 INTRODUCTION

The exponential rise in the production of both liquid and solid wastes over the last few years may be directly attributed to the expansion of the human population and the expansion of human activities. Even while 99.9% of wastewater is water, the remaining 0.1% is the source of most of the problems. The remainder of the 0.1% consists of substances that are seldom seen in wastewater. Nonetheless, the water supply may be contaminated with dangerously high levels of these substances, posing threats to human health and the natural world. However, in underdeveloped nations, only a fraction of effluent from sewered neighborhoods is really cleaned. The lack of funding is a common source of wastewater treatment problems, but the lack of knowledge about low-cost wastewater treatment technologies and the financial advantages of recycling treated wastewater also play a role. Untreated sewage from farms, homes, and factories pollutes over half of the world's waterways, making them unfit for human consumption because of the abundance of fecal germs present. Human health and the health of the ecosystem are negatively impacted by this untreated effluent. Therefore, wastewater treatment is necessary to reduce the spread of excreta-related infections, as well as to mitigate water pollution and its resultant harm to aquatic biota. In light of the importance of protecting both the environment and human health, proper wastewater treatment must be seen as a need. Numerous methods are used for wastewater treatment, each with its own set of pros and cons. However, in underdeveloped nations, wastewater treatment concentrates on sludge treatment and disposal rather than water treatment. Wastewater treatment nowadays often employs eco-friendly methods. The parameters of the wastewater being treated guide the selection of both non-traditional and traditional treatment technologies. Biotechnology, which was previously unknown, is now recognized as an integral aspect of wastewater treatment. Isolated microbes are selected, manipulated, and genetically altered to perform optimally in wastewater bioremediation. Both on-site and off-site treatment options for polluted wastewater are possible with this setup. These biotechnology-based wastewater treatment methods include bioreactors, phytoremediation, bioremediation, and bioaugmentation. Wastewater may be treated using either aerobic processes or anaerobic processes, or a combination of the two.

6.2 SCARCITY OF WATER AND WASTEWATER TREATMENT

Because of water scarcity and the need of treating wastewater, as a result of its scarcity, water is one of the planet's most precious resources. In many third-world countries, metropolitan centers and rural communities alike continue to struggle with water shortages and pollution. With a growing global population, this is anticipated to become more prevalent, particularly in nations where urbanization is progressing more rapidly. Water contamination became a global issue as industrialization spreads rapidly. Large amounts of freshwater are used by these businesses, and the wastewater they produce is then released into the ecosystem Water shortages are a problem in many nations due to a combination of reasons, including climate change, the development of agricultural operations, urbanization, population increase, and other industrial activities. Many cities in industrialized nations are struggling to cope with rapid population growth and its attendant social and environmental concerns. There is a shortage of potable water because of issues, including inadequate housing, unsanitary living conditions, and weak power infrastructure. Major environmental issues are raised by the wastewater created from diverse sources due to the presence of numerous hazardous chemicals, and the discharge of wastewater into the ecosystem leads to multiple detrimental consequences on plants and animals. The release of wastewater into the ecosystem cannot be prevented, but its negative impacts may be mitigated by a variety of treatment strategies. Although traditional treatment approaches have been used effectively in the past, they are not without drawbacks.[6] Biological treatments are gaining popularity as a potential alternative to traditional wastewater treatment methods because to its low environmental impact, low cost, and low energy consumption. The primary goal of biological wastewater treatment is to eliminate the most dangerous pollutants from various wastewaters, making them safe to dispose of in a manner that does not endanger human or environmental health.

6.3 ECO-FRIENDLY METHODS OF WASTEWATER TREATMENT

"Green technology of wastewater" refers to environmentally friendly and sustainable methods used for wastewater purification. Biotechnology, green chemistry, and other emerging technologies may help reduce negative impacts on the environment. Catalysts, adsorption, nanocomposite, photo catalysis, an electrochemical approach, photo-Fenton, magnetic nanoparticles, and mesoporous materials are only some of the green ways for wastewater treatment. In order to help society, fulfil its needs, green wastewater technology promotes the growth and improvement of methods, processes, and applications that upgrade or replace the present technology. On the other hand, in the face of resource shortages and ecological hazards, these indigenous-developed technologies significantly lessen human influence on the planet. Improvements in human production and spending habits to reduce pollution caused by waste discharge and contamination, as well as the development of new, cleaner technologies to replace those with harmful effects on health and the environment.

6.4 THE IMPORTANCE OF BIOTECHNOLOGY IN LIMITING POLLUTION TO THE ECOSYSTEM

Wastewater treatment with biotechnology has been on the rise for decades. Water and sewage treatment might benefit greatly from the manipulation of microbial populations and nucleic constituents using this method. Many of the major wastewater treatment processes may be manipulated using techniques made available by biotechnology. Moreover, "biotechnology" is "the science and technology utilization of the capacities of microbes and cultured tissue cells through the integration of microbiology, engineering sciences and biochemistry."

Bioaugmentation: is the process of introducing microbes into a system in order to increase the concentration or accumulation of a target component. You may do this by supplying the plant with more air, water, and nutrients.

Bioremediation: refers to the process of using microbes, either those that already exist in the environment or those that have been changed genetically, to degrade or metabolize harmful compounds.

Bioreactor: The purpose of a bioreactor is to either generate or eliminate a certain species of bacteria by maintaining them in an enclosed container with carefully calibrated environmental conditions.

Phytoremediation: refers to the process of using plant life to cleanup contaminated environments.

6.5 UTILIZING BIOTECHNOLOGY FOR WASTEWATER TREATMENT IN UNDER DEVELOPED COUNTRIES

Even while wastewater treatment has become more effective in underdeveloped nations, it remains a major priority and a long-term sustainability problem. Activated sludge (AS), oxidation ponds, trickling filters, bio filters, and anaerobic treatment are all examples of biotechnology-based wastewater treatment systems. Environmental engineers have already put biotechnology to use in a variety of ways, such as biotrickling filters, solid waste composting methods, and biosorption. In order to effectively complete the treatment procedures, it is crucial in all approaches to identify the bacteria responsible for the breakdown of organic compounds.

- **Anaerobic treatment:** After nitrogen (N), phosphorus (P), and disease-causing microorganisms (IPM) have been removed, organic carbon is the main target of anaerobic treatment of wastewater. Cost-effective and environmentally benign, anaerobic bioprocesses are seen as preferable to physicochemical methods. Sludge treatment via anaerobic digestion has a lot going for it. Its benefits include significant decreases in chemical oxygen demand, NO_3^-, solids, biochemical oxygen demand as well as the biogas production and biofertilizer. The approach offers a low cost of operation. Wastewater high in sulfur may be treated in up flow anaerobic sludge blanket reactor. This method reduces biological surplus sludge while increasing treatment efficiency, decreasing capital costs, using no oxygen, producing methane, and needing minimal nutrients. Since anaerobic digestion is not efficient below 30°C, it must be coupled with low-value heat recovery methods. The 10-day COD elimination efficiency of 85% with acclimatized exogenous granular sludge compared well with 37% with indigenous anaerobic sludge in the treatment of swine wastewater. Exogenous granular sludge that has been acclimatized to a swine wastewater environment performs better because it is better able to tolerate ammonia inhibition and break down volatile fatty acids.
- **Biocatalyst:** The term "biocatalyst" refers to enzyme-based catalysts. Significant interest was shown in them when they were shown to be more effective than traditional approaches at cleaning up wastewater. Since this is the case, it may be used as either an adjunct or main method of care. In comparison to chemical

degradation methods, enzymatic biodegradation has several advantageous features, including high selectivity, specificity, and catalytic activity, moderate reaction mechanism, and low by-product production. Nontoxic, versatile, biodegradable protein molecules like lipase, amylase, and protease are used commercially as biocatalysts in a wide variety of industries, including the textile, paper, detergent, biofuel, leather, pharmaceutical, food and beverage, and environmental sectors. Biocatalysts are often produced by cultivating specific microorganisms with action against specified substrates. Therefore, the creation of multifunctional biocatalysts is currently essential for processing complicated substrates. Biomass anaerobic digestion, which might include food and agricultural processing wastes, could produce biocatalysts. A close association between biocatalyst-mediated microenvironmental improvement and anaerobic dye degradation has been observed. Since municipal waste AS contains a wide variety of complex organic compounds, pretreating it with a multihydrolytic biocatalyst was undertaken to boost the sludge's industrial viability. The sludge's enzymatic pretreatment under better circumstances increased the COD by 32%, the proteins by 53.57%, and the carbohydrates by 76.71%. Thus, organic solid wastes may be efficiently processed into a multifunctional biocatalyst that could be employed to improve the efficacy of AS for the treatment of municipal waste.

- **Activated sludge:** Among the many current methods for treating wastewater, AS is the most extensively utilized biotechnique due to its cheap cost. There is a significant adsorption capability of contaminants by the zoogloea on AS flocs, which is mostly made up of extremely hydrophilic polar material. In the presence of AS flocs, contaminants, such as heavy metals and organic pollutants are absorbed. Adsorption is based on microbial properties and the pollutant's affinity for living things. Electrostatic adsorption, ion exchange, coordination, complex formation, oxide reduction, nonpolar microprecipitation, and enzymatic mechanism, are only a few of the physical and chemical processes involved in this adsorption procedure. Heavy metal, nanoparticles, antibiotic, hepatic gas, dye and nanoparticles exposures all showed significant improvements after AS treatment. Microorganism adsorption was shown to be enhanced by the presence of extracellular polymers in sludge and by ions in sewage.

- **Membrane filters:** Over the last several decades, membrane filtration methods have been widely used for treating household wastewater. The wastewater runs downhill through a layer of microbial biofilm that forms on a bed of gravel, pebbles, or plastic media in these systems. Adsorption processes of organic and certain inorganic substances, such as nitrates by a microbial biofilm layer under aerobic circumstances are included in the pollutant removal from the wastewater stream. Thick biofilm layers eventually shed into the effluent and combine with primary sludge to generate secondary sludge. The removal of VOCs (volatile organic compounds) by biotrickling filtration has been found to be both cost-effective and efficient. Because bacteria, fungi, and viruses (or "microbes") constitute the foundation of BTFs, the selection of certain strains has a considerable impact on their efficacy. When it comes to biogas purification and N removal, the two-stage trickling filter may be used as a cost-effective desulfurization technique.
- **Membrane bioreactor:** Effective removal of COD and other minor pollutants is achieved using a membrane bioreactor. Wastewater from a variety of commercial and residential sources may be treated effectively using a membrane bioreactor. Membrane bioreactor systems are an intelligent option for wastewater treatment in the near future despite the fact that certain current water treatment solutions are more cost and energy-effective than membrane bioreactor technology. AS was used in almost every study that sought to employ a membrane bioreactor to treat contaminated surface water, and this resulted in membrane fouling. As a solution to the membrane fouling issue, a tethered growth. the immobilized microorganisms on the carrier of a membrane bioreactor provide a potential technical option, as they help to retain biomass inside the reactor, hence enhancing the biodegradation rate. The use of a sequencing-batch membrane bioreactor has the ability to lessen the ecological danger posed by antibiotics in swine wastewater by removing up to 60% of the organic matter and nutrients present in the water.

6.6 THE APPLICATION OF GENETICALLY MODIFIED MICROBES AND THE ELIMINATION OF POTENTIALLY HARMFUL SUBSTANCES

Many methods exist for enhancing wastewater treatment and eliminating xenobiotic compounds, with genetic engineering of microorganisms being one of them. Xenobiotics are compounds that are released into the environment from all sources due to human activity and thus take place at concentrations higher than natural.

6.7 DISCUSSION

6.7.1 CASE STUDIES

Artificial neural networks were done by Alsulaili and Refaie[9] to estimate the efficacy of wastewater treatment facilities in Kuwait, using input parameters including conductivity, pH, water temperature, and biological oxygen demand. Results revealed that adding more than three parameters to the model had little influence on the artificial neural networks, and that the initial concentration of conductivity and biological oxygen demand had the greatest impact on the effluent quality of water. Sensitivity analysis indicated that the initial concentration of chemical oxygen demand was the most important factor in determining the biological oxygen demand in the effluent. Several important indicators of water quality were utilized by the authors, including total suspended solids, total phosphorus, nitrates, total nitrogen, ammonia, total algae counts, and turbidity. While the total nitrogen and total phosphorus readings were within the permitted ranges, the turbidity rates were found to be higher than those considered safe for human consumption. Authors argue that a decent estimate of water quality may be obtained by evaluation and monitoring utilizing a mixture of in situ measurements and remote sensing. The authors have also developed a system for continuous monitoring of water quality in Olushandja Dam using remote sensing (Namibia). Predictions of heavy metal removal from synthetic wastewater using a rotating biological contactor are also made using artificial neural networks in another research. This time the metals of interest were Pb (II), Cd (II), and Cu (II).[17] After collecting 90 samples from the rotating biological contactor over the course of 3 months, the authors fed them into a feed-forward back propagation neural network and utilized the resulting model to make predictions about the metal and chemical oxygen demand removal rates. Authors determined a network design of 2-12-2 and hydraulic retention time as the most responsive parameter for toxic metal and chemical oxygen demand removal in the rotating biological contactor after extensive trial and error during training and testing and by properly optimizing the internal network parameters.

In a wine-producing area of France,[12] surveyed residents to see whether they would support using treated wastewater for agricultural irrigation or landscaping purposes. After interviewing 845 prospective customers face-to-face, the authors expected to find some intriguing findings. Farmers were against the idea of wastewater reuse, working women were less receptive than men and with the development of adequate communication, farmers might increase the social acceptance of wastewater reuse, but at the cost of losing their wine customers. The writers of this study do an excellent job of laying out the perspectives of people who purchase their fresh produce, meat, and wine from this region on the topic of a wastewater reuse project. According to the authors, noted that it was challenging to persuade the hesitant group, who are fewer medium friendly and wary of health concerns, to support wastewater reuse initiatives in the wine-growing area. Vasudevan and Natarajan (2021) performed a survey using questionnaires to assess people's knowledge of the need for better rainwater collecting systems in Tamil Nadu, India. The authors developed 45 survey questions using a multiple-choice format, dividing them into subcategories, such as demographic information (11%), information about water sources (11%), loans (20%), quality of water and treatment awareness (22%), technical knowledge of operation, solving wastewater problems and cleanup procedures (36%). This research found that although some individuals prefer direct groundwater replenishment, the vast majority of people utilize gathered water for household uses, either from roofs or from porches and gardens. In addition, this research brought to light a significant public concern, the need to treat rainwater stored in tanks before it is used for human use, and to enhance its quality.

Algae and waste AS harvested from a brewery effluent treatment system were evaluated for their potential as a fertilizer by Taylor et al. (2021). This research examined the effects of WAS and algal fertilization to those of a traditional inorganic fertilizer on soil quality and agricultural yields. Whether the soil was fertilized with fertilizer, WAS, or algae, the output of Swiss chard was found to be comparable, and the salt absorption ratio of the soil was raised in all three cases. The pH of soil, and electrical conductivity, and exchangeable acidity were all comparable across all treatments of fertilizer, whereas the inorganic fertilizer had the highest quantities of P and K.

A resource-based method to reducing agricultural non-point source pollution via the use of biological agents was proposed by Kan and Duan,[15] who also underlined the significance of non-point source pollution. Authors state that this novel method efficiently used household organic waste, wastewaters, and animal waste to produce biological agents to remove toxic chemical agrochemicals that lead to non-point source pollution. Since 2010, the suggested method has been used in Xichuan County and Xixia County, demonstrating its efficacy in reducing pollution from non-point sources. In order to ensure the long-term success of this strategy, the authors suggested two changes: (i) enhancing environmental safety infrastructure in developed and undeveloped areas to better ensure clean water, and (ii) implementing an organization and business activity of rural pollution control by enlisting the help of local stakeholders and the farming communities.

6.8 CONCLUSIONS

Particularly in underdeveloped regions, a viable answer is needed to the problem of inadequate water supply in the face of rising water pollution and water shortages. Sustainable methods for treating wastewater are the future, and their adoption is a step in that direction. One possible approach is to use biotechnology. Utilizing the capabilities of the microorganism and cultivated tissue cells requires a multidisciplinary approach that incorporates microbiology, engineering, and biochemistry. AS is one of the most popular and cost-effective biotechnological wastewater treatment technologies which effectively adsorb contaminants in wastewater. The low cost and low environmental impact of anaerobic treatment make it a viable alternative to physicochemical methods. Biotechnological approaches, include media filters, biocatalysts, and membrane bioreactors, rotating biological contactors, microbial biosorption, and many more. To further enhance wastewater treatment and get rid of the xenobiotic chemical, genetically engineered microbes are utilized. There are a variety of biotechnological approaches to wastewater treatment, but there is still room for innovation and adaptation.

KEYWORDS

- **wastewater**
- **anaerobic treatment**
- **biosorption**
- **usage of membrane filters**
- **application of biocatalysts**
- **usage of hydroponic system**
- **genetic engineering**

REFERENCES

1. Devi, R. Evaluation of Low-Cost Bio-Technology for Community-Based Domestic Wastewater Treatment. In: *Biotechnology: Prospects and Applications*; Salar, R. K. G.; Siwach, S. K.; Duhan, P. J. S., Eds., Vol. 17; Springer, 2013; pp 227–234.
2. Zhang, D.; Jinadasa, Q.; Gersberg, K. B. S. N.; Liu, R. M.; Ng, Y.; Tan, W. J. et al. Application of Constructed Wetlands for Wastewater Treatment in Developing Countries. A Review of Recent Developments (2000–2013). *J. Environ. Manag.* **2014,** *141*, 116–131.
3. Crini, G.; Lichtfouse, E. Advantages and Disadvantages of Techniques Used for Wastewater Treatment. *Environ. Chem. Lett.* **2019,** *17*, 145–155.
4. Gallego-Schmid, A.; Tarpan, R. R. Z. Life Cycle Assessment of Wastewater Treatment in Developing Countries: A Review. *Water Res.* **2019,** *153*, 63–79.
5. Hodaifa, G.; Paladino, O.; Malvis, A.; Seyedsalehi, M.; Neviani, M. Green Techniques for Wastewaters. *Interface Sci. Technol.* **2019,** *30*, 217–240.
6. Singh, R. L.; Singh, R. P. Advances in Biological Treatment of Industrial Waste Water and Their Recycling for a Sustainable Future. In *Applied Environmental Science and Engineering for a Sustainable Future*; Springer, 2019.
7. Magwaza, S.; Magwaza, T.; Odindo, L. S.; Mditshwa, A. O. Hydroponic Technology as Decentralized System for Domestic Wastewater Treatment and Vegetable Production in Urban Agriculture: A Review. *Sci. Total Environ.* **2020,** *698*, 134154.
8. Owa, F. D. Water Pollution: Sources, Effects, Control and Management. *Mediterranean J. Soc. Sci.* **2013,** *4* (8), 65–68.

9. Alsulaili, A.; Refaie, A. Artificial Neural Network Modeling Approach for the Prediction of Five-Day Biological Oxygen Demand and Wastewater Treatment Plant Performance. *Water Supply* **2021,** *21* (5), 1861–1877.
10. Blanco, G. C.; Stablein, M. J.; Tommaso, G. Cultivation of Chlorella Vulgaris in Anaerobically Digested Gelatin Industry Wastewater. *Water Supply* **2021,** *21* (5), 1953–1965.
11. Chen, H.; Wang, Z. An Inexact Multi-Stage Interval-Parameter Partial Information Programming Model for Water Resources Management Under Uncertainties. *Water Supply* **2021,** *21* (5), 2139–2157.
12. Garin, P.; Montginoul, M.; Noury, B. Waste Water Reuse in France—Social Perception of an Unfamiliar Practice. *Water Supply* **2021,** *21* (5), 1913–1926.
13. Gong, X.-J.; Dong, Y.-Q.; Li, W.-G. Mesoporous Structure Regulation of Activated Carbons and Effects on the Synergistic Efficacy of Synchronous Adsorption and Bio-Degradation in Biological Activated Carbon Process. *Water Supply* **2021,** *21* (5), 2058–2074.
14. Jin-Xi, W.; Singh, R. P. Experimental Study on the Performance of Modified Flocculant (PEI-T) for Removal of Turbidity and Heavy Metals from Aqueous Solution. *Water Supply* **2021,** *21* (5), 2075–2083.
15. Kan, D.; Duan, P. Resource-Based Agricultural Non-Point Source Pollution Control Using Biological Agents. *Water Supply* **2021,** *21* (5), 2035–2046.
16. Kapalanga, T. S., Hoko, Z., Gumindoga, W.; Chikwiramakomo, L. 2021 Remote-Sensing-Based Algorithms for Water Quality Monitoring in Olushandja Dam, North-Central Namibia. *Water Supply* **2021,** *21* (5), 1878–1894.
17. Kiran, M. G., Das, R., Behera, S. K., Pakshirajan, K.; Das, G. 2021 Modelling a Rotating Biological Contactor Treating Heavy Metal Contaminated Wastewater Using Artificial Neural Network. *Water Supply* **2021,** *21* (5), 1895–1912.
18. Darweesh, M.; Zedan, A. M.; El-Banna, A.; Elbasiuny, H.; Elbehiry, F. Biotechnology for Green Future of Wastewater Treatment, 2021.

CHAPTER 7

Assessment of Bioremediation Techniques for Wastewater Management

SUBBULAKSHMI GANESAN, G. PADMAPRIYA, AGAMPODI SANDUNI, and ANUPAMA DE ZOYSA

Department of Chemistry, Jain University, Bangalore, Karnataka, India

ABSTRACT

Environmental pollution is increasing daily due to the toxins which are released to the environment by the anthropogenic activities. Bioremediation is a technique used to reduce the toxicity of certain harmful chemicals that are released into the environment by using plants, animals, or microbes. Depending on many factors like cost, type, and concentration of the pollutants which are present, bioremediation is carried out as ex situ as well as in situ. Moreover, the engineered bioremediation techniques, such as biosparging, bioventing, bioslurping, biostimulation, bioaugmentation, and natural attenuation; slurry phase bioremediation, such as bioreactor; and solid phase bioremediation, such as biopiling, land farming, composting, and bio filter are used. The bioremediation techniques can be used as ecofriendly methods for wastewater treatment.

7.1 INTRODUCTION

Environmental pollution is occurring due to the anthropogenic activities. Bioremediation is a successful technique which is used to reduce the environmental pollution occurred due to the toxins released to the

Sustainable Water Treatment and Ecosystem Protection Strategies. Hossein Hariri Asli, Tamara Tatrishvili, Ann Rose Abraham, & A. K. Haghi (Eds.)

environment.[1] Plants, animals, fungi, algae, and bacteria are mainly used in reducing pollution. Moreover, bioremediation is a technique which is highly involved in immobilization, eradication, detoxification of the toxins, as well as with degradation. Here, the toxins are converted into lesser toxic forms. Bioremediation consists of different methods, such as bioattenuation, biopiles, bioaugmentation, biostimulation, bioventing, and biopiles. It is a technique which is very much effective, ecofriendly, economical, and can be easily applied in order to prevent pollution. Wastewater contamination creates huge environmental issues. In any country, aquatic systems get affected first for pollution. Pollution can happen in two ways—point and nonpoint sources of pollution.[2] If the pollutant is emitted directly to the environment, it is called as point sources. The most common point of sources of pollution are the municipal as well as industrial waste water, where the runoff from the industrial sites, drainage, and solid disposal sites are included here. The flow of waste water from the agricultural lands, orchards, and urban runoff which comes from unsewered areas is included as nonpoint sources. The pollution of water highly affects not only the aquatic organisms, but also birds and terrestrial animals.[3] Contaminated water can destroy aquatic lives as well as terrestrial animals. It reduces the ability to reproduce too. Finally, polluted water becomes unsuitable for consumption in day-to-day life of man, as well as for domestic usage.

7.2 THE IMPACT OF THE ANTHROPOGENIC ACTIVITIES ON THE QUALITY OF WATER

A severe effect of water pollution is due to pathogens that causes waterborne diseases; about 2 million people every year in the under developed countries are adversely affected because of this. Sewage treatment is the best solution that can be taken for this. Pollutants such as organics, heavy metals, plant nutrients, like N and P, causes pollution of water.

7.2.1 WATER POLLUTANTS

The waste that demands oxygen is a pollutant that is extremely important for the eco systems. When the depth increases in any healthy water body of water, the dissolved oxygen concentration decreases rapidly. "Oxycline" occurs as a result of exposure to the air which is situated in the upper eutrophic zone, where the oxygenic phototrophs such as cyanobacteria and algae are present that split the water by releasing the oxygen gas as a byproduct in photosynthesis. The oxygen amount which is dissolved in the water is used by the aquatic living organisms for their cellular respiration.[4] Archaea and aerobic heterotrophic bacteria are capable of decomposing the dead organic matter into carbon dioxide, oxygen, and water. When the decaying organic matter present in water increases, it will remove oxygen from the water where it will create a deficiency in oxygen for the aquatics as well as for the marine animals.[5] BOD is the level of oxygen which is used by the aerobes in water bodies. An unpolluted water body with respect to oxygen is a turbulent river that flows through a natural forest. Turbulence helps the water to come in contact with the atmosphere where the oxygen level is restored. The dissolved amount of oxygen level is thus restored. The dissolved amount of oxygen in such a river usually ranges from 10 to 14 ppm, where the BOD value is low with clean water fish.

Whereas, a stagnant deep lake present in an urban area with combined sewage system is a polluted water body with respect to the oxygen level. This kind of a water system may contain dead organic carbon in larger amounts because of which there is very less chance for the water to circulate and come in contact with the atmosphere.[6,7] In this kind of a lake, the amount of oxygen which is dissolved is less than or equal to 5 ppm, where the BOD is high with lower levels of oxygen.

7.2.2 EUTROPHICATION AND ALGAL BLOOMS

Nitrogen as well as phosphorous are gathered in water bodies overtime that demand higher oxygen levels. As nitrogen and phosphorous are the main requirements for the growth of cells and are present in the lower levels of waters which are not polluted. Higher levels of nitrogen and phosphorous from the anthropogenic activities can cause eutrophication that will lead into algal blooms or the explosive growth of algae as well as cyanobacteria, which will form a thick mat that floats while creating water pollution, causing damage for the eco system due to the clogging of the gills of fish, blocking the sunlight, and by damaging the lake aesthetics.

Some of the species of algae and cyanobacteria and algae have the ability to kill fish, birds, and the mammals and also cause illness in humans.[8] The marine environments contain certain species like dinoflagellates that can cause algal blooms referred as "red tides." These particular algae have the ability to secrete toxins that can kill fish as well as the other organisms and they can accumulate in the bodies of shell fish.

When humans and the other animals consume these, they will get paralytic shell fish poisoning. This is an intoxication that is caused due to the ingestion of dinoflagellates toxin that can't be inactivated by cooking. This is a fatal case. Cyanobacteria too can form blooms in the marine and also in the fresh water ecosystems and some are capable of production of the toxins. They are known as microcystins that form allergic reactions in the body and also causes damage in the liver. After contaminating the water, when the algal bloom dies, it creates a dead zone or a hypoxic zone that becomes an oxygen-demanding zone.

7.2.3 OIL SPILLS IN THE OCEAN

Oil spills in the ocean are considered to be an alternative method of causing organic pollution. When the oil spills on the water surface, as this is less dense than the water, the light-weighted hydrocarbons is removed by evaporation which reduces the amount of oil spill.[9] These hydrocarbon-degrading organisms decomposes the oil, which is a time-taking process. After sometime, only 15% of the volume of oil remains but it is present as a thick asphalt lump, which is very harmful to the aquatic life.

7.2.4 PATHOGENS

Pathogens enter water from the human faecal waste matter because of the inadequacy of the sewage system.[10] In many of the underdeveloped countries, waste water are released into the waterways, which are not treated. According to the WHO, 880 million people in the world didn't have drinking water in 2008. And, also about 2.6 billion people, who had access to public sewage system, lived without improved sanitary facilities; there were simple pit latrine and septic tanks. Every year about 1.7 million people in the world die due to diarrhoea caused by polluted drinking water, poor sanitation, and hygiene. About 90% of the deaths are reported among the children under the age of 5.

7.2.5 BIOREMEDIATION OF OIL SPILLS

Marine bacteria have the ability to degrade the oil that exists in the world's oceans. Most of the bacteria are aerobes. Some of the hydrocarbons in the oil particles are broken down into very small particles. For example, *Alcanivorax borkumensis* is a bacterium that can produce surfactants that can solubilize oil. Some have the ability to form carbon dioxide by degrading the oil. Moreover, this bacterium is genetically modified to increase degradation efficacy. The problem here is that the amount of oil spill is greater than the bloom of the bacteria. Therefore, lots of bacterial species are needed to be generated from the genetic engineering technology. By using the bioremediation agents, we can reduce the environmental cost as well as the final cost.[11] Seeding of the waste water, which is contaminated with the competent microflora that is able to degrade the hazardous waste, is a popular method. The microbes which are used in this process may be the naturally occurring organisms or the prepared microbes in the lab to attack the particular waste compound.

7.3 THE ROLE OF MICROBES IN THE PROCESS OF BIOREMEDIATION

Microbial bioremediation is a process that can destroy or immobilize contaminants, and thereby protect the peoples' health and the environment. Moreover, this is a cost-effective method. Hence, microbes that are genetically modified also can be used in this process.[12] This bioremediation activity by using the microbes can be stimulated by supplementing the nutrients such, as nitrogen, phosphate, and oxygen as the electron acceptor, and the substrates, such as methane, toluene, and phenol, or by the introduction of microbes having desired catalytic capabilities. Achromobacter, Arthrobacter, Alcaligens, Acinetobacter, Flavobacterium, Corynebacterium, Bacillus, Nocardia, Micrococcus, Mycobacterium, Pseudomonas, Rhodococcus, Vibrio, and Sphingomonas species are commonly used microbes in bioremediation. Effective wastewater treatment can be done by using the lactic acid bacteria called as Lactobacillus plantanum, Streptococcus lacti, Lactobacillus casei, and the photosynthetic bacteria such as Rhodobacter spaeroide and Rhodopseudomonas palustrus ect.[13] Arthrobacter has the ability to reduce nitrogen by the bioremediation of polychlorinated biphenyls in the sediments of the river and dechlorination of trichloroethylene and chloroform. The microbe Enterobactercloacae has the capability of utilizing trinitrotoluene as a source of nitrogen.[14] Pentaery thriol tetra nitrate reductase is an enzyme that involves in the degradation of the esters of nitrates that are capable of reducing the TNT aromatic ring that causes the liberation of the nitrite.

Endosulfin, which is an extremely toxic pesticide that causes toxicity for the fish and aquatic invertebrates, binds to the soil particles and can persist for a very long time with a half-life that consists of 60–800 days. About 57–90% of the toxin alpha endosulfun as well as 74–94% beta endosulfun within a period of 7 days were degraded by bacteria, such as Pseudomonas and Arthrobacter species. Moreover, *Pseudomonas putida* has the ability to degrade petroleum. According to most of the researches, *Pseudomonas* species has the capability of degrading 42 different types of pollutants such as black liquor from the kraft pulp, tannery effluent, paper mill effluent, and steel industrial effluent.[15] Aerobic bacteria like *Pseudomonas, Sphingomonas, Alcaligens, Mycobacterium,* and *Rhodococcus* are the microbes that are capable of degrading the toxic pollutants. They have the ability to degrade pesticides as well as hydrocarbons. Phenol degradation and also the PAHs in a batch culture are degraded by *Exiguobacterium aurantiacum.* Phosphate removal can be successfully done by *Acinetobacter* species.

Not only that, *Lampropedia* species, *Aeromonas* species, *Vibrio* species, *Pseudomonas species*, and *Tetraphoera* species have the ability to remove phosphates in the water. *Pseudomonas* species and *Brew bacillus* species are nitrate-reducing bacterial starins which are isolated from the petroleum-contaminated soil.

7.3.1 BIOAUGMENTATION

Bioaugmentation is a method in biotechnology to remove the recalcitrants in the industrial waste water.[16] The activated sludge obtained from waste water contains the microbes which occur naturally and can biodegrade a lot of pollutants in water. But there are some toxins that are biodegradation resistant. Bioaugmentation can overcome the challenges mentioned above. This method can address the higher number of emerging pollutants that are available in higher concentrations. Table 7.1 shows the examples of bioaugmentation of waste water released from industries for bioremediation.

TABLE 7.1 Bioaugmentation Treatment.

Pollutant	Medium for setting up	Medium needed in bioaugmentation	Bacteria which is augmented
4-Fluoroaniline	Batch reactor	A medium with inorganic salt	*Acinetobacter*
Quinoline	Sequential batch reactor	Waste water from petroleum refinery	*Bacillus*
Bromoamine	MBR 10 L	SW	*Sphingomonas xenophaga*
Cyanide	Cokes with full scale of wastewater treatment	Waste water with cokes	*Cryptococcus humicola*
Nicotine	SBR	Tobacco waste water which is diluted in tap water	*Pseudomonas*
Naphthalene	MBR	Coal gasification of waste water	*Streptomyces*
Phenol mixture, pyridine, Quinoline, naphthalene, carbazole	A system of sequential anaerobic reactor which is having volume 4.9 L	Wastewater coaking	Consortium of six bacteria with *paracoccus denitrificans* and five starins of the species *pseudomonas*
Phenol	Biological way of oxidation reactor contamination	Gasification of coal	A mix of bacteria which can degrade phenol

Moreover, the bacterial species which are capable of hydrocarbon degradation are *Bacillus, Corynebacterium, Staphylococcus, Streptococcus, Alcaligens, Acinetobacter, Shigella, and Klebsiella.* The bacterial species which are capable of degrading the aromatic hydrocarbons are mainly extracted from the soil.[17] These are normally gram negative bacteria and most of them belong to the genera of *Pseudomonas.* Biodegradable channels also can be found in the *Mycobacterium* gene, *Rhodococcus, Bacillus, Aeromonas,* and *Corynebacterium.* Though many bacteria have the ability to metabolize the organic pollutants, there is one type of bacteria which has enzyme potential to degrade the organic compounds.

7.3.2 BIOREMEDIATION OF INDUSTRIAL EFFLUENTS

Industrialization and urbanization are causing enhanced levels of organic contaminants in the environment. *Cyan bacterium* is used for treating the industrial effluents, bioremediation of the aquatic as well as the terrestrial habitats, and detoxification of the effluents and biofertilizer. *Oscillatoria, Synococcus,* and *Cyanothece* species are cyan bacterial species with higher biodegradation as well as biosorption of the effluents. The activated sludge treating of the municipal sewage comprises of the microbes like *Pseudomonas, Alcaligens*, *Flavobacterium*, and *Zoo lea* species.

7.3.3 BIOSPARGING

When a particular place is filled with mineral oil or compounds with aromatic substances, such as benzene, xylene, toluene, and naphthalene, those toxins can be extracted by using a biological method called as biosparging. Aerobic bacteria, which are capable of breaking down the mineral oils as well as BTEXN, are used here. Most of the times, the deeper layers of the soil as well as the ground water are anaerobic. In order to obtain the bacteria which uses oxygen, we should provide this into the soil as well as ground water with the filters known as injection filters. The supply of oxygen directly in these bacteria improves the capacity for degradation of the minerals called as oils and also as BTEXIN. This technique is suitable for in situ remediation of both the soil as well as groundwater and it's the best method for the removal of mineral oil. This method is used to remove higher amount of mineral oil too.

Normally, the duration of a project with remediation will vary several years and it will depend majorly on its concentration—the different types of contamination present and the extent to which the contaminants present in the soil contain oxygen. After the process of remediation of mineral oil, some of the residual contaminations consist of immobile oil fractions, which are very heavy without the risk for diffusion when compared with the physical methods of in situ techniques. The amount of the residual physical modes is predominantly targeting the volatile as well as fraction of oil that are soluble. Moreover, the biological ways can remove heavy weight and the less mobile fractions. Therefore, biosparging can be considered as a technique which is more effective as well as sustainable and very cheap.

7.3.4 BIOVENTING

Bioventing is an in situ remediation technique which uses the microbes for biodegrading the organic constituents which are adsorbed on the soil in the zone of unsaturation. Bioventing can enhance the activity of the indigenous bacteria which are having the capability to stimulate the natural in situ biodegradation of the hydrocarbons. In this process, oxygen is supplied by the injection of air directly to the residual contamination. Bioventing helps in the degradation of the volatile compounds which are organic along with the vapors which move very slowly. This rate of natural degradation is limited normally due to the lack of the nutrients. In the systems with conventional bioventing, the amount of oxygen will be gained by an electrical power for the subsurface wells. While comparing with the soil vapor, vacuum extraction, this bioventing method uses a low rate of airflow in order to provide enough amount of oxygen in order to sustain the activity of the microbes. Passive bioventing systems are capable of using the air exchange naturally in order to transfer oxygen for the sub surface by the bioventing wells. One valve will allow air to enter the well when the pressure which is in the well is less than the pressure of the atmosphere. When the pressure of the atmosphere reduces, the pressure within the subsurface, the valve will close by the trapping of air inside soil by surrounding that well. This method is normally used in the bioremediation of soil. But if we reduce the toxins that are present in the soil, it is easy to remove the toxins that are available in water too. This is because ground water can get polluted if the surface water is toxic.

7.3.5 NATURAL ATTENUATION

Natural attenuation can be used to clean up the contaminated ground water into which the municipal waste water is released in the soil-cleared water systems. Leachates can cause a series of range of redox reactions that normally occur in the subsurface. Intrinsic bioremediation is a method which is causing degradation of the contaminants by the indigenous microorganisms under the prevailing conditions.

Aqueous and mineral intrinsic biodegradation assessment can be used to asses several aspects of intrinsic bioremediation, such as,

a) Redox processes—Redox reactions caused by the microbes.

b) Plume finger print
c) Expressed capacity—The amount of the hydrocarbon that is destroyed by a redox process.
d) Rate of decay—Contaminant source as well as the plume provides necessary parameters for the modeling.

Natural attenuation contains the methods of precipitation, dispersion, volatization, sorption, advection, intrinsic biodegradation, and hydrolysis. Natural attenuation should be responsibly as well as effectively applied in order to be established as a leachate management option. Natural attenuation must be quantifiable. If natural attenuation has the ability to eliminate the need for the active method of remediation, the landfill cost would be significantly reduced. Requirements that are included in the intrinsic biodegradation of the organic compounds include the presence of the organic substrates, electron acceptors, microbial consortium, as well as the absence of the factors that will inhibit the growth of the microbes.

Few studies that concern the landfill leachate exist and such studies have primarily dealt with the sandy, aerobic aquifers and complicating factors, such as possibility of the presence of different plumes and inability for the usage of production of chlorides as an indicator. Redox zones are dependent on local conditions that will develop after the entry of the leachate and also such conditions will form in a way that the methanogenic conditions are noted which are near to the source in the oxidizing conditions along with the perimeter.

7.4 BIOREMEDIATION OF AQUATIC WASTE WATER

Aqua culture is one of the fastest growing food production fields because of its traditional polyculture. This is based on optimum utilization of farm resources, farm wastes, etc. Modern farming technologies use fertilizers and chemicals in higher amounts and due to this the aqua culture is now considered as one of the potential polluters in the aquatic environment.

7.4.1 WASTEWATER PRODUCTION IN AQUA CULTURE

This waste which is present in aqua culture farms is considered as:

1. Residual food matter and the faecal particles in fish

2. By-products of the process of metabolisis
3. Residual present in the biocides as well as the biostats
4. The waste derived from fertilizers
5. The waste produced in the period of moulting
6. Collapsing of the algal blooms

The usage of the bioremediation will result in a very lower accumulation of organic matter in to water resource. Better oxygen would thus be present in the system and finally leading to form a better aquatic environment. A successful method of bioremediation involves:

1. Optimization of the nitrification rate for keeping the low concentration of ammonia.
2. Increasing the rates of denitrification for eliminating excess of nitrogen from aquatic systems as the nitrogen gas.
3. Improving the oxidation of sulphide in order to minimize the sludge accumulation.
4. Improving the primary productivity of shrimp and also the secondary crops.
5. Maintenance of diverse and stable aquatic community where the species which are undesirable do not become dominant.

7.4.2 BIOREMEDIATION OF ORGANIC DETRITUS

The organic matter which is dissolved and suspended in water mainly contains the microbes that are capable of clearing the carbonaceous wastes effectively from water. Species such as *Bacillus subtilis, Bacillus licheniformis, Bacillus coagulants,* and *Phenibacillus polymyxa* are the examples of bacteria that are suitable for bioremediation of the organic detritus.

Usually these microbes are not present in necessary numbers in a water column. If we add these species in relevant amounts for the plants where it needs, it is possible to carry out bioremediation with the help of these microbes.

Apart from all which comes under bioaugmentation, the *Bacillus* species are produced and they can be mixed with sand or clay and can be deposited under the titration of an aquatic system.

Lactobacillus too is another species which can be used along with *Bacillus* in order to degrade the organic matter into smaller parts like detritus. These bacteria have the ability to produce various enzymes that can break down the proteins and the starch into smaller molecules which are taken up as the sources of energy by the other microbes. This removal of larger organic compounds is capable of reducing the turbidity in water.

7.4.3 BIOREMEDIATION OF NITROGENOUS COMPOUNDS

Excess of nitrogen compounds in aquatic systems leads to the deterioration of the quality of water through the gathering of the chemical compounds with nitrogen, such as ammonia and nitrate, causing toxicity for fishes as well as shrimps. The principal ammonia sources are the flux which is obtained from the mineralization of the organic matter as well as the diffusion of molecules and the sediments which are reduced;, even though the nitrogen fixation done by cyanobacteria and the atmospheric deposition are also important.

7.4.4 USAGE OF DIFFERENT BIOREMEDIATION TECHNIQUES FOR WASTEWATER TREATMENT

"Treatment technology" is a series of unit operations which can change the components of a hazardous chemical substance through the chemical, physical, or the biological practices and therefore it will lower the toxicity, volume of the material which is contaminated, and the mobility.

Bioremediation is a method used in restoration of the damage of the eco systems through the increase in different kinds of living beings such as fungi, plants, as well as bacterial species which are capable of removing, degrading, or transforming of the toxic organic substances into lesser toxic metabolic products. The most common types of bioremediation techniques are phytoremediation, chelation leaching, precipitation, electrobioremediation, and methylation. In phytoremediation, the plants that have microbes associated in the rhizosphere are used for the recovery of the contaminated water. This is an in situ method where there are no expensive costs and this is environmentally as well as ecologically friendly. Phytoremediation consists of phytoextraction, phytovolatization, phytodegradation, and phytostabilization.

Plants that are used in phytoremediation should have capability of higher biomass production, tolerance to environment of the ecosystem, preferably the local species, and should be easily harvestable. The technique of phytoremediation of the heavy metals is useful enough because of the ability of the plants for the absorption as well as accumulation of the contaminants of inorganic substances. Plants differ in their capability for the absorption as well as accumulation of the toxic metals obtained from water, sludge, and soil. The plants have different mechanisms against the availability of the metals that are present in their environment based on the resistance for the processing of the efficient mechanisms needed in the metal exclusive that will restrict their transport for the aerial parts, as the nontoxic chemical species. The plants species used for the decontamination of the heavy metals are *Calendula officinals, Alcearosea, Impatiens balsa mina, and Zea mays.* Some of the above-mentioned plant species falls under the hyper accumulators due to their capability of accumulating a higher concentration of the heavy metals present in the tissues of the plants. Rhizofiltration is similar to the process of phytoextraction and it can be used in decontaminating the waste water by the filtration of the weight of roots of the plants which are grown hydroponically and therefore the contaminants which are dissolved can be absorbed and they will be collected. In studies conducted in Mexico with the help of hydroponic plants such as *Helianthus annuus* and *Mentha deriva*, such plants were not successfully able to remove the heavy metals but they were able to remove nitrates of 60% and phosphates of 40%. Moreover, the usage of the macrophytes for the wastewater removal has better results.

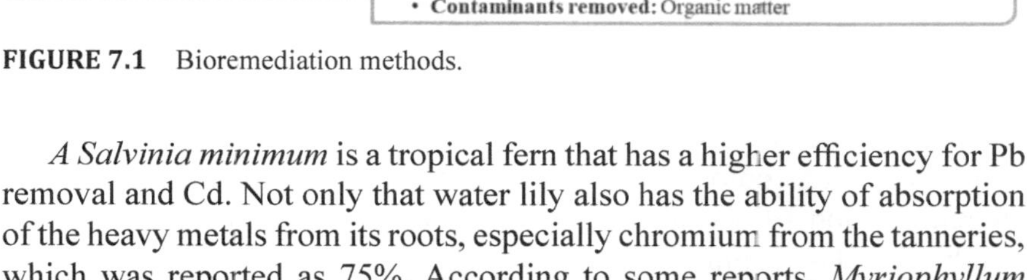

FIGURE 7.1 Bioremediation methods.

A Salvinia minimum is a tropical fern that has a higher efficiency for Pb removal and Cd. Not only that water lily also has the ability of absorption of the heavy metals from its roots, especially chromium from the tanneries, which was reported as 75%. According to some reports, *Myriophyllum aquaticum* has the capability of removal of chromium and copper. Moreover, several successful studies too were done based on *Eichhornia crassipes*, which is believed to have the ability to remove the heavy metals. From the experiments conducted based on *Eichhornia crassipes*, it was found that it possesses the ability of removal of chromium and copper and it has the ability to absorb metals with different concentrations of $NiCl_2$. The accumulation ability of metals by *E. crassipes* is based on the plant's age, light intensity, temperature, concentration of the metals, which are toxic, as well as due to the competition among the other ions. It is also very efficient in the removal of cadmium than copper as well as iron from the wastewater that comes from the textiles, metallurgy, and pharmaceutical industries.

Additionally, the aquatic plants have the ability to remove radioactive metals like the ability to remove radioactive metals like uranium. The aquatic plants, which can remove uranium, are *Callitriche stagnalis, Potamogeton peetinatus*, and *Potamogen natans.* These plants are found in central Portugal and have a higher biomass accumulation and they are efficiently utilized for the phytofiltration of uranium from the contaminated water for gaining higher reductions in the uranium amount.

The plants that are applied in the phytoremediation are known for the tolerance under the extreme amounts of both organic as well as inorganic compounds as well as from their physiological mechanisms that are included with resistance, survivability in the contaminated water, as well as their tolerance. Trees are also used in the phytoremediation of waste water.

Willow plants are also used in treating waste water having a content of $N\text{-}NH^{4+}$ that can achieve higher decontamination efficiency due to higher evapotranspiration rate as well as due to the nutrient retention.

Nitrogen and phosphorous from waste water can be removed by microalgae also. *Chlorella* and *Scedesmus* are some of the microalgae's that are used in removing the heavy metal substances. Use of *Chlorella vulgaris* in tertiary wastewater treatment reduces nitrogen and phosphorous content; whereas, *Scenedesmus dimorphous* can remove NH^{4+} and phosphorous in the secondary treatments. *C. vulgaris* can remove higher amount of Cd than *Scedemus acutus.* But both the species have the ability to remove Cu to upto 80%. *Chlorella sp.* too gives better results in the bioremediation of the aqua culture in waste water. Azolla–Anabaena symbiotic relationship also provide some of the benefits in the bioremediation of the waste water.

7.5 CONCLUSIONS

Bioremediation of waste water helps to clean and reuse waste water. It is a cheaper technology which is not needed in complex infrastructure and also it is used for the treatment of large volumes of the waste water. Bioremediation is very useful for the removal and transformation of degrading the different kinds of the contaminants that are seen in waste water. It is useful for the detoxification as well as reduction of the toxin levels caused by several types of bacteria, fungi, plant, etc. This is an efficient as well as an ecofriendly method that can be used in the removal of toxins from waste water.

KEYWORDS

- **biosparging**
- **bioventing**
- **bioslurping**
- **biostimulation**
- **bioaugmentation**
- **natural attenuation and biopiling**

REFERENCES

1. Azeez, R. A. A Study on the Effect of Temperature on the Treatment of Industrial Wastewater Using *Chlorella vulgaris*. *Alga. Eng. Tech. J.* **2010,** *28* (4), 1–8.
2. Bankar, A.; Nagaraja, G. Recent Trends in Biosorption of Heavy Metals by Actinobacteria. In *New and Future Developments in Microbial Biotechnology and Bioengineering*; Elsevier, 2018; pp 257–275.
3. Kim, S. J.; Chung, J. H.; Kim, T. Y.; Cho, S. Y. Biosorption of Heavy Metals and Cyanide Complexes on Biomass. *Stud. Surface Sci. Catal.* **2006,** 141–144.
4. Kulshreshtha, A.; Agrawal, R.; Barar, M.; Saxena, S. A Review on Bioremediation of Heavy Metals in Contaminated Water. *IOSR J. Environ. Sci. Toxicol. Food Technol. (IOSR-JESTFT)*. **2014,** *8* (7), 44–50.
5. Newman, L. A.; Reynolds, C. M. Phytodegradation of Organic Compounds. *Curr. Opin. Biotechnol.* **2004,** *15* (3), 225–230.

6. Pandey, R. S.; Upadhyay. B. V. Pseudomonas Fluorescens Can Be Used for Bioremediation of Textile Effluent Direct Orange-102. *Trop. Ecol.* **2010,** *51* (2). 397–403.
7. Paul, D.; Pandey, G.; Pandey, J.; Jain, R. K. Accessing Microbial Diversity for Bioremediation and Environmental Restoration. *Trends Biotechnol.* **2005,** *23*, 135–142.
8. Rizwan, M.; Singh, M.; Mitra, C. K.; Morve, R. K. Ecofriendly Application of Nanomaterials: Nanobioremediation. *J. Nanopart.* **2014,** 1–7.
9. Shamim, S. Biosorption of Heavy Metals. *Biosorption* **2018,** *2*, 21–49.
10. Sharma, G. K.; Khan, S. A. Bioremediation of Sewage Wastewater Using Selective Algae for Manure Production. *Int. J. Environ. Eng. Manag.* **2013,** *4* (6), 573–580.
11. Singh, R.; Singh, P.; Sharma, R. Microorganism as a Tool of Bioremediation Technology for Cleaning Environment: A Review. *Int. Acad. Ecol. Environ. Sci.* **2014,** *4* (1), 1–6.
12. Smith, E.; Thavamani, P.; Ramadass, K.; Naidu, R.; Srivastava, P.; Megharaj, M. Remediation Trials for Hydrocarbon-Contaminated Soils in Arid Environments: Evaluation of Bioslurry and Biopiling Techniques. *Integr. J. Biodeterior. Biodegrad.* **2015,** *101*, 56–65.
13. Suman, J.; Uhlik, O.; Viktorova, J.; Macek, T. Phytoextraction of Heavy Metals: A Promising Tool for Clean-up of Polluted Environment? *Front. Plant Sci.* **2018,** *9*, 1476.
14. Verma, R.; Dwivedi, P. Heavy Metal Water Pollution—A Case Study. *Recent Res. Sci. Technol.* **2013,** *5* (5).
15. Vidali. M. Bioremediation. An Overview. *Pure Appl. Chem.* **2001,** *73* (7), 1163–1172.
16. Wang, W. X. Bioaccumulation and Biomonitoring. In *Marine Ecotoxicology*; Academic Press, 2016; pp 99–119.
17. Youcai, Z.; Ziyang, L. General Structure of Sanitary Landfill. In *Pollution Control and Resource Recovery: Municipal Solid Wastes at Landfill*; Elsevier: Amsterdam, 2017; pp 1–10.

CHAPTER 8

Biofilters: An Organic Approach to Wastewater Treatment

SUBBULAKSHMI GANESAN and MRUDULA P.

Department of Chemistry, Jain University, Bangalore, Karnataka, India

ABSTRACT

Biofilters are one of the most effective and commonly used devices that are suitable for treatment of wastewater. These biofilters are made up of minute and distinct biofilms that are equipped with the task of reducing the impurities in the water and causing them to settle down. The bio films also help to trap the unwanted solids present in water and in turn result in the passage of clean water. Due to the easy set up and usage, biofilms serve as a sustainable method of treatment of wastewater. Biofilters are also suitable instruments to get rid or remove any dissolved gases that may be present in the wastewater. The dissolved gases, such as hydrogen sulphide, in wastewater majorly contribute to its bad smell; hence, biofilters help to remove this bad odor from the wastewater. In certain commercial industries, the wastewater is usually concentrated with volatile compounds, such as toluene and xylene. These compounds are present in commercial products such as detergents, and hence are a major polluting component of the large-scale manufacture of detergents. This chapter elaborates on the method of using biofilters to remove dissolved gases, volatile compounds, and other effluents that are present in wastewater, thereby reducing the major health risks that occur when human beings and other organisms, such as fish, are exposed to this wastewater. As bio filters are easy to set-up and handle, they are a suitable option for filtration and purification of

Sustainable Water Treatment and Ecosystem Protection Strategies. Hossein Hariri Asli, Tamara Tatrishvili, Ann Rose Abraham, & A. K. Haghi (Eds.)

wastewater at a small-scale level. Additionally, biofilters are cost-effective solutions to the rising demand for the treatment of wastewater.

8.1 INTRODUCTION

Biofilters are instruments that utilize organisms to remove the effluents or compounds that are present in wastewater. These organisms are generally referred to as the biomass of the biofilter and they vary according to the type of impurities that are needed to be removed from the wastewater. A common example of biofilters is the use of a trickling filter, which is for the specialized use of wastewater. Other examples of biofilters include sand filters, rock filters, and so on. Bio filters were first made using materials like slag as the medium for filtration. However, nowadays, the range of medium being used is vast, with even plastic being used as the medium for filtration. The working of a biofilter is based on the principle of breakdown of impurities and other pollutants in the wastewater by the microbes present on the surface medium of the biofilter. The need for the use of biofilters increased greatly after it was discovered that microorganisms can grow effectively within the same pipes that are used to carry this water to and from the various households and industries in a city.

The layer of biomass in these pipes can pose many health risks in people who consume or come in contact with this water. Further, if the concentration of chlorine molecules in wastewater increases due to action of harmful microbes on it, the wastewater becomes carcinogenic and even more hazardous. For treating this type of wastewater from these pipes, a particular biofilter, known as GAC biofilter, is best suited. The filtration medium in such a biofilter is mainly occupied by bacterial colonies which breakdown the organic compounds present in the wastewater and use it as their source of nutrition.

As a result of the biofiltration by GAC filter, the overall quality of the wastewater improves with improvement further observed in the smell and taste of the wastewater.

8.2 PRIMARY PROCESSES INVOLVED IN WORKING OF BIO FILTERS

The working efficiency of a biofilter is directly related to the concentration and type of microorganisms that are grown on its medium for filtration.

1. Adherence of the microbes to the filtration medium

The first step of the process of growth of microbes in the biofilter is effectively making the microbe colonies to adhere to the filtration medium. This process first requires the microbes to be carefully moved from their site of initial growth to the vicinity of the biofilter. The second step involves the attachment of the microbes to the surface medium, followed by the process of allowing the microbes to grow and thrive on the surface of the filtration medium. The methods to move the microbes from their in situ position include sedimentation, convection, and Brownian motion. Once the microbes have been moved and allowed to get attached to the medium, it is necessary to check whether this attachment is permanent or temporary.

The reaction of attachment of the microbes with the medium will depend on the total electrostatic force in combination with the van der Waals force of attraction. There are various other factors that affect the attachment process. These include, features of the medium, characters of the chosen microbes, whether they can grow suitably in a water medium, angle at which the microbes have been introduced to the medium, and so on. However, some of these features cannot be manually controlled and can only be chosen or eliminated after trials with different types of microbes are done with different materials of biofilters.

2. Use of medium and growth of microorganisms

Once the microorganisms start growing on the medium, they form layers of colonies or a thin film of biomass (biofilm). There are many factors that affect the rate at which the microorganisms consume their nutrition from the medium. These factors include:

- Transport of the medium within the microbial colonies.
- Rate of absorption of the medium by the microbial colonies.
- The dynamics of absorption and use of the media by the microbial colonies.

3. Release of the microbial layer

The crucial aspect which directly contributes to the working efficiency of a biofilter is the preservation and nurture of the microbial colonies attached to the surface medium of the biofilter. Release of the microbial layers from the biofilter should be avoided at all costs.

The release or removal of microbial layers can occur due to the following factors:

- Abrasion—This occurs when the microbial layers are disturbed or removed due to their rough or vigorous contact with external sources.
- Erosion—This occurs when the rate of flow of water is such that it erodes or chips off the microbial layers from the filter medium.

In some cases, the release of microbial layers can also occur when the flow of water is reversed. This reversed force of water has the ability to severely remove the microbial layers present.

8.3 MODELS TO CHECK THE APPLICABILITY OF A BIOFILTER

- McCarty and Rittman (1980)—Their model was based on a biofilm whose dynamics were explained using the second law of Fick as well as the equation given by Monod.
- The assumption made was that the minimum amount of media concentration required in mass amount is also the same required to ensure that a balanced condition of microbial layers is present in the filter surface.
- Rittman and Chang[15]—They worked on a model that elaborated on the physical dynamics of growing the microbial layers on carbon that was activated.
- They based their work on the concept of breakdown of nutrients by the microorganisms, release of the nutrient media, and growth of the microbial layers.
- Billen (1992)—He developed the model known as "CHABROL." The main function of this model was to work on finding out the amount of organic carbon present that was dissolved and could be broken down by the microorganisms.

  The model further explains three forms of reactions and communications which are responsible for the mechanism with which the microbes grow and thrive on the filter media, they are:

  1. Reaction with the support, which is mainly solid.
  2. Reaction with the organic compounds, present in the dissolved state.
  3. The life span of the microorganisms being grown.

The model further goes on to elaborate how the large-scale functioning of biofilters is in turn dependent on the various processes taking place on the microscopic level. It also helps to find out the microbial layers that have fully adhered to the filter surface as well as the amount of organic compounds present in the outgoing water based on the features of the incoming polluted wastewaters.

The working of the above-mentioned model was experimentally validated in Choisy-le-Roi and Neuilly-sur-Marne wastewater treatment plants located in France.

- Bouwer and Hozalski (2001)—They developed a model known as "BIOFILT" which was numerical based.

  This model was capable of mimicking or replicating the imbalanced condition of microbially lined biofilters, which are mainly used to treat water that is consumed.

  The model has the ability to replicate the nutrient media broken down by the microbes and the microbial layers (adhered and free) in the form of a function of time.

  The model can further detect the release or removal of microbial film owing to reverse flow of water and other reasons.

  However, this model has a few disadvantages, such as:

  1. It makes the assumption that the constituents of water remain constant, despite the changes brought about due to the reverse flow of water.
  2. The breakdown of only one specific form of nutrient media can be measured using this model.
  3. On-site accurate data about the various conditions, such as temperature and pressure, need to be obtained for effective replication/mimicking.

8.4 FEATURES OF THE BIOFILTER SETUP THAT AFFECT ITS FUNCTIONING

1. **Filtration medium:** The choosing of the filter medium depends on different parameters, such as the estimate of the amounts of pollutants present in the wastewater. The conditions will further differ based on the type of wastewater being treated, such as primary and tertiary wastewater. For example, a suitable choice

to treat primary water would be synthetic materials. For tertiary water, coal, sand, or anthracite should be preferred. Since this type of water is going to be mainly used for consumption purposes, the removal of dissolved organic components from this water is of the main concern. The filtration medium should be able to provide the following conditions: sufficient area for the microorganisms to grow on. Its quality should be such that the microorganisms can successfully adhere to it.

2. **Time of contact:** The time of contact is a small but crucial factor that affects the working of the biofilter. It is a significant factor that is directly proportional to the rate at which organic components are removed.
3. **Reverse flow of water:** The microbial layers present in the surface medium of the biofilter should be able to withstand the backwards flow of water in the biofilter. Hence, there is a need to devise effective methods to allow the backflow of water in the biofilter without affecting the adherence of the microbes to the filtration surface. Backflow of water needs to be conducted every 60–90 h of working of the biofilter. It should also be ensured that the microbes are not affected during these backflow sessions.
4. **Temperature:** The rate of microbial activity showed an exponential increase when the temperature was increased gradually. The microbial colonies showed maximum increase in their growth rate when their temperature was increased to about 20 °C higher than their optimum growth temperature.

8.5 SIGNIFICANT COMPONENTS OF BIO FILTERS AND THEIR ESTIMATION

1. **Medium used:** The applications of biofilters are not limited to only primary forms of wastewater. Instead, they can be used to treat potable and tertiary water too. Hence, the components of the biofilter being used will also vary with the type of wastewater that is being treated. For example, for the treatment of primary wastewater, factors such as Biological Oxygen Demand need to be considered. For tertiary wastewater, the measurement of dissolved organic compounds in water needs to be used as the basis of measurement. The main function of biofilters in the case

of potable wastewater is to decrease the amount of requirement of chlorine in the water and to prevent the probability of formation of by-products. Hence, for potable water, the basis of measurement is considered to be AOC and BDOC. These parameters require specific methods and measurements.

The methods used to calculate the above-mentioned parameters include:

1. Werner technique
2. Kemmy technique
3. Van der Kooij technique
4. Jago Stanfield technique.

The units used in the measurement of AOC in technique number 2 are the Colony Forming Units in unit volume of the water solution. In techniques 1 and 4, the content is measured using ATP or adenosine triphosphate as the basis of measurement. All these techniques give precise values that remain constant over a period of time.

2. **Growth of the microbial colonies:** As mentioned previously, the effectiveness of a biofilter is directly proportional to the amount of microbial colonies effectively growing on the surface medium. However, the preservation and support of microbial life on the filtration medium depends on the features of its surface. There arises a need for effective tools to measure the microbial content on the filter surface medium owing to the microscopic size of these colonies. One method to calculate the growth on microbes on the surface is the technique of Heterotrophic Plate Counting technique, for primarily bacteria that is heterotrophic. In the above technique, the enumeration of bacterial colonies was done using an agar called R2A. The colonies were inoculated using the spread plate method. The period of incubation in the above-mentioned method is for 7 days at 20 °C. The standard values for other components in this technique are considered for experimentation. Another technique includes the use of phospholipids to enumerate the microbial growth on the filtration medium. In this technique, about 1 g of the microbial colonies were removed and separated along with the filtration medium. After this, it was washed under tap water, from which the Chlorine content was removed. The washing of the filtration medium with the attached microbial growth was mainly

done to successfully get rid of any deposited suspended solids. Once the suspended solids have been washed off thoroughly, the microbial colonies can be accurately measured without any error caused by the presence of other particles. The first step of the technique involves the isolation of Phosphorus present along with the microbial colonies. This phosphorus is broken down by the microbial colonies and further consumed, causing the formation of the inorganic form of phosphate. The amount of inorganic form of phosphate can be accurately calculated using the colorimeter device, using the Vanadate Molybdate reagent. The concentration of the microbial growth in the above technique is measured in terms of the nano moles of the phospholipids present in per gram of the filtration medium on drying it. With regard to the microbial colonies that were made to grow on carbon that was activated and used as the filtration surface medium, enumeration of bacterial colonies was difficult. The main reason for this difficulty was the major irregularities on the surface area of the carbon medium. Hence, a different technique was devised for these types of filtration media. In this technique, the activity of the bacterial colonies are first observed and enumerated under the standard parameters. Once the measurement is done according to standard parameters, these values are related to the activity of the bacterial colonies on the filtration medium. The glucose respiration rate is used for the above-mentioned co-relations. The unit used for measurement in terms of nano moles of the respired glucose in every hour. For the measurement of microbial colonies on synthetic form of wastewater, the dry mass method is used. This technique of measurement is suitable for microbial colonies present in higher quantities.

The highest amounts of microbial colonies were found to be about 0.1 g per gram of the filtration medium. This quantity was observed after 42 days of use of the biofilter.

The lowest amounts of microbial colonies were found to be about 0.04 g per gram of the filtration medium. This quantity was observed after 63 days of use of the biofilter. A different set of technique showed the quantity of microbial colonies to be about 0.1 g per gram of the filtration medium over a time period of 1 month. The concentration of the microbial colonies is hence directly proportional to the concentration of organic compounds in the water along with the rate of hydraulic loading mechanism.

8.6 USES OF BIOFILTERS

The main use of biofilters is in the process of wastewater treatment. Biofilters can either be used in the process of primary or secondary units of wastewater treatment. The use of biofilters usually are applied in between the processes of pre-treatment and sedimentation. Based on the type of wastewater being treated, there are different customizations available for the shapes and materials to be used for making the filtration medium. Bio filters have also commonly be used in the process of secondary treatment as a form of trickling filter. They are mainly used in this case to treat wastewater from households, otherwise known as domestic wastewater. Biofilters can be used in association with other treatment units to enhance the quality of the effluent water. The main materials used in the construction of a biofilter used as a trickling filter include stones, sand, plastic, or slag. However, the use of biofilters in the form of a trickling filter varies with different parameters, such as, the type of media used, the target objectives of the treatment process, as well as the working of the other instruments in the treatment units.

The combination of two or more biofilter units can also be applied for attaining advanced mechanisms, such as, the process of roughing, nitrification, and different forms of oxidation.

There are many advantages of using biofilters in the form of trickling filters:

1. Less cost of operation
2. Less amount of space occupied by the setup
3. Easy stabilization of the sludge without any floatation issues.
4. Use of biofilters as trickling filters prevents the formation of bulk amounts of the sludge.

8.7 CONCLUSIONS

Biofilters are suitable instruments that can be used in wastewater treatment for obtaining highly treated effluent water. The reasons that make biofilters applicable in wastewater treatment are:

1. Easy to use and operate.
2. They can be used for a long period of time without any wear and tear.
3. High efficiency rate of removal of TOC.

The use of microorganisms ensures that the concentration of organic compounds present in the wastewater obtained after treatment was a constant value over a period of time.

The reverse flow of water that is required to be done every day to loosen the filtration medium, does not negatively impact the microbial colony layers as well as their rate of growth. Hence, there is not affect on the composition of the resulting effluent obtained.

However, from the points mentioned previously, there are a few factors that could directly have an impact on the functioning of the biofilter. These factors include:

1. Rate of filtration.
2. Concentration of organic compounds in the incoming wastewater.
3. Concentration of the influent.

These issues can be resolved by ensuring that the operation of the biofilter is carried out at the standardized range of optimum conditions for effective functioning.

KEYWORDS

- **pollution**
- **waste water**
- **treatment**
- **biofilm**
- **bio filters**

REFERENCES

1. Arora, S.; Rajpal, A.; Bhargava, R.; Pruthi, V.; Bhatia, A.; Kazmi, A. A. Antibacterial and Enzymatic Activity of Microbial Community During Wastewater Treatment by Pilot Scale Vermifiltration System. *Bioresour. Technol.* **2014,** *166*, 132–141.
2. Arora, S.; Rajpal, A.; Kumar, T.; Bhargava, R.; Kazmi, A. A. A Comparative Study for Pathogen Removal Using Different Filter Media During Vermifiltration. *Water Sci. Technol.* **2014,** *70*, 996–1003.
3. Flemming, H. C.; Wingender, J. The Biofilm Matrix. *Nat. Rev. Microbiol.* **2010,** *8*, 623–633.

4. Hammes, F.; Velten, S.; Egli, T.; Juhna, T. *Comprehensive Biotechnology*, 2nd ed.; Elsevier, B.V, 2011.
5. Kumar, T.; Bhargava, R.; Prasad, K. S. H.; Pruthi, V. Evaluation of Vermifiltration Process Using Natural Ingredients for Effective Wastewater Treatment. *Ecol. Eng.* **2015,** *75*, 370–377.
6. Lewandowski, Z.; Boltz, J. P. Biofilms in Water and Wastewater Treatment. *Treatise Water Sci.* **2010,** *4*, 529–570.
7. Naz, I.; Saroj, D. P.; Mumtaz, S.; Ali, N.; Ahmed, S. Assessment of Biological Trickling Filter Systems with Various Packing Materials for Improved Wastewater Treatment. *Environ. Technol.* **2015,** *36* (1–4), 424–434.
8. Prakash, B.; Veeregowda, B. M.; Krishnappa, G. Biofilms: A Survival Strategy of Bacteria. *Curr. Sci.* **2003,** *85* (9), 1299–1307.
9. Rippy, M. A. Meeting the Criteria: Linking Biofilter Design to Fecal Indicator Bacteria Removal. *Wiley Interdiscip. Rev. Water.* **2015,** *2*, 577–592.
10. Schifman, L. A.; Kasaraneni, V. K.; Sullivan, R. K.; Oyanedel-Craver, V.; Boving, T. B. Bacteria Removal from Stormwater Runoff Using Tree Filters: A Comparison of a Conventional and an Innovative System. *Water (Switz.)* **2016,** *8*.
11. Verma, S.; Daverey, A.; Sharma, A. Slow Sand Filtration for Water and Wastewater Treatment—A Review. *Environ. Technol. Rev.* **2017,** *6*, 47–58.
12. Xiong Y, Liu, Y. Biological Control of Microbial Attachment: A Promising Alternative for Mitigating Membrane Biofouling. *Appl. Environ. Microbiol.* **2010,** *86*, 825–837.
13. Bakke, R.; Characklis, W. G.; Turakhia, M. H.; Yeh, A.-I. Modelling a Monopopulation Biofilm System: Pseudomonasaeruginosa. In *Bio Films*. Characklis, W. G., Marshall, K. C.; Eds.; Wiley: New York, 1990.
14. Bouwer, E. J. and Crowe, P. B. Biological Processes in Drinking Water Treatment. *J. AWWA* **1988,** *80* (9), *82*.
15. Chang, H. T.; Rittman, B. E. Mathematical Modelling of Biofilm on Activated Carbon. *Environ. Sci. Tech.* **1987,** *21* (3), 273.
16. Chaudhary, D. S. Adsorption-Filtration Hybrid Systems in Wastewater Treatment and Reuse. PhD Dissertation, University of Technology, Sydney, Australia, 2003.
17. Hozalski, R. M. Removal of Biodegradable Organic Matter in Drinking Water Bio Filters: Experimental Studies and Model Development. PhD dissertation, Johns Hopkins University, Baltimore, 1996.
18. Huck, P. M. Measurement of Biodegradable Organic Matter and Bacterial Growth Potential in Drinking Water. *J. AWWA* **1990,** *82* (7), 78.
19. Lu, P.; Huck, P. M. Evaluation of the Methods for Measuring Biomass and Biofilm Thickness in Biological Drinking Water Treatment. *Proceedings AWWA WQTC*, Miami, 1993.
20. Marquette, P.; Prevost, M.; Servais, P.; Beaudet, J. F.; Coallier, J.; Lafrance, P. Shutdown of BAC Filters: Effects on Water Quality. *J. AWWA* **1998,** *90* (12), 53.

CHAPTER 9

Bioremediation: A Renewable Approach for Wastewater Treatment

SUBBULAKSHMI GANESAN, G. PADMAPRIYA, and J. H. THARUDINI

Department of Chemistry, Jain University, Bangalore, Karnataka, India

ABSTRACT

Environmental biotechnology is one of the major applications in biotechnology. The main objective of environmental biotechnology is to safeguard against the damage caused by the deteriorating environment. Therefore, there are various applications of biotechnology that can be used to remove pollutants from the environment. So that we can apply different techniques of biotechnology to treat wastewater. Pollution due to wastewater is a critical problem that negatively affects human life. Wastewater can lead to a variety of problems such as soil infertility, an increase of pathogens, ecological succession, loss of aquatic life, an increase in infections, loss of groundwater quality, and loss of ecosystems. So that we can apply different techniques in biotechnology to treat wastewater like primary, secondary, and tertiary treatments. By using the applications of biotechnology, we are able to find solutions for the problems stated above. To create a clean and sustainable environment, biotechnology applications can be practiced especially in wastewater treatment.

9.1 INTRODUCTION

Water pollution is a critical problem that has been faced by people globally because of different reasons, causing a variety of problems to

Sustainable Water Treatment and Ecosystem Protection Strategies. Hossein Hariri Asli, Tamara Tatrishvili, Ann Rose Abraham, & A. K. Haghi (Eds.)

humankind and all living organisms. The scarcity of fresh drinking water is the major problem caused due to water pollution. Water pollution can be caused by different reasons. Most of them are anthropogenic reasons. As a result of Industrial Revolution, different kinds of industries were started, and because of that many synthetic chemicals came into to picture. An increase in world population leads to an increase in more requirements so that the production of synthetic products increased due to an increase in demand for those products for a cheap price. The release of these synthetic products into the environment is the main reason for the pollution of the environment including water pollution. The main synthetic product that pollutes the water is plastic. Synthetic plastic causes huge damage to water and also to aquatic organisms. According to environmentalists, there will be more plastic than fish in the ocean after few years, not only in water but also in soil. The main problem of synthetic plastic is they are nonbiodegradable. Synthetic plastic and other disposal matter can stay in the environment for decades and hundreds of years without degrading.

Material	Number of years taken to degrade
Cigarette butt	10
Plastic grocery bag	20
Styrofoam cup	50
Aluminum cans	200
Plastic beverage holder	400
Disposal diaper	450
Plastic bottle	450
Fishing line	600

When these materials are added to water, they deposit in the base of the water bodies, and also aquatic animals consume them. As a result, those animals die because they are very toxic to living organisms. Therefore, these synthetic products can lead to so many problems. The other main reason for water pollution is the release of industrial waste to water bodies without treating them properly. The addition of human and animal feces to water is one of the other reasons for water pollution.[1] Wastewater can cause many problems such as soil infertility, ecological succession, loss of aquatic life, increase in pathogens, the spread of diseases, loss of ecosystems, loss of groundwater quality, and increased growth of weeds. Wastewater contains a high amount of nutrients and contaminants. Increasing nutrients like nitrates, phosphates, and phosphorous leads to the growth of algal blooms that can cause eutrophication. Eutrophication is caused by to excessive growth of blue–green algae in water. It leads to problems such as reduction of the clarity of the water, decrease in dissolved oxygen in water, invasion of new species, etc. Sunlight is an important factor for all aquatic organisms. When eutrophication occurs growth of blue–green algae on the surface of water prevents sunlight from penetrating into the depth of the water body which leads to the reduction of photosynthesis of aquatic plants. Also, it can cause the death of aquatic organisms including both plants and animals.[2] As a result of an increase in the organic content of water and degradation of these organic compounds microbes occur. Because of that, dissolved oxygen content in water decreases and causes water pollution. It gives a bad odor and the spread of disease carriers such as mosquitoes, flies, and other insects occurs. Apart from these, change in microclimate is one of the major problems that happened due to water pollution. It leads to other critical environmental issues such as global warming and the greenhouse effect. So, we can see that there are a lot of problems created due to wastewater. Therefore, we should find a solution to these problems. There are a lot of applications of biotechnology that can be applied to the problems created due to water pollution. Therefore, we can use those applications to find solutions for water pollution. Due to the aforementioned wastes, dissolved oxygen levels in the water are increased, and the addition of nutrients leads to excessive development and a rise in suspended particles or sediments in water bodies.

The primary cause of water pollution is wastewater. Water waste, which is gathered in cities and towns and is unfit for human consumption, includes sewage waste and industrial liquid waste. The primary remedy for the issues brought on by wastewater is wastewater treatment. The process of turning wastewater that is no longer required into effluent that can be returned to the water cycle with little environmental impact is known as wastewater treatment. Suspended solids are pollutants in wastewater. Harmful bacteria, biodegradable organic material, and nutrients such as nitrogen and phosphorus. Wastewater can come from residential areas (human and animal excrement and water used for cooking, bathing, and cleaning), business facilities, dairy operations, and other sources.

9.2 METHODOLOGY

The primary goals of wastewater treatment are the elimination of infections, harmful materials, and contaminants. To release discharged water into bodies of water or onto agricultural land with better water quality. Wastewater treatment primarily lowers eutrophication, chemical oxygen demand (COD), and biological oxygen demand (BOD).

9.2.1 PRELIMINARY TREATMENT

The screening process is a crucial part of the initial treatment. In this step, wastewater is processed through screens, grit chambers, and skimming tanks to remove debris, sediments, oil, and grease. Prechlorination is a procedure that kills disease spores and gets rid of the unpleasant wastewater odor. Normally, this area uses 10–15% chlorine.[3]

9.2.2 PRIMARY TREATMENT

Sixty percent of the suspended solids are eliminated in this stage. About 30% of COD, 35% of BOD, 10% of phosphorous, and 20% of total nitrogen make up COD. There are a few steps in primary treatment.

9.2.2.1 SEDIMENTATION

About 50% of suspended solids can be removed by this step under gravitational settling.

9.2.2.2 MECHANICAL FLOCCULATION AND COAGULATION

Wastewater is clarified and coagulants such as alum and poly electrolytes are used to remove fine suspended solids and colloidal particles.

9.2.2.3 NEUTRALIZATION

Highly acidic and alkaline wastewater is neutralized by lime slurry or NaOH and H_2SO_4 or CO_2, respectively.

9.2.3 SECONDARY TREATMENT

It is also referred to as a biological treatment. In this process, aerobic or anaerobic techniques are used to remove the dissolved and colloidal organic materials in sewage. In systems like aerated lagoons, trickling filters, activated sludge units, oxidation ponds, traditional biofilters, high-efficiency biofilters, membrane bioreactors, etc., the effluent from the primary sedimentation tank is first treated to aerobic oxidation. In the sludge digester, the sludge produced in the first sedimentation tank is exposed to anaerobic digestion.[4] About 80% COD; 90% BOD; 30% phosphorous; 50% total nitrogen; and oil, grease, phenol, grit, and scum are removed during secondary treatment.

9.2.3.1 ACTIVATED SLUDGE METHOD

This is a method used in the secondary treatment of water. Activated sludge is a mixture of bacteria, fungi, protozoa, and rotifers maintained in suspension by aeration and mixing. In this process, the biomass of aerobic organisms is grown in large aerated tanks. These organisms degrade the waste and use it as their food for their growth.

9.2.3.2 AERATION TANK

Sewage water is treated in the aeration tank directly during the procedure. It is an aerobic process. Extended aeration is more effective at removing BOD than activated sludge, which makes it particularly appealing to use in situations where it will be followed by tertiary treatment for reuse.

9.2.3.3 AERATED LAGOONS

Between algal ponds and activated sludge systems are aerated lagoons. Aeration is used to supply oxygen. Compared to algal ponds, this unit requires much less ground space and is around 3–5 m deeper. Wastewater can be settled and biologically improved further by being kept in big man-made ponds or lagoons. These lagoons have a high aerobic capacity and are populated by local macrophytes. By eliminating particles, small filter-feeding invertebrates such as the Rotifera and Daphinia species substantially aid in treatment.

9.2.3.4 WASTE STABILIZATION PONDS

The stabilization ponds are of open flow through basins that were specifically designed to treat biodegradable industrial wastes and sewage. They offer lengthy detention periods that last anything between a few and several days. Aerated lagoons are pond systems where oxygen is produced mechanically rather than through algae photosynthesis. Maturation ponds are lightly loaded ponds used as the final stage of wastewater treatment to polish secondary effluents and remove bacteria.

9.2.3.5 TRICKLING FILTERS

It is an attached growth procedure where wastewater treatment microorganisms are attached to inert packaging material. Rock, gravel, slag, sand, redwood, and a variety of plastic and other synthetic materials are exploited as packing materials in the attached growing process.

9.2.3.6 ROTATING BIODISCS

A form of secondary biological treatment is a rotating biological contactor (RBC). It is composed of a variety of tightly packed, parallel discs that are mounted on a revolving shaft that is supported just above the wastewater's surface.

9.2.3.7 MEMBRANE FILTRATION

There are filters different types of membrane filters according to the pore size of the membrane. There are different classifications of membranes. According to the separation mechanism, there are three types of mechanisms. They are the sieve effect, solution–diffusion mechanism, and electrochemical effect. There are two main classes of membranes based on the separation mechanism namely, porous membranes and nonporous membranes.

Porous membranes- In these membranes fixed pores are present and these pores can be further divided into three types,

i. Macro pores: Pore size is larger than 50 nm.
ii. Mesopores: Pore size is in the range of 20–50 nm.
iii. Micro pores: Pore size is less than 2 nm.

Membranes with dense media are nonporous membranes. The free volume that exists in between the membrane material's macromolecular chains is where species diffusion occurs.

While nanofiltration could be categorized in a class between porous and nonporous membranes, microfiltration, ultrafiltration, and both are porous membranes.

9.2.4 NANOFILTRATION

Due to its specific attribute of distinct charged-based repulsion, it is a fast-developing membrane separation technique used for wastewater treatment as well as the separation of medicines and antibiotics.[5] Membrane softening or low-pressure reverse osmosis (RO) are other names for this procedure. In nanofiltration, the operating pressure ranges 0.5–1.5 MPa.

9.2.5 ULTRAFILTRATION

It is a membrane operation for clarifying, disinfection, and disinfection. Because ultrafiltration membranes are porous, they can only reject solutes that are coarser. By using this technique, it is possible to eradicate all forms of germs, including bacteria, viruses, and particles. The operating pressure ranges 50–500 kPa.

9.2.6 MICROFILTRATION

Microfiltration membranes have pores that are at least 0.1 μm in size. It is typically used to remove particle debris. Similar to ultrafiltration, this method also uses pressure.

9.2.7 MEMBRANE BIOREACTOR (MBR)

Membrane bioreactor technology is based on biological treatment followed by membrane separation. The system is made up of a vigorous activated sludge process, and the biomass separation stage is completed by membrane cassettes that are located outside of the aeration tank in a separate membrane tank. In typical activated-sludge systems, the settling stage is replaced with membranes, effectively revolutionizing the procedure.[6] Membrane-based separation of biomass from treated water results in filtered, high-quality final effluent with potential for reuse. It enables the bioreactor to create very high biomass mixed liquid suspended solids (MLSS) concentrations without the negative side effects typically connected with conventional settlement methods.

Membrane fibers in a membrane bioreactor have countless tiny pores on their surface. While permitting clean water molecules, the pores act as a barrier to contaminants.

Membrane fibers in a membrane bioreactor have countless tiny pores on their surface. While permitting clean water molecules to pass, the pores create a barrier to contaminants. The water is then gently suctioned through the pores.

9.2.8 ADVANTAGES OF MBR

i. High-quality effluent
ii. High loading rate capacity
iii. Reducing tertiary filtration and secondary clarifiers
iv. Reducing plant budget
v. Easy operation
vi. Easy modulation

9.2.9 DISADVANTAGES OF MBR

i. Increase capital cost
ii. Relatively expensive to install and operate
iii. The membrane may be sensitive to some chemicals
iv. Require pre-treatment
v. Frequent membrane monitoring and maintenance
vi. Membrane fouling

9.2.10 TERTIARY TREATMENT

The process of treating sewage ends here. It is also known as the enhanced treatment and removal of organic content, which is done to remove nutrients from sewage and destroy dangerous microorganisms.[7] The primary goal of tertiary treatment is to offer a last step of treatment to further enhance the effluent quality prior to release into the environment, such as the sea, rivers, lakes, wetlands, or for possible re-use in agriculture. Tertiary therapies come in a variety of forms. Some of these include biowave therapy, duckweed pond treatment, reed bed treatment, water hyacinth pond treatment, and algal fish ponds.

9.2.11 WATER HYACINTH POND TREATMENT

The water hyacinth plant (*Eichornia crassipies)* is planted in the water body as part of this treatment.[8] Their lengthy, fibrous roots are covered in microbial biofilms. They function as hyperaccumulators and can take in other contaminants as well as heavy metals including arsenic, cadmium, and lead.

9.2.12 DUCK WEED PONDS

Growing on the water's surface, duckweed is a tiny plant with a fibrous root structure and a high rate of development and reproduction.[9] *Wolfia*, *Lemna*, and *Pistia* are a few of the plants that are utilized in duckweed ponds.

9.2.13 REED BEDS

In this method, grasses, shrubs, and trees that rely on pollutants are grown near water bodies.

Examples: Tape grass, *Phragmoides australis*, *Muntigia calabura*, Calotropis, Castor, Eucalyptus, and sunflower.

9.2.14 ALGAE AND FISH PONDS

In this process, algae with the potential to eliminate toxins, such as *Chlorella*, *Nostoc*, *Chlamydomonas*, *Spirogyra*, and *Volvox,* are produced in ponds. Growth of fish that can survive harsh circumstances in wastewater. These fish include catfish, carp, Catla, Rohu, and others.

FIGURE 9.1 Wastewater advanced treatment.

9.2.15 BIOREMEDIATION

Bioremediation is a natural and genetically engineered technology that involves in cleaning up polluted air, water, and soil using various biological forms of life such as bacteria, fungi, fishes, animals, and plants.[10] There are different types of bioremediations based on the site of cleaning, biological origin, and based on the organism.

Bioremediation is a very good solution for the treatment of wastewater.

9.2.16 DIFFERENT METHODS OF BIOREMEDIATION

- Natural attenuation: Cleaning of the environment by the mentioned in the following.

Based on the site of cleaning	**In situ: Cleaning up at the site** **Ex situ: Cleaning away from the site**
Based on the biological origin	Bacterial bioremediation
	Mycoremediation
	Pisciremediation
	Phycoremediation
	Zoo remediation
	Phytoremediation
Based on the organism	Aerobic bioremediation
	Anaerobic bioremediation

- Bioaugmentation: Practice of adding cultured microorganisms into the subsurface for the purpose of biodegrading specific ground-water contaminants.
- Biostimulation: It is the process of stimulating one organism by some other organism by releasing some chemical compounds like plant growth substances (e.g., rhizobium helps in N_2 fixation).
- Biofiltration: Roots of some plants help to absorb toxic compounds in water and soil.
- Bioaccumulation: It is a process where the organism first utilizes toxic materials and they use those toxic materials for its growth and development, then the excess things are accumulated as granules within its cells.
- Biostabilization: It is the advanced process of bioaccumulation.
- Biodegradation It is the process where the organisms convert toxic materials to fewer toxic substances.
- Biovolatilization: Process where conversion of any toxic material to less toxic volatile substances.
- Biohyper accumulation: It is the process of bringing all the above-mentioned methods.
- Biosparging: The process where the microorganisms in water or air influx air to remove pollutants.

9.3 CONCLUSIONS

The world has a long history of recycling wastewater, but growing urbanization, particularly in developing nations, is creating new problems for the management of water resources. Reusing water is encouraged by mounting demands on freshwater supplies including considerable engagement in nations such as the United States, Mexico, Mediterranean nations, the Middle East, Australia, South Africa, Japan, China, and Singapore. However, the lack of a suitable infrastructure to offer effective treatment and delivery could obstruct development. We can conserve and use more water by prioritizing using recovered water while maintaining control over its quantity and quality. Issues with regard to trace, there are still long-term health repercussions emergent microbes, and chemical pollutants. However, for the development of sustainable water resources, recovering and reusing the water is essential rather than releasing cleansed wastewater for disposal to the environment. The amount of water reused and health is tightly related, to security, technical viability, popular acceptance, and the community's perception of the water's value. As a society, we can no longer afford to utilize water only once, but it is our duty to manage and ensure water supplies.

KEYWORDS

- **environmental biotechnology**
- **pollutants**
- **ecological succession**
- **wastewater treatment**

REFERENCES

1. Bengani-Lutz, P.; Zaf, R. D.; Culfaz-Emecen, P. Z. et al. Extremely Fouling Resistant Zwitterionic Copolymer Membranes with Similar to 1 nm Pore Size for Treating Municipal, Oily and Textile Wastewater Streams. *J. Member. Sci.* **2017,** *543*, 184–194.
2. Escapa, A.; Mateos, R.; Martínez, E. J. et al. Microbial Electrolysis Cells: An Emerging Technology for Wastewater Treatment and Energy Recovery: From Laboratory to Pilot Plant and Beyond. *Renew. Sustain. Energy Rev.* **2016,** *55*, 942–956.
3. Sarkar, B.; Mandal, S.; Tsang, Y. F. et al. Designer Carbon Nanotubes for Contaminant Removal in Water and Wastewater: A Critical Review. *Sci. Total Environ.* **2018,** *612*, 561–581.

4. Wang, Y.; Wang, D.; Liu, Y. et al. Triclocarban Enhances Short-Chain Fatty Acids Production from Anaerobic Fermentation of Waste Activated Sludge. *Water Res.* **2017,** *127*, 150–161.
5. Wu, H.; Lai, C.; Zeng, G. et al. The Interactions of Composting and Biochar and Their Implications for Soil Amendment and Pollution Remediation: A Review. *Crit Rev Biotechnol.* **2017,** *37*, 754–764.
6. Yi, H.; Huang, D.; Qin, L. et al. Selective Prepared Carbon Nanomaterials for Advanced Photocatalytic Application in Environmental Pollutant Treatment and Hydrogen Production. *Appl. Catal. B Environ.* **2018,** *239*, 408–424.
7. Yi, H.; Qin, L.; Huang, D. Nano-Structured Bismuth Tungstate with Controlled Morphology: Fabrication, Modification, Environmental Application and Mechanism Insight. *Chem. Eng. J.* **2019,** *358*, 480–496.
8. Yi, H.; Yan, M.; Huang, D. et al. Synergistic Effect of Artificial Enzyme and 2D Nano-Structured $B_{i2}W_{O6}$ for Eco-Friendly and Efficient Biomimetic Photo Catalysis. *Appl. Catal. B: Environ.* **2019,** *250*, 52–62.
9. Yi, H.; Zeng, G.; Lai, C. Environment-Friendly Fullerene Separation Methods. *Chem Eng J.* **2017,** *330*, 134–145.
10. Zhang, Q. H.; Jin, P. K.; Ngo, H. H. et al. Transformation and Utilization of Slowly Biodegradable Organic Matters in Biological Sewage Treatment of Anaerobic Anoxic Oxic Systems. *Bioresour. Technol.* **2016,** *218*, 53–61.

CHAPTER 10

Bioremediation: A Green Approach for Environmental Contaminants

SUBBULAKSHMI GANESAN, AGAMPODI SANDUNI, ANUPAMA DE ZOYSA, and SNEHA S.

Department of Chemistry, Jain University, Bangalore, Karnataka, India

ABSTRACT

Bioremediation is one of the eco-friendly methods which is used to reduce the toxicity of the toxic substances using plants animals and microorganisms. This method is preferable because it is cost efficient and eco-friendly and does not use any chemicals or hazardous substances. It is a sustainable mode of remediation technique which is used to clean the environment. The most commonly used microorganisms are *Pseudomonas putida*, *Dechloromonas aromatica*, *Methylibium petroleiphilum* involved in the bioremediation process. They degrade the pollutants and some microbes can even digest those pollutants to attain their nutrition through it. Some of the microbes which can degrade heavy metals are *Flavobacterium*, *Enterobacter*, and *Bacillus* species. Phytoremediation is also a type of bioremediation in which plants such as mosses are used in the removal of pesticides and other harmful substances in the soil as well as in the environment. *Alyssum lesbiacum* is used to remove copper. Not only this, zoo remediation is a technique in which animals such as oysters, muscles, calms, and sponges are used to clean up the environment. This article revolves around the above bioremediation techniques which lead to a pollution-free environment.

Sustainable Water Treatment and Ecosystem Protection Strategies. Hossein Hariri Asli, Tamara Tatrishvili, Ann Rose Abraham, & A. K. Haghi (Eds.)

10.1 INTRODUCTION

As there is a huge increase in the population the need for agriculture and industries is developing in a very high rate and pollution, on the contrary, also keeps on growing intensely. The toxins released by the industries and the chemicals present in the pesticides and some of the heavy metals extensively affect the environment and result in soil pollution, water pollution; and heavy metal poisoning and they in turn cause serious effects on human health.[1] As technology is increasing day by day, researchers are also coming up with better ideas which does not involve any harm to human life and biologically trying to reduce pollution in a cost-efficient way.

The chemicals such as polyesters, polychlorinated biphenyl (PCB), polyvinyl chloride (PVC), and hydrocarbon are released into the environment through many industries which are looking just for their income source and are least bothered for the environment.[2] These chemicals are hazardous and persist for a longer time even for a few decades, removing these toxic pollutants is not so easy because they cannot be degradable easily and they are carcinogenic in nature.

Remediation is a term used to solve any problem and bioremediation is biologically solving the problem. However, this method is not applicable for a higher concentration of heavy metals since they are carcinogenic to humans and also to microorganisms and lead to death. We all come across many microorganisms such as fungi, bacteria, and algae on a daily basis some are useful and some are harmful but scientists and researchers have come across certain methods and have been invented in bioremediation in which microorganisms are used to clean up the environment by degrading the toxic materials to live a pollution free life.[3]

Over the years, the use of synthetic matter and fertilizers has been increased; now, alternatives for these substances are being developed using biological means and microorganisms.

This is a sustainable method and eco-friendly method. This uses nitrogen-rich compounds because it enhances the bioremediation process.

There are two types of bioremediation—in situ bioremediation and ex situ bioremediation.

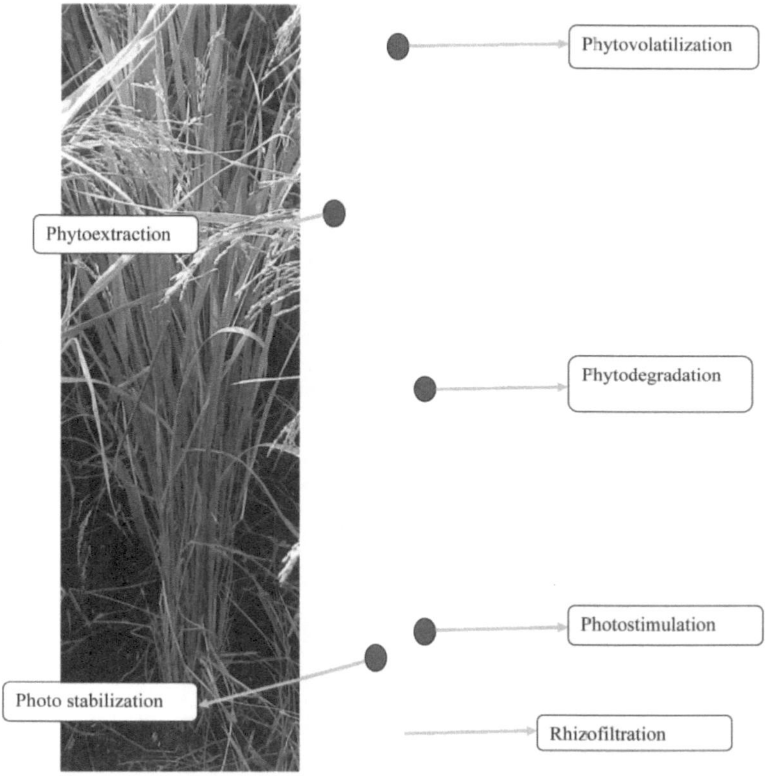

FIGURE 10.1 Bioremediation process.

10.2 IN SITU BIOREMEDIATION

In situ bioremediation is a method in which the pollutants are not excavated but treated in their original place.[4] The steps involved in this are sparging, bioventing, percolation, pump and treatment.

10.3 EX SITU BIOREMEDIATION

Ex situ bioremediation is the method in which the contaminants are taken out and then treated with microorganisms. The steps involved in this process are land farming, biopiles, and compost pile irrigation.

10.4 BASED ON MICROORGANISMS

Bioremediation is classified according to the microorganism used for the remediation.

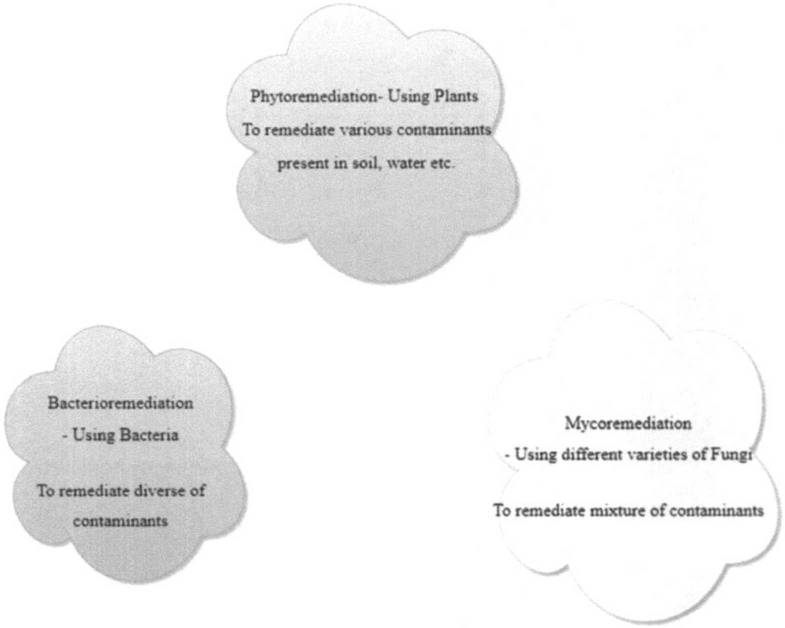

FIGURE 10.2 Microorganism used for the remediation.

10.5 METHODOLOGY

The remediation process involves phytotransformation, phytoextraction, phytovolatization, and phytostabilization with the help of transformation, adsorption, mineralization and the degradation of the pollutants and nitrogen fixation is done for plant growth.[5]

The term bioremediation was first coined by Robinson while he was working on sewage water treatment. Further researchers made recent developments to obtain better results and now we come across genetically modified stains and some new variety of plants to clean up the environment. Various methods such as biostimulation, bioaugmentation, biopiling, bioattenuation, and biopiles are also used to remediate the contaminants.

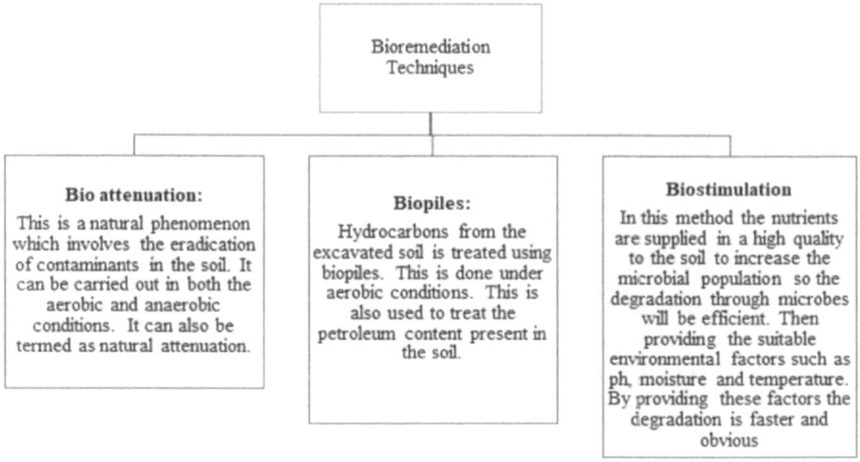

FIGURE 10.3 Bioremediation techniques.

10.6 FACTORS AFFECTING THE GROWTH OF MICROBES

Mainly, the environmental factors affect the growth of microorganisms.

10.7 ENVIRONMENTAL FACTORS

The major environmental factors which affect the growth of microorganisms are pH, moisture, temperature, climate, and humus.[6] The interactions of microorganisms are established successfully if the environmental factors are favorable to the microbes and also the site of interactions.

Even the redox potential and soil texture determine the growth and development of microorganisms. These factors in turn determine the degradation time of microbes. A pH of about 6.7–8.6 is required for the degradation.

10.8 AVAILABILITY OF NUTRIENTS

Although there are nutrients present in the soil on certain days, those nutrients are not enough for the microorganisms and also for the plants.[7,8] For the bioremediation to be more effective there must be a balance in the nutrients and it can be brought by supplying the additional nutrients. It also increases the metabolic activity of the microorganisms.

10.9 TEMPERATURE

Temperature is the most important factor which determines the survival of the microorganisms. The region with low temperature is not preferable because the growth and degradation time is longer. The temperature differs for each organism some may require high temperature, and some may require low temperature. When there is an increase in temperature the growth of microorganisms also increases and vice versa.

10.10 OXYGEN LEVEL

Oxygen level differs for each organism. Some organisms require a high level and some may require concentration or can even grow in the absence of oxygen.[9]

Mainly hydrocarbon mechanism is influenced by the presence of oxygen.

10.11 MOISTURE CONTENT

An adequate water supply is required for the growth of the microbed.

10.12 pH

pH is the measure of the acidity or alkalinity nature of the compound. Even a slight pH change can affect the growth of microbes.

10.13 METAL IONS

For bacteria and fungi, the metal ions can be required in smaller amounts.[10]

Toxic compounds: Even though the microorganisms can degrade the contaminants, higher concentrations of pollutants cannot be degraded by them.

10.14 RESULT AND DISCUSSION

As per the studies and surveys, we came across very useful and nutrient-rich manure, hircus manure and mosses, which can treat the contaminated air and the water hyacinth which can clean up the environment. Although microorganisms such as bacteria, fungi, and yeasts are used in this biological treatment this is not sufficient for the whole waste to be removed there are also some factors that affect the growth of microbes.[11] We must analyze these factors and then we can build suitable conditions for their growth. Biodegradation is also similar to bioremediation in which the organic toxins are degraded to fewer toxic compounds. It can take place either in the presence or absence of oxygen.

The growth of the microorganisms is influenced by the environmental factors and some of the enzymes such as ligases and hydrolyses. The microorganisms must attack enzymatic allies to degrade the pollutants in less time. It is known that the hircus manure is very useful for the bioremediation process. The hircus manure is available in large quantities in Nigeria commercially.[12] These indirectly help the remediation process by enhancement of microbial activities. This manure contains high amounts of nitrogen, carbon, and other nutrients which are basically required for life in turn stimulates the growth of microorganisms to obtain energy.

The major types of bioremediations involve biopiling, bioaugmentation, biosparging, bioventing, and biopiles.[13,14] Inorganic substances of phosphorus, nitrogen, and hydrocarbons, chlorinated compounds are the major contaminants that are present in the soil and water.

Due to the development of industries and big infrastructure, the nation may keep on developing but the environment is facing many issues and most of them are not even aware of it.[15] To clean up the environment and live a better life, we must somehow come across certain discoveries which are made day by day one such as bioremediation which is a conventional method and cost-efficient method so every individual will be able to do this.

10.15 CONCLUSIONS

Bioremediation is a very useful and effective method to clean up the environment without causing any harm to the environment and mankind. This method is majorly based on microorganisms, plants, and animals and is an eco-friendly process. The growth of microorganisms is also affected by certain factors such as temperature, humidity, moisture, pH of the soil. It is also similar to biodegradation but biodegradation is a natural method and it cannot degrade the toxins thereby we came across a new method which is bioremediation in which the microorganisms are able to reduce the toxicity or some of them can even degrade to obtain their nutrition. Bioremediation was not so effective for the contaminants of higher concentration. Once the process was successful, we were also able to degrade the toxins with the help of plants which is termed as phytoremediation. As per our studies, it is known that hircus manure is considered as a nutrient-rich media for the growth of microorganisms. This manure is produced in high quantities in Nigeria. Since there is a high usage of pesticides in the agricultural field the contaminants then settle on the soil and then it affects the property of the soil, to degrade such contaminants this hircus manure can be used.

Since pollution keeps on increasing it is hard for us to survive, each time we go out we inhale all the polluted air and face many health issues. To reduce such contaminants present in the air, we can use plants that can take up the gases or other contaminants in the air and give out the oxygen which can be brought out by phytoremediation. It is also a natural method and can be practiced easily.

KEYWORDS

- **bioremediation**
- **zoo remediation**
- **phytoremediation**
- **microorganisms**
- **environment**
- **pollutants**

REFERENCES

1. Dados, M. O.; Demetriou, K.; Papastephanou, C.; Ioannides, I. M. Rapid Remediation of Soil Heavily Contaminated with Hydrocarbons: A Comparison of Different Approaches. *Ann. Microbiol.* **2015,** *65* (1), 241–251.

2. Iyagba, G.; Offor, U. S. Effect of Crude Oil and Biostimulant (Bioremediation) on Growth Extract of Maize (Zea mays (l.) and Cowpea (Vignaunguiculata (l.) Walp). *Eur. Sci. J.* **2014,** *10* (6), 284–291.
3. Chikere, B.; Azubuike, C. C. Catechol-2,3-dioxygenase Screening in Putative Hydrocarbon Utilizing Bacteria. *Int. Res. J. Microbiol.* **2013,** *4* (1), 1–6.
4. Beolchini, F.; Rocchetti, L.; Regoli, F.; Dell'Anno, A. Bioremediation of Marine Sediments Contaminated by Hydrocarbons: Experimental Analysis and Kinetic Modelling. *J. Hazard. Mater.* **2010,** *182* (1–3), 403–407.
5. Gavrilescu, M. Overview of In Situ Remediation Technologies for Sites and Groundwater. *Environ. Eng. Manag. J.* **2006,** *5* (1), 79–114.
6. Ayotamuno, M. J.; Kogbara, R. B.; Ogaji, S. O. T.; Probert, S. D. Bioremediation of a Crude-Oil Polluted Agricultural-Soil at Port Harcourt, Nigeria. *Appl. Energy* **2006,** *83* (11), 1249–**1257**.
7. Mirlahiji, S. G.; Eisazadeh, K. Bioremediation of Uranium by Geobacterspp. *J. Res. Dev.* **2014,** *1*, 52–58.
8. Abioye, P. O.; Abdul Aziz, A.; Agamuthu, P. Enhanced Biodegradation of Used Engine Oil in Soil Amended with Organic Wastes. *Water, Air, Soil Pollut.* **2010,** *209* (1), 173–179.
9. Adebusoye, S. A.; Ilori, M. O.; Amund, O. O.; Teniola, O. D.; Olatope, S. O. Microbial Degradation of Petroleum Hydrocarbons in a Polluted Tropical Stream. *World J. Microbiol. Biotechnol.* **2007,** *23* (8), 1149–**1159**.
10. Agarry, S. A.; Owabor, C. N.; Yusuf, R. O. Bioremediation of Soil Artificially Contaminated with Petroleum Hydrocarbon Mixtures: Evaluation of the Use of Animal Manure and Chemical Fertilizer. *Bioremed. J.* **2010,** *14* (4), 189–195.
11. Agarry, S. E.; Owabor, C. N.; Yusuf, R. O. Enhanced Bioremediation of Soil Artificially Contaminated with Kerosene: Optimization of Biostimulation Agents Through Statistical Experimental Design. *J. Petrol. Environ. Biotechnol.* **2012,** *3* (3), 1–8.
12. Bamforth, S. M.; Singleton, I. Bioremediation of Polycyclic Aromatic Hydrocarbons: Currant Knowledge and Future Directions. *J. Chem. Technol. Biotechnol.* **2005,** *80* (7), 723–736.
13. Sinha, S. N.; Biswas, M.; Paul, D.; Rahaman, S. Biodegradation Potential of Bacterial Isolates from Tannery Effluent with Special Reference to Hexavalent Chromium. *Biotechnol. Bioinf. Bioeng.* **2011,** *1*, 381–386.
14. Makadia, T. H.; Adetutu, E. M.; Simons, K. L.; Jardine, D.; Sheppard, P. J.; Ball, A. S. Re-use of Remediated Soils for the Bioremediation of Waste Oil Sludge. *J. Environ. Manag.* **2011,** *92* (3), 866–871.
15. Ijah, U. J. J.; Antai, S. P. The Potential Use of Chicken-Drop Micro-Organisms for Oil Spill Remediation. *Environmentalist* **2003,** *23* (1), 89–95.

CHAPTER 11

Promising Technologies in Wastewater Treatment: The Current Standpoint

SUBBULAKSHMI GANESAN

Department of Chemistry, Jain University, Bangalore, Karnataka, India

ABSTRACT

Wastewater treatment can be done in three main ways such as primary treatment, secondary treatment and tertiary treatment. Primary treatment of wastewater includes the removal of floating materials on wastewater as well as sedimentation. Secondary treatment is carried out by using activated sludge, trickling filters, anaerobic lagoons, aerated lagoons, and stabilization. Tertiary treatment involves the further purification of wastewater as well as recycling. Adsorption, chemical coagulation and flocculation, chemical oxidation, desalination as well as oxidation ponds are the methods that are used in the tertiary treatment of water. The wastewater that is recycled can be used in irrigation, for the nonfood crops such as grasslands and playgrounds.

11.1 INTRODUCTION

Wastewater treatment is divided into three categories namely, primary, secondary, and tertiary treatment methods. These methods majorly aim for the removal of biochemical oxygen demand (BOD), nitrogen (N), phosphorous (P), different solids, and bacteria.[1] The composition of sewage is a very complex process that differs depending on the sources and the type of treatment.

Sustainable Water Treatment and Ecosystem Protection Strategies. Hossein Hariri Asli, Tamara Tatrishvili, Ann Rose Abraham, & A. K. Haghi (Eds.)

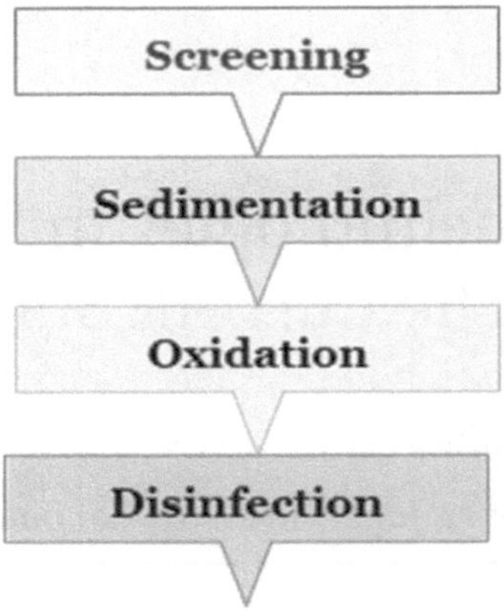

Here, the waste matter which was on the surface of the water is screened and those particles were sediment by adding different chemicals such as aluminum sulfate. Then the sludge that formed was oxidized and after that, it was disinfected by adding chlorine into it. The addition of chlorine will remove the microbes that are infectious as well as it will remove the bad odor that is released from it.

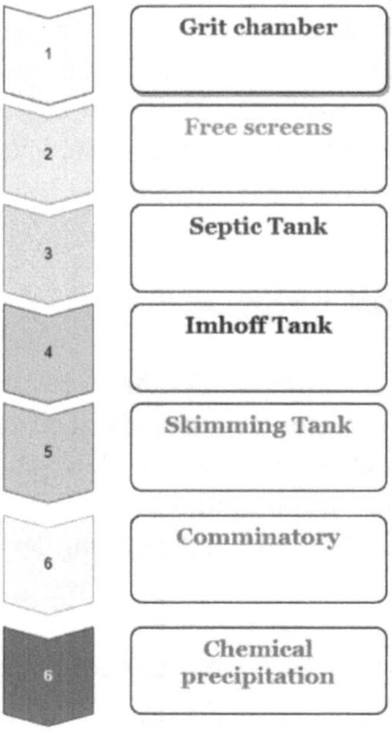

11.2 PRIMARY TREATMENT

It is a method which consists of the removal of floating and suspended solids by the mechanical means. More than half of the suspended solids are removed by the primary treatment. In this, the large solids are removed and grease and scam are also removed. The screening of these substances is done through the metal bars which are first spaced 25–50 mm apart and it may range down to 0.8-mm openings. Sand and the other coarse materials too are removed with the help of the grit chambers. After the screening as well as the removal of the grit, the wastewater is directly run into settling or the sedimentation tanks, and this process of which the solids that are suspended are removed by the method of gravitational settling.[2]

Primary sludge is considered a huge problem as it is bulky and should be removed. It also contains 94–99% of water. In most of the cases, sludge is dried in the beds along with some water removed by the filtration. According to the report of the American Chemical Society, it indicates that the primary treatment has the capability to reduce 60% of the suspended solids, 35% BOD, 30% chemical oxygen demand (COD), 10% phosphorous, and 20% nitrogen.

Moreover, biological oxidation methods fall under the following primary treatment methods:

1. Carbonaceous oxidation.
2. Nitrification.
3. Ammonification (e.g., biofilms, zoogleal slime layers).
4. Activated sludge process.
5. Removal of nitrogen, phosphorous, and potassium (NPK).

During the carbonaceous oxidation, the organic compounds are converted into carbon dioxide, and during the nitrification process, any nitrogen-containing compound is converted into nitrates and nitrites. This can be done by some aquatic plants such as water hyacinth. Ammonification is the conversion of nitrogen gas present in the atmosphere into ammonia. This process of ammonification is executed by using biofilms or zoogleal slime layers. A biofilm is an assembly of surface-associated microbes that are enclosed in an extracellular polymeric substance matrix. Mineral crystals, clay/particles of silt, corrosion particles, and blood components depend on the environment in which the biofilm has developed and it may also be found in the biofilm matrix. The microbes which are present in the biofilm may differ from the freely suspended counter parts with the genes that are transcribed. Biofilms normally form on different varieties of surfaces, that include the living tissues, industrial water system piping, indwelling medical devices, and natural aquatic systems. Water system biofilms are highly complex with corrosion products, clay particles, fresh-water diatoms as well as filamentous bacteria. The intersurface between the solid–liquid layers of a surface and an aqueous medium such as water? Blood provides an environment for the attachment of the microbes. Hydrodynamics of the aqueous medium, properties displayed by the cell surface, conditioning films formed, and effects of substratum are taken into account for the attachment.[3]

Zoogleal slime layers are a complex population of organisms that are capable of forming a slime layer on a media that can break down the organic matter in the wastewater into simpler compounds as well as into nontoxic forms. These microbes live on the wastewater, slit, and the dead organisms. Activated sludge is a flocculant culture of microbes developed in the aeration tanks under artificially provided conditions. Activated sludge is brown in color and it is also called the waste activated biosolids or the waste activated solids. This activated sludge contains a mixed blend of the microbes of 95% which are mostly aerobic species of bacteria. Activated sludge contains fungi and protozoa and the higher forms of the invertebrates other than bacteria. The basic function of this activated sludge process is the maintenance of viable population microbes.

11.2.1 AERATION TANK

This provides a detention time that is required and also it ensures the activated sludge and the influent wastewater are mixed thoroughly.

11.2.2 MECHANICAL OR DIFFUSED AERATION

Agitators or the mixers are used to mix air and liquor in the mechanical aeration. Pressurized air is released through the diffusers near the bottom of an aeration tank in the diffused aeration systems.

11.2.3 SETTLING TANK

Plain settling tanks are made to provide roughly 2–4 h of the hydraulic detection time in the activated sludge system.

11.2.4 ACTIVATED SLUDGE AND RETURN SLUDGE

Return sludge consists of a pump, timer, and variable speed drive as well as a flow measurement device.

11.2.5 ACTIVATED SLUDGE PROCESS

This is a process that is designed to speed up the rate of the decomposition of the waste material in the water and it was first developed in 1914. This process was first tested at Manchester–Davyhulme wastewater treatment plant. Most of the parts of the pilot plant unit were made up of wood, but most of the characteristic features of the activated sludge process are used these days. This consists of separate clarifiers with a continuous flow arrangement fill and draw arrangement and the activated sludge recycle this is called a sequential batch reactor (SBR) Activated sludge process treats both municipal water and industrial water. This activated sludge process is a multichamber reactor that uses highly concentrated microbes that are capable of degrading the organics and removal of the nutrients from the wastewater, which produces effluent with high quality.[4] The goal of this activated sludge process is to maintain aerobic conditions. To achieve this, a continuous as well as timed supply of oxygen is needed. Loads of bacteria are suspended and they are mixed in the wastewater with an aerated tank. These bacteria use organic pollutants for their growth and this is transformed into energy, carbon dioxide, water as well as new cell material. These flocs are removed in the secondary clarifier with the help of the gravity settling and some of this sludge will be recycled from the clarifier and back to the reactor. Finally, the effluent is discharged for polishing.

Activated sludge process consists of different designs. Generally, all the activated sludge process consists of three major components, namely, an aeration tank which serves as a bioreactor, return activated sludge, a settling tank for the separation of activated sludge solids.[5]

The concentration of biodegradable components in an influent is decreased in all these activated sludge plants because of the biological processes that occur in the particular aeration tank. Hydraulic residence time present in the aeration tank is obtained by dividing the volume of the aeration tank by the flow rate and it controls the efficiency of removal of the sludge. Here oxygen is injected into the wastewater where the solids are smashed during this process. The sewage liquor will be discharged into a chamber with the activated sludge and the live bacteria sink onto the bottom of the tank where the dead bacteria start floating on the surface. The live bacterium returns to the digestion chamber and then the clean water is discharged into a water course.

11.2.6 ADVANTAGES

1. The procedure is less complicated.
2. The method can be odorless.
3. Costs are very low.
4. The quantity of the unwanted sludge is very low.
5. Beneficial bacteria reseed in the sewage treatment plants.
6. It allows for larger emptying intervals.

11.2.7 DISADVANTAGES OF ACTIVATED SLUDGE TREATMENT PROCESS

1. This should be built by the professionals.
2. Experts are needed to operate as well as for the maintenance.
3. Electricity should be provided continuously where it increases the energy consumption.

11.2.8 REMOVAL OF NPK

Both N and P in higher concentration in water bodies will lead to eutrophication. Nitrate is causing risk for the human health as it is leading for the diseases such as infant methemoglobinemia. After the activated sludge process, a posttreatment process is done to remove N and P from the effluent which is collected. These effluents contain very small amounts of organic compounds where the sources of carbon (C) are used to remove N. Conventional methods in posttreatment denitrification employ methanol to remove nitrate. As the effluents are having very small number of organic compounds, C sources are added to remove N. Conventional methods for the posttreatment of denitrification use methanol for the removal of nitrate. *Desulfovibrio* species CMX is a sulfate-reducing bacteria that can remove NO present in iron or ethylene diamines tetra acetic acid.[6] Sulfate-reducing bacteria along with wood play an important role in the removal of nitrogen. In the iron polarization ferrous ions are generated and combine with the phosphate ions to produce vivianite and also the other ferrous phosphates by the removal of phosphorous. Moreover, Potassium is removed by reverse osmosis.

The primary wastewater treatment is also done by the following methods:

1. Bar screen method
2. Grit chamber
3. Primary clarifier
4. Centrifugation

11.2.9 BAR SCREEN METHOD

During this method, larger objects such as wood and stones are removed prior to the grit chamber.

11.2.10 GRIT CHAMBER

It is a sedimentation tank that is designed to slow down the flow of water in the wastewater treatment plants. Heavy solid particles, sand, and grit settle in the pipes as well as for the other equipment.

11.2.11 PRIMARY CLARIFIER

This is a tank that provides a longer hydraulic aeration time for the quiescent zone for the solids to settle from an influent stream. The solids which are settled are removed from the tank and it is sent to anaerobic biological treatment, where the clarified wastewater is directed into secondary biological treatment. These clarifiers are used to remove oils and grease as well as other floating materials.

11.2.12 CENTRIFUGATION

It is the use of centrifugal force to promote the accelerated settling of the particles present in a solid–liquid mixture.

11.3 SECONDARY TREATMENT

In this treatment, microbes are used as a biological method for the removal of the contaminants in the wastewater. Secondary treatment can be aerobic or anaerobic where each process is utilizing a different community of bacteria.

Most of the instances, aerobic biological treatment methods are most common in the municipal wastewater treatment. In an aerobic process, the organic contaminants are first converted into carbon dioxide, additional microbes and water. Aerobic lagoons are typically large with a shallow basin that provide the adequate time for the bacteria and algae to treat naturally. In the activated sludge method, the aerobic microbes are used to remove the carbonaceous matter from the wastewater.

Rotating biological contactor involves in an attached growth process of which consists of a series of closely arranged, parallel discs that are mounted by keeping it on a rotating shaft which is a partially submerged substance where the wastewater is treated. In the usage of trickling filter, the wastewater is distributed on the fixed bed of media like the rocks, plastic, gravel as well as the substrate. The wastewater is flowing downwards over the surface of the media of which the microbes are forming a layer of biomass and the contaminants in the water are consumed.

Aerobic biological treatment employs the organisms that are functioning in the absence of molecular oxygen. During the anaerobic process, organic contaminants are converted into biofuel gas comprised of carbon dioxide and methane mainly, along with the other end products. Anaerobic processes are usually used to treat the high-strength wastewater where the biogas is produced and where it is practical to utilize the aerobic processes.[7] Aerobic processes use lower amount of energy than the mechanical aeration. The main objective of the wastewater treatment is for the removal of organic matter which is biodegradable and also suspended solids by a process called as aeration and filtration and here the effluent is further treated from the primary treatment to remove the organic residues and the particles which are suspended.

11.3.1 TRICKLING FILTERS

Trickling filters are low in the initial cost, lower generating cost, maintenance cost also is very low and also it is a very simple process.

Normally, the trickling filters have a depth of 5–7 ft for the low rates and for the high rates; they have a depth of 10–40 ft for the columnar media. Polypropylene media filters are designed with the depth ranges of low as well as intermediate rate filters. Trickling filters which are deeper are having the ability to improve the nitrification process and it can be used as the second stage of a two-stage biological system which is designed for the nitrification process. Recirculation is a recommended method used to increase the BOD removal efficiency of the high-rate trickling filter processes.

There are two major types of the secondary treatment namely aerobic treatment units and biofiltration systems. These systems are available commercially as the pre manufactured units which are designed for treating the wastewater which are produced by some of the residences and also in some cases the nonresidential facilities.

11.3.2 AEROBIC TREATMENT UNITS

The aerobic units are called as the package sewage treatment plants that will inject the air into wastewater to create the aerobic microbes that will break down the contaminants of organic wastewater. Aerobic treatment units consist of a tank with three compartments where the first compartment functions like a septic tank and the solids are settled out there where the middle compartment contains air which is injected with the help of a blower and the third which provides the additional removal of the solids.[8] A septic tank is used mainly along the aerobic treatment unit to enhance the performance of the system. Secondary treatment method is used to overcome the difficulties such as the low permeability. In higher permeability in sand, the shallow depth for the groundwater as well as the bedrock. In some instances, the usage of the secondary treatment methods is needed under the onsite of the wastewater management systems regulation like when the disposal fields are given permission for the preparation of provisional parks, etc.

There is different kind of secondary treatment systems are as follows:

1. Aerobic systems
   1) White water ATU
   2) Pro flow aerobic systems
   3) Wastewater system which is fast
2. Biofiltration systems
   1) Water biofilter
   2) AdvanTex biofilter
   3) Puraflo peat biofilter
   4) Ecoflo biofilter
3. Combined treatment
   1) Enviro septic
   2) Eljen GSF

The alternative methods used in the secondary treatment are as follows:

1) Activated sludge
2) Aerobic lagoon
3) Trickling filter

The secondary treatment is designed to degrade the biological components in the sewage that are derived from the waste of humans, food, detergents, and soaps. This is a biological process where the aerobic digestion includes processes such as the trickling filters, aerobic granulation, RBC or the anaerobic processes such as the aerated lagoons, membrane bioreactors, and the constructed wetlands.

11.3.3 ACTIVATED SLUDGE

It is a process used for treating the sewage as well as the industrial wastewater by using the air which is composed of the bacterial and protozoan population.

11.3.4 AERATED LAGOON

Aerated lagoon is also called as an aerated basin which is with artificial treating the pond which is provided biological oxidation of the wastewater.

There are lots of other biological processes that are used in the treatment of the wastewater (e.g., trickling filters, rotating biological contractors, and biofilter).

All of them are having common use of the oxygen and also the action of the microbes for biotreating of the pollutants within the wastewaters.

11.3.5 TRICKLING FILTER

It is made up of large cylindrical concrete slabs which are having the diameter of 8–16 m and also the depth of 2–3 m. This tank is filled with the porous bed with the crushed stones, rocks and the clinkers. Voids are present to prevent clogging. Here the sewage enters through the inlet and it will reach to the sprinkler. Because of its arms which rotate, sewage is sprayed on to a porous bed where a network of drain is present below it.

The sewage will trickle to the bottom and will be collected in a depression called as slump from where a single main outlet will carry the treated sewage effluent.[9,10] The main rotating arms in the sprinkler will be spraying the sewage drops along with oxygen. As the sewage is intermittent, sewage is getting enough space and also time for percolating and the system will maintain the aerobic conditions. When the sewage trickles, the microbes that present in the sewage will colonize the surface and will form a microbial film. When the surface is covered with an aerobic population of microbes that comprises of the bacterial species which includes *Sphaerotilus notans*, *Flavobacterium*, *Beggiota*, Microalgae. etc., slime-producing bacteria such as *Zooglea ramigera* plays a very important role and since it is a microbe, the film is called as Zooglea film. When the time is increasing, the film too becomes thicker. After some of the time, the nutrients as well as the oxygen cannot reach the innermost cells and they will die when the film will be getting slogged off. Due to this, the pores will be blocked and the untreated sewage will start to accumulate on the top where the porous bed has to be changed.

11.3.6 ADVANTAGES OF TRICKLING FILTER

1. Lowers the BOD level of sewage by 75–85%.
2. Efficiency is increased further by the usage of a series of filters.
3. Cheaper in price.

4. Advanced than one oxidation pond.
5. Easy operation.

11.3.7 DISADVANTAGES

1. Because of clogging, the porous bed needs to be changed.
2. Efficiency is affected by the environment.
3. Due to the spray of aerosols, spread of pathogens will take place.

11.4 TERTIARY TREATMENT

In the tertiary treatment, mostly the toxic substances, different chemicals are removed. Her the methodologies such as advanced oxidation, membrane processes as well as the filtration processes are widely used.[11] Most of the time, the toxicity of the chemicals will be removed or the toxicity will be reduced.

11.5 CONCLUSIONS

Wastewater treatment is done with the help of three methods namely, primary, secondary, and tertiary. These are eco-friendly methods of wastewater treatment which are very effective in usage. The treated water can be used to in agriculture and also can be used in purposes such as fire extinguisher. So, these methods can be concluded as effective methodologies in the wastewater treatment.

KEYWORDS

- **flocculation**
- **oxidation**
- **desalination**
- **sedimentation**
- **wastewater**

REFERENCES

1. Aho, I. M.; Lagasi, J. E. A New Water Treatment System Using *Moringa oleifera* Seed. *Am. J. Sci. Ind. Res.* **2012,** 3 (6), pp. 487–492.
2. Khangembam, C. D. Effect of Temperature on Nitrifyingmicrobes, Emphasizing on Ammonia Oxidizing Archaea Andbacteria. *J. Biol. Sci. Med.* **2016,** *2* (2), 7–14.
3. Murphy, F.; Ewins, C.; Carbonnier, F.; Quinn, B. Wastewater Treatment Works (WwTW) as a Source of Microplastics in the Aquatic Environment. *Environ. Sci. Technol.* **2016,** *50* (11), 5800–5808.
4. Farhaoui, M.; Derraz, M. Review on Optimization of Drinking Water Treatment Process. *J. Water Resour. Protect.* **2016,** *8*, 777–786.
5. Farhaoui, M. Managing and Modeling of the Drinking Water Treatment Sludge. *Int. J. Res.—Granthaalayah* **2017,** *5* (2), 168–179.
6. You, J.; Das, A.; Dolan, E. M.; Hu, Z. Ammonia-Oxidizing Archaea Involved in Nitrogen Removal. *Water Res.* **2009,** *43* (7), 1801–1809.
7. Munno, K.; Helm, P. A.; Jackson, D. A.; Rochman, C.; Sims, A. Impacts of Temperature and Selected Chemical Digestion Methods on Microplastic Particles. *Environ. Toxicol. Chem.* **2018,** *37* (1), 91–98.
8. Kyncl, M. Opportunities for Water Treatment Sludge Reuse. *Geo Sci. Eng.* **2008,** *LIV 2* (1), 11–22.
9. Mukheled, A. A Novel Water Pretreatment Approach for Turbidity Removal Using Date Seeds and Pollen Sheath. *J. Water Resour. Protect.* **2012,** *4*, 79–92.
10. Mintenig, S. M.; Bauerlein, P. S.; Koelmans, A. A.; Dekker, S. C.; van Wezel, A. P.Closing the Gap Between Small and Smaller: Towards a Framework to Analyse Nano- and Microplastics in Aqueous Environmental Samples. *Environ. Sci.: Nano* **2018,** *5*, 1640–1649.
11. Shen, Y. H. Treatment of Low Turbidity Water by Sweep Coagulation Using Bentonite. *J. Chem. Technol. Biot.* **2005,** *80*, 581–586.

PART III
Case Studies

CHAPTER 12

Synthesizing Monodisperse Polydiene Latexes in an Emulsifier-Free Heterogeneous Monomer–Water System

N. G. GRIGORYAN, A. A. HOVHANNISYAN, A. G. NADARYAN, and G. K. GRIGORYAN

Scientific and Technological Center for Organic and Pharmaceutical Chemistry of the National Academy of Sciences of Armenia, Yerevan, Armenia

ABSTRACT

Latexes are successfully used in immunological studies in based on the phenomenon of agglutination reactions.

An investigation on the possibility of synthesizing monodisperse polydiene latexes in a emulsifier-free heterogeneous monomer–water system, in the presence of sucrose is demonstrated in the present work. The aim here being is to obtain latexes with unsaturated chemical bonds and hydroxyl groups on the surface of latex particles. Such an approach is interesting as it enlarges the field of application of immunolatexes. Chloroprene was chosen as the monomer.

12.1 INTRODUCTION

The passive hemagglutination test in immunodiagnosis is usually used to diagnose diseases caused by bacteria and viruses. This reaction is also used

Sustainable Water Treatment and Ecosystem Protection Strategies. Hossein Hariri Asli, Tamara Tatrishvili, Ann Rose Abraham, & A. K. Haghi (Eds.)

to determine certain hormones, to identify the patient's hypersensitivity to drugs and hormones.

In biotechnology, for the production of immunodiagnostic tests, monodisperse latex particles are used, on the surface of which antigens (bacterial, viral, tissue) or antibodies are immobilized. Agglutination of latex particles occurs by adding appropriate sera or antigens to the latex.

The sorption properties of a latex particle are determined by their surface chemical composition. For this purpose, it is essential that latex particles must have certain functional groups on their surface. Desides, the main requirement for immunological latex is to be monodisperse with specified diameter of particles. Only under these conditions can the process of agglutination occur.

Usually emulsion polymerization (EP) is considered to be the main method of latexes synthesis. Nevertheless, the mechanisms of formation of latex particles during EP are still largely unclear, and the synthesis of monodisperse latexes with a given particle diameter is carried out rather intuitively than according to pre-programmed recipes.

EP is mainly carried out in micellar emulsions and the surface of the synthesized latex particles is coated with emulsifier molecules. This factor severely limits the possibility of expanding the range of immunological latexes.

The synthesis of monodisperse latexes is most often possible in heterogeneous static monomer–water systems without using emulsifiers.[1–3] Dispersion process in a heterogeneous system (formation of a new interface), in a static condition, occurs because of an increase in the free energy of the system due to the heat of polymerization.[1,3]

In a heterogeneous monomer–water system, polymerization reactions proceed in the immediate vicinity of the phase boundary, where monomer microdroplets are formed. It is further assumed that the density of the particles increases as the polymerization progresses and the particles settle to the bottom of the aqueous phase. In the absence of emulsifier, the surface of the latex particles is covered with hydrophilic fragments of the monomer molecule. This makes it possible to synthesize qualitatively different immunolatexes by changing the nature of the monomer and using special additives. In this work, sucrose was used as an additive.

12.2 EXPERIMENTAL

Polymerization of chloroprene was carried out in a heterogeneous static system of the monomer—an aqueous solution of potassium persulfate.

Methods for carrying out static heterophase polymerization are given in Refs. [1, 4, 5]. In thermostated tubes, the monomer is carefully layered onto the aqueous phase. Initially, sucrose was dissolved in the water phase. Polymerization was initiated with potassium persulfate at 60°C. Turbidity of the aqueous phase, indicating the appearance of dispersed particles in the system, is observed after a certain period of time. Photos of the dynamics of the turbidity in the aqueous phase of chloroprene—an aqueous solution of potassium persulfate and sucrose is shown in Figure 12.1.

FIGURE 12.1 Photographs of the aqueous phase in the heterogeneous system chloroprene-aqueous solution of potassium persulfate and sucrose. (1) at the beginning (before polymerization), (2) after 3 h, and (3) after 6 h of polymerization.

Turbidity is observed throughout the entire volume of the aqueous phase, at a sucrose concentration of less than 4% in the water phase (Figure 12.2 test tube 2). With a sucrose concentration of more than 4%, turbidity is observed at the interface and slowly spreads throughout the entire volume of the aqueous phase over time (Figure 12.2, test tube 3).

The synthesized emulsifier-free polychloroprene latex was highly stable due to the presence of sulfate and hydroxyl groups and unsaturated bonds on the particle surface. Latex stability was determined by centrifugation.[6]

To obtain monodisperse latexes, polymerization was carried out in three-phase system chloroprene–water–sucrose crystals (Figure 12.3). In such a system, a density gradient is created along the height of the water phase and dispersed particles are localized in the zone of their nucleation. Such localization contributes to the complete formation of particle sizes and narrowing of their distribution over the diameter.[1,5] The concentration of the resulting latex was determined by the gravimetric method and by measuring the density of the aqueous phase. In a static system, the polymerization process can be considered complete when the amount of the solid residue remains unchanged. The dry residue of the final latex in this series of experiments was 2%. An electron microscope photo of the obtained latex is shown in Figure 12.4.

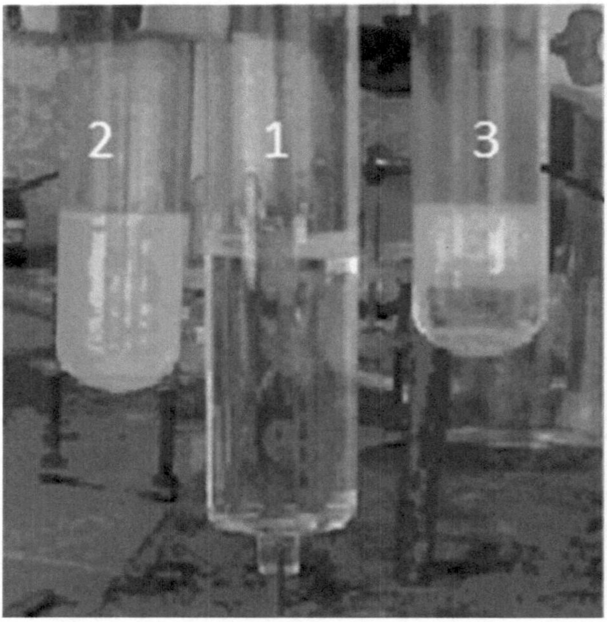

FIGURE 12.2 Polymerization of systems chloroprene—aqueous solution of sucrose. (1) The system before polymerization. The concentration of sucrose in the aqueous phase: in a test tube. (2) 0.3%; (3) 0.4%. The concentration of potassium persulfate is 0.2%.

12.3 CONCLUSIONS

It is shown that emulsifier-free emulsion polymerization of chloroprene in the heterogeneous system monomer–aqueous solution of potassium persulfate–sucrose crystals leads to the production of monodisperse latexes, with hydroxyl groups and unsaturated double bonds on the surface of their particles.

FIGURE 12.3 Polychloroprene latex synthesized in a three-phase system monomer–water–sucrose crystals.

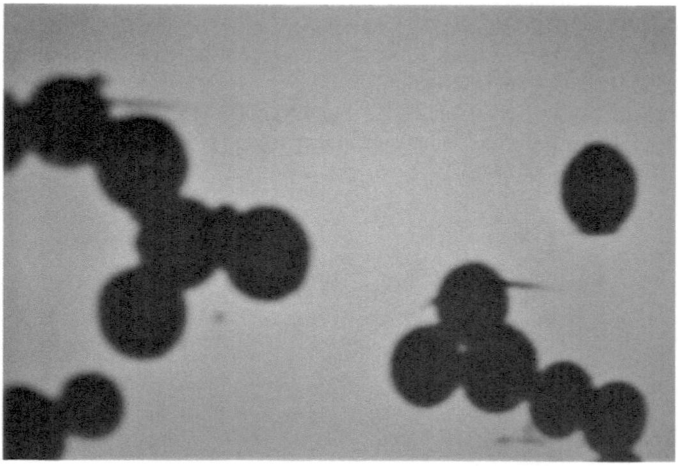

FIGURE 12.4 Electron microscopic photo of polychloroprene latex obtained in the presence of sucrose.

KEYWORDS

- **polymerization**
- **static system**
- **latexes**
- **emulsion polymerization**
- **microdroplet**

REFERENCES

1. Oganesyan, A. A., Freeradical polymerization and phase formation in heterogeneous monomer/water systems, Doctoral (Chem.) Dissertation, Moscow: Moscow Inst. of Fine Chemical Technology, 1986.
2. Oganesyan, A. A.; Khaddazh, M.; Gritskova, I. A.; Gubin, S. P.; Grigoryan, G. K.; Muradyan, G. M.; Nadaryan A. G. Polymerization in the Static Heterogeneous System Styrene–Water in the Presence of Methanol. *Theoret. Foundations Chem. Eng.* **2013,** *47* (5), 600–603.
3. Hovhannisyan, A. A.; Khaddazh, M. *On the Theory of Emulsion Polymerization*; Lambert Academic Publishing: Germ, 2016.
4. Hovhannisyan, A. A. Polymerization in Monomer—Water Heterogeneous Systems. *J. Chem. Sci.* **2018,** *9*, 60–65.
5. Oganesyan, A. A.; Gukasyan, A. V.; Airapetyan, K. S.; Boyajyan, V. G.; Matsoyan, S. G. Methods of Synthesis of Monodisperse Polychloroprene Lattices. Patent N 4090394/05, 20.05.86. (in Russian)
6. Oganesyan, A. A.; Grigoryan, G. K.; Muradyan, G. M.; Nadaryan, A. G. Stability of Non-Emulsifier Polychloroprene Monodisperse Latexes. *Chem. G. Armenia* **2011,** *64* (4), 575–579. (in Russian)

CHAPTER 13

Characterization of Fulvic Acids, Isolated from Natural Waters of Georgia

T. MAKHARADZE[1], M. SOSELIA[1], G. MAKHARADZE[2], and S. ROGAVA[2]

[1]*Ivane Javakhishvili Tbilisi State University R. Agladze Institute of Inorganic Chemistry and Electrochemistry, Tbilisi, Georgia*

[2]*Ivane Javakhishvili Tbilisi State University, Tbilisi, Georgia*

ABSTRACT

Pure samples of fulvic acids (ashing 0.50–0.82%) were isolated from natural waters of Georgia by the adsorption-chromatographic method (sorbent-activated charcoal). It was determined elemental composition of fulvic acids: C- changes from 48.72 to 53.75%; H- from 3.80 to 5.30%; N-from 0.89 to 2.50%; S- from 0.51 to 0.98%, and O- from 39.64 to 44.45%.

Acid–base properties of fulvic acids were studied by the direct and back potentiometric titration, $pK_{H,COOH}$ changes from 4.0 to 4.4; $pK_{H,Ph-OH}$ changes from 9.9 to 10.5.

The containing of carboxyl groups changes from 3.2 to 3.6 mmol/g and phenolic group from 1.5 to 1.9 mmol/g.

13.1 INTRODUCTION

Fulvic acids (FA)—macromolecular organic substances are natural organic complexing compounds, found in the environment that originate from chemically and microbially modified plant and animal matter. They

Sustainable Water Treatment and Ecosystem Protection Strategies. Hossein Hariri Asli, Tamara Tatrishvili, Ann Rose Abraham, & A. K. Haghi (Eds.)

are major organic substances dissolved in natural waters (70–80%).[1–6] FA form stable complexes with heavy metals and radionuclides and stipulate migration forms thereof in natural waters and soils.[1,3,7–14] In spite of numerous researches, data about the physical–chemical characteristics of FA are heterogeneous. Chemical structure of FA has not been finalized yet. Elemental analysis is one of the most simple and important way of characterizing FA. According to the results of elemental analysis, the containing of carbon in FA changes from 28.1 to 59.7%; H- from 0.4 to 8.9%; N- from 0.3 to 8.2%; S- from 0.1 to 4.3%, and O- from 35 to 58%.[1,15–21] The reason for such scattering results is the hygroscopicity of fulvic acids, high ashing of samples and the most importantly is using different methods for the isolation of FA.

The data are also different on one of the fundamental characteristic of fulvic acids-dissociation constants: $pK_{H,COOH}$ and $pK_{H,Ph-OH}$. $pK_{H,COOH}$ changes from 1.6 to 7.0 and $pK_{H,Ph-OH}$—from 8.7 to 10.6.[1,4,5,22–27]

Our objective was to study physical–chemical properties such as elemental composition, dissociation constants, and quantity of carboxyl and phenolic-OH groups of fulvic acids, isolated from natural waters of Georgia.

13.2 EXPERIMENTAL METHOD

Pure samples of fulvic acids were isolated from natural waters of Georgia by the adsorption-chromatographic method.[1,4,10] After filtration through membrane filters (0.45 μm pore size), the water samples were concentrated by the frozen method. The concentrated water samples were acidified with 6 M HCl to pH 2 and were put for 2 h on water bath at 60°C for coagulation of humin acids. Then the solution was centrifuged for 10 min at 8000 rpm (Centrifuge T-23). For the isolation of FA centrifugate was used the adsorption-chromatographic method. Charcoal was used as a sorbent. Desorption of amino acids and carbohydrates were performed by means of 0.1 M HCl. For desorption of polyphenols was used 90% acetone water solution. Elution of the fraction of FA was performed with 0.1 N NaOH solutions. Obtained alkaline solution of FA, for the purification was passed through a cation-exchanger (KU-2-8). The part of the solution was dried under vacuum until the constant weight was obtained. Dried samples of FA were analyzed.

Elemental composition of fulvic acids was determined using C, H, N, and S elementary analyzer ("Elementar," Germany). The percentage contents of carbon, hydrogen, nitrogen, and sulfur were directly obtained by measurement and the percentage of oxygen was calculated by subtracting the percentages of carbon, hydrogen, nitrogen, and sulfur from 100.

For the determination of dissociation constants of fulvic acids was used potentiometric titration (direct and back titration). The 50.0 mg of fulvic acids was solved in 20 mL of 0.05 M NaOH. The end volume of solution by bidistilled water was reduced to 25 mL (μ = 0.1 KCl). The titration of alkaline solution of fulvic acids was performed by 0.106 M HCl. Just after finishing of inverse titration (pH = 2.98), direct titration was started with 0.095 M NaOH. The measurement was made on pH meter (pH meter(pH-673), after adding of each portion of reagent (0.1 mL). Both direct and back titrations were carried out in nitrate area.

13.3 RESULTS AND DISCUSSIONS

As it is shown from Table 13.1, elemental composition of fulvic isolated from natural water by the same method, slightly differs from each other. C changes within 48.72–53.75%; H within 3.80–5.30%; O within 39.64–44.45%; N within 0.89–2.50%, and S within 0.51–0.98%.

Acid–base properties of fulvic acids, isolated from natural waters were studied by the potentiometric titration method. Potentiometric titration curves of fulvic acids, isolated from different natural objects doesn't differ from each other sharply, that's why for the illustration, let us discuss the potentiometric titration results of fulvic acids isolated from samples of Lake Paravani. The one peak is shown obviously on the differential curve of the direct potentiometric titration (Fig. 13.1). According to Ref. [1], it is caused by the titration of carboxyl and partly phenolic hydroxyl groups, because it couldn't be possible to fix the latter as separate peak. If we consider, that only such phenolic hydroxyl groups have high enough acidity which is close to electron-withdrawing groups, for example, nitro groups, then we can conclude that the given peak can be caused only by the titration of carboxyl groups. The correctness of this opinion is confirmed with the data obtained by the back titration. Two peaks are clearly observed on differential curve, which correspond to the titration of carboxyl and phenolic hydroxyl groups (Figure 13.1).

TABLE 13.1 Elemental Composition of Fulvic Acids, Isolated from Natural Waters of Georgia.

Research object	%						Groups, mmol/g		Dissociation constants	
	Ash	C	H	O	N	S	COOH	phenol-OH	$pK_{,COOH}$	$pK_{,Ph-OH}$
Lake Paravani	0.50	53.75	4.29	40.56	0.89	0.51	3.4	1.7	4.4	10.2
Lake Paliastomi	0.58	51.58	5.30	39.64	2.50	0.98	3.6	1.8	4.0	9.9
Sioni Reservoir	0.82	48.72	4.54	44.45	1.74	0.55	3.2	1.9	4.1	10.5
River Mtkvari	0.79	51.78	3.80	42.28	1.58	0.56	3.6	1.5	4.3	10.0

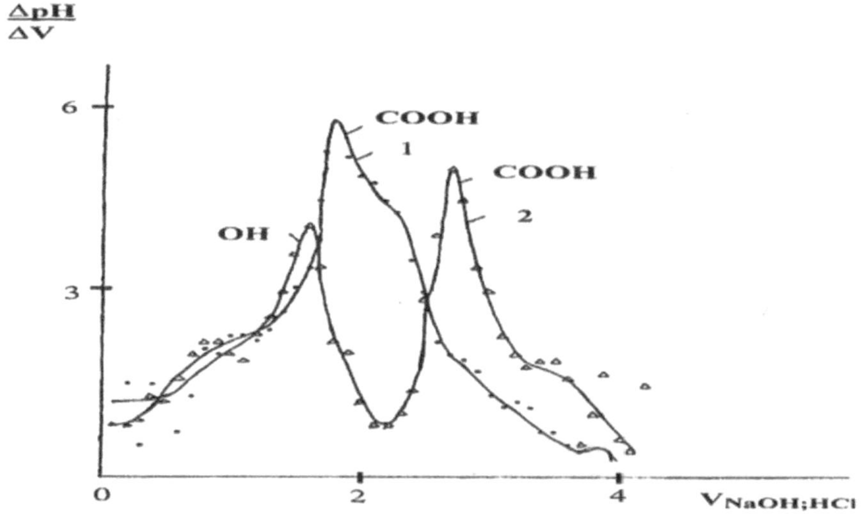

FIGURE 13.1 Differential curve of potentiometric titration of fulvic acids isolated from Lake Paravani. (1) the direct titration, (2) back titration; $V_1 = 25$ mL; $C_{NaOH} = 0.095$ M; $C_{HCl} = 0.106$ M; $\mu=0.1$ (KCl)

Such nature of peaks gives us the opportunity to discuss fulvic acids as monofunctional substances in the given area of pH and to use the graphical option of Henderson–Hasselbalch method for the calculation of dissociation constant. The connection between the concentration of H^+ ions, dissociation degree, and indicator of acidic dissociation constant is expressed in the equation: $pH = pK\alpha - \lg [(1-\alpha)/\alpha]$.

Dissociation degree α = Ci/Cmax, where Ci is the concentration of base (acid), mmol, which was spent on titration of fulvic acids by the before given pH and Cmax is the quantity ob base(acid) which was spent on titration of fulvic acids, on potentiometric curve, before the pH of the corresponding the maqsimum of the peak. Each concrete value of pH corresponds to determined dissociation degree of given functional group. Graphically, the Henderson–Hasselbalch equation could be shown in coordinates: $pH - \lg [(1-\alpha)/\alpha]$.

When $\lg [(1-\alpha)/\alpha] = 0$, $pH = pK_a$. Because, during the determination of pKα by the graphical method, great importance is given to the straight line of tilt angle, for the creating, the graphic was used the least square method and was calculated acidic dissociation constants of fulvic acids isolated from natural waters (Table 13.1). $pK_{H,COOH}$ changes from 4.0 to 4.4, and $pK_{H,Ph-OH}$ changes from 9.9 to 10.5.

According to potentiometric data was calculated the quantity of carboxyl and phenolic-OH groups. For the calculation of the quantity of phen.-OH, back titration data were processed by Gran method.[28] The containing of carboxyl groups changes from 3.2 to 3.6 mmol/g and phenolic group from 1.5 to 1.9 mmol/g (Table 13.1).

13.4 CONCLUSIONS

On the basis of received results, we can conclude, that contrary to the widespread assumption in the literature, fulvic acids isolated from different samples of natural waters don't differ from each other neither elemental composition nor acid–base properties.

ACKNOWLEDGMENTS

This research was supported by Shota Rustaveli National Science Foundation of Georgia (SRNSFG) [Grant Number №YS-21-3728].

KEYWORDS

- **fulvic acids**
- **elemental composition**
- **dissociation constant**

REFERENCES

1. Varshal, G. M. Migration Forms of Fulvic Acids and Metals in Natural Waters, *Dissertation.* Vernadsky Institute of Geochemistry and Analytical Chemistry of Russian Academy of Sciences, 1994.
2. Dulaquas, G.; Waeles, M.; Gerringa, L. A.; Midag, R.; Rijkenberg, M.; Riso, R. G. The Biogeochemistry of Electroactive Humic Substances and Its Connection to Iron

Chemistry in the North East Atlantic and the Western Mediterranean Sea. *J. Geophys. Res.: Oceans* **2018,** *123*, 5481–5499.
3. Osadchyy V.; Nabyvanets, B.; Linnik, P.; Osadcha, N.; Nabyvanets, Y. *Processes Determining Surface Water Chemistry*; Springer International Publishing: Switzerland, 2016.
4. Makharadze, G. A.; Supatashvili, G. D.; Varshal, G. M. Humic Acids in Surface Waters of Georgia. *Hydrochem. Mater.***1989,** *106*, 22–30.
5. Makharadze, G.; Goliadze, N.; Khaiauri, A.; Makharadze, T.; Supatashvili, G. Fulvic and Humin Acids in Surface Waters of Georgia. In *High-performas Polymers for Engineering-based Composites*. Apple Academic Press: Waretown, 2016; pp. 167–179.
6. Pisarek, I.; Glowacki, M. Quality of Groundwater and Aquatic Humic Substances from Main Reservoire of Ground Water No. 333. *J. Ecol. Eng.* **2015,** *16*, 46–53.
7. Moiseenko, T. I.; Diny, M. I.; Gashkina, N. A.; Kremlevaa, T. A. Occurrence Forms of Metals in Natural Waters Depending on Water Chemistry. *Water Resour*. **2013,** *40*, 407–416.
8. Adusei-Gyamfi, J.; Ouddane, B.; Rietveld, L.; Cornard, J.; Criquet, J. Natural Organic Matter-Cations Complexation and Its Impact on Water Treatment: A Critical Review. *Water Res*. **2019,** *160*, 130–147.
9. Makharadze, G. A.; Supatashvili, G. D.; Varshal, G. M. The Research of the Forms of Copper in Surface Waters. *Hydrochem. Mater*. **1988,** *103*, 3–16.
10. Makharadze, T.; Makharadze, G. Investigation of Complex Formation Process of Copper with Macromolecular Organic Substances, Isolated from Natural Water. *Organ. Chem. Plus* **2020,** *1*, 1–5.
11. Maccarthy, P.; O'Cinneide, S. Fulvic Acids: Interactions with Metal Ions. *Eur. J. Soil Sci*. **2006,** *25*, 429–437.
12. DE OLIVEIRA VAZ, D.; FERNANDE, S. A. N.; SZPOGANIC, Z. B. Complexations of Divalent Metallic Ions with Fulvic Acids Eclética química. **2018,** *43*, 54–58.
13. Makharadze, T.; Makharadze, G. Measurement of Complex Formation Process of Nickel (II) with Freshwater Fulvic Acids Using the Solubility Method. *Fine Chem. Eng.* **2021,** *2* (2), 54–61.
14. Zhu, B.; Ryan, D. K. Characterizing the Interaction Between Uranul Ion and Fulvic Acid Using Regional Integration Analysis (RIA) and Fluorescence Quenching. *J. Environ. Radioact*. **2016,** *153*, 97–103.
15. Rice, J. A.; MacCarthy, P. Statistical Evaluation of the Elemental Composition of Humic Substances. *Organ. Geochem*. **1991,** *17*, 635–648.
16. Aiken, G. R.; Mcknight, D. M.; R. A. Harnish R. A. Chemical Characteristics of Aquatic Fulvic Acid Isolated from Lake Fryxell, Antarctica. *Analy. Chim. Acta* **1999,** *395*, 323–334.
17. Kovács, K.; Gáspár, A.; Sajgó, C.; Schmitt-Kopplin, P.; Tombácz, E. Comparative Study on Humic Substances Isolated in Thermal Groundwaters from Deep Aquifers Below 700. *Geochem. J.* **2012,** *46*, 211–224.
18. Ma, H.; Allen, H. E.; Yin, Y. Characterization of Isolated Fractions of Dissolved Organic Matter from Natural Waters and a Wastewater Effluent. *Water Res*. **2001,** *35*, 985–996.
19. Kucerik, J; Bursakova, P.; Prusova, A; Grebikova, L.; Schaumann, G. E. Hydration of Humic and Fulvic Acids Studied by DSC. *J. Therm. Analy. Calorimetry*. **2012,** *110*, 451–459.

20. Zhang, Y.; Gong, G.; Zheng, H.; Yuan, X.; Xu, L. Synergistic Extraction and Characterization of Fulvic Acid by Microwave and Hydrogen Peroxide–Glacial Acetic Acid to Oxidize Low-Rank Lignite. *ACS Omega* **2020,** *5* (12), 6389–6394.
21. Fujitake, N.; Kodama, H.; Seiya Nagao, S.; Tsuda, K.; Yonebayashi, K. Chemical Properties of Aquatic Fulvic Acids Isolated from Lake Biwa, a Clear Water System in Japan. *Humic Substances Res.* **2009,** *5/6*, 45–53.
22. Rey-Castro, C.; Mongin, S.; Huidobro, C.; David, C.; Salvador, J.; Garces, J.; Galceran, J.; Mas, F.; Puy, J. Effective Affinity Distribution for the Binding of Metal Ions to a Generic Fulvic Acid in Natural Waters. *Environ. Sci. Technol.* **2009,** *43*, 7184–7191.
23. Lenoir, T.; Manceau, A. Number of Independent Parameters in the Potentiometric Titration of Humic Substances. *Langmuir* **2010,** *26*, 3998–4003.
24. Leenheer, J. A.; Wershaw, R. L.; Brown, G. K.; Reddy, M. M. Characterization and Diagenesis of Strong-Acid Carboxyl Groups in Humic Substances. *Appl. Geochem.* **2003,** *18*, 471–482.
25. Ephreim, J. H.; Boren, H.; Arsenie, I.; Pettersson, C.; Allard, B. A Combination of Acid-Base Titrations and Derivatization for Functional Group Determinations of an Aquatic Fulvic Acid. *Sci. Total Environ.* **1989,** *81/82*, 615–624.
26. Ritchie, J. D.; Perdue, E. M. Proton-Binding Study of Standard and Reference Fulvic Acids, Humic Acids, and Natural Organic Matter. *Geochim. Cosmochim. Acta* **2003,** *67*, 85–96.
27. Song, F; Wu, F.; Guo, F.; Wang, H.; Fenga, W.; Zhou, M.; Denga, Y.; Bai, Y.; Xingc, B. P.; Giesy, J. P. Interactions Between Stepwise-Eluted Sub-Fractions of Fulvic Acids and Protons Revealed by Fluorescence Titration Combined with EEM-PARAFAC. *Sci. Total Environ.* **2017,** *605–606*, 58–65.
28. Masini, J. C. The Use of Linear Potentiometric Titration Curves in the Determination of Alkalinity and Acid-Base Properties of Diluted Solutions of Humic Substances. *Talanta* **1994,** *41*, 1383–1389.

CHAPTER 14

Biotechnology in the Pharmaceutical and Polymer Industries: Environmental Burden Reduction Measures and Management of Natural Resources

PRZEMYSŁAW GNATOWSKI and MARTA RAZIMOWICZ

Department of Polymer Technology, Faculty of Chemistry, Gdansk University of Technology, Gdansk, Poland

ABSTRACT

In this chapter it is shown why there is a great interest in the field of biotechnology usage in pharmaceutical and polymer industries. It is also emphasized that in the pharmaceutical industries, biotechnology is being used widely in practical scale for production of antibiotics, vitamins, or enzymes.

Among modern technologies, the most popular is biotechnology, which has found wide application in many industries. The biggest revolution is being seen in medicine, starting with drug design, synthesis, and the introduction of innovative treatments for a wide variety of diseases.[1] Biotechnology as a technique is highly regarded, primarily in the production of biopharmaceuticals. In addition to providing innovative ways to manufacture well-known products, the field is taking an active role in the production of entirely new products.[2]

There are two main divisions of biotechnology: traditional and modern. The first is based on naturally occurring organisms, which, through

Sustainable Water Treatment and Ecosystem Protection Strategies. Hossein Hariri Asli, Tamara Tatrishvili, Ann Rose Abraham, & A. K. Haghi (Eds.)

artificial selection, are used for numerous production processes. We are talking about bioindicators—wild strains of bacteria that have not been modified in any way. In addition to bacteria, traditional biotechnology also uses small chemical molecules that can form intermediates or end products of metabolic processes—so-called primary metabolites. This group includes ethyl alcohol, sugars, polysaccharides, amino acids, organic acids, and some vitamins, such as B_2 and B_{12}. Among the basic biotechnological processes is the enzymatic and at the same time anaerobic process of transformation of organic compounds with the participation of microorganisms—fermentation. Fermentation is divided into lactic (lactic bacteria) and ethanol (occurs with the participation of yeast). These processes are used to produce dairy products, alcoholic beverages such as beer or wine, cakes, or sourdough bread.[3,4]

The second type of biotechnology is innovative technology based on biocatalysts and enzymes that have been modified through conscious and deliberate human interference with genetic material and molecular biology. Through genetic modification, it is possible to improve biocatalysts so that they enable technological processes to be carried out in the intended manner and allow the presumed effects to be achieved. In short, we modify the genome so as to later create an organism from it that possesses certain characteristics. The processes of modern biotechnology compared to traditional biotechnology are carried out on a smaller scale. As a result of the introduction of modern techniques, it was possible to introduce an immuno-vaccine, insulin, or growth hormone to the pharmaceutical market. Chemicals obtained by innovative biotechnology methods are used in the production of detergents and biopesticides. These methods include: pharmacogenomics—which studies the impact of the human genome in response to pharmacotherapy, gene expression, protein engineering, and peptide engineering. In addition, tissue engineering, whose main focus is on tissue formation, cell fusion, gene therapy or bio: -catalysis, -leaching, -filtration, and many others.[3–5]

14.1 BIOTECHNOLOGY IN PHARMACEUTICAL INDUSTRY

This chapter will present the aspect of the use of biotechnology in the pharmaceutical industry, as biotechnologically produced medicinal substances are gaining more and more value in the pharmaceutical market over the years.[1]

One of the most widely used biotechnological processes is microbial fermentation. It is one of the few methods used in the commercial production of some large-scale manufactured products. The total pharmaceutical market is growing at an accelerating rate. In addition, the group of active pharmaceutical substances produced by microorganisms can include blood cholesterol-lowering agents, statins, various types of enzyme inhibitors, immunosuppressants, or anticancer compounds. In the production of most of these drugs are biotechnological processes. New fermentation methods are constantly being developed for known active substances and drug precursors. One example is the production of Lovastatin, a drug used to normalize blood lipid levels, which is one of the main ingredients actively involved in the manufacture of other semisynthetic statins. Increasingly, companies are turning to modern strategies to improve the efficiency of bioprocesses aimed at increasing the productivity and efficiency of those various processes.[6–8]

Semi-synthetic penicillin and cephalosporin antibiotics derived from 6-aminopenicillic acid (6-APA) and 7-aminocephalosporic acid (7-ACA), respectively, are obtained with the involvement of enzymatic processes.[9,10] Cephalexin is a semisynthetic beta-lactam antibiotic derived from the cephalosporin C.[10] Polfa Tarchomin S.A. is considered the main manufacturer of cephalosporins. Traditional chemical production of cephalexin had up to 10 steps. The conventional process provided up to 50 kg of waste per kg of antibiotic. Only after technological changes were made was this reduced to about 15 kg. Over time, a four-step process involving enzymes was developed, which further reduced the amount of waste and the consumption of most resources. Nevertheless, a better option than the enzymatic route proved to be the direct fermentation route.[11,12]

Standard vitamin production using organic chemistry is increasingly being replaced by innovative technological processes. Examples include DSM Nutritional Products and BASF (Germany), which have phased out the six-step process for producing vitamin B_2.[13] They have replaced it with a one-step fermentation process involving Bacillus subtilis bacteria, on the grounds that it shows less environmental impact compared to classical production. The scheme to produce vitamin B_2 by different methods is shown in Figure 14.1. The production by fermentation was made possible by genetic engineering of the bacteria, which increased the vitamin yield by about 300,000 times compared to what was achieved with the wild strain. Thus, the one-step fermentation process significantly reduced production costs by 50% compared to the previously used mechanism (Figure 14.1).[12,14]

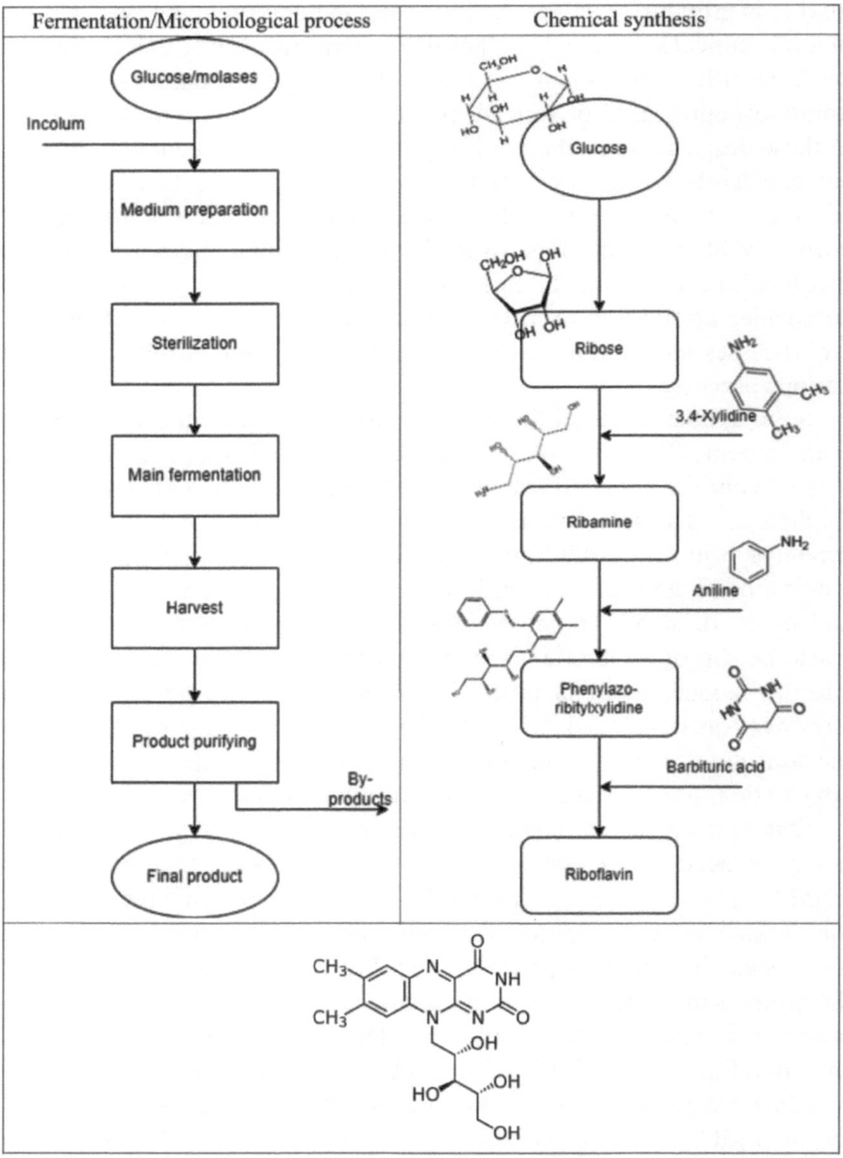

FIGURE 14.1 Production of vitamin B_2 by various methods.

Synonymous with advances in biotechnology is the prevention and therapy of new types of drugs called biopharmaceuticals. Therapeutic substances produced by biotechnological methods currently play the greatest role in the pharmacotherapy of diseases in which standard chemical drugs have proved ineffective. Biopharmaceuticals are a different group of drugs than small-molecule compounds, that is, antibiotics. This class of compounds includes tissue plasminogen activator (tPA), insulin, and recombinant hepatitis B vaccine. Oncology is among the major fields in which biopharmaceuticals play a significant and most important role. As of today, biotechnology drugs are considered safer and very well tolerated. In contrast, typical immune reactions for this group of compounds are sometimes rare. With the gradual expiration of patent protection for the first-generation biotech active substances in recent years, one can expect the gradual emergence of biotech drugs that are the equivalent of classic drugs in the medical field.[15] Biopharmaceuticals belong to a group of drugs created through biotechnological processes, including genetic engineering methods using bacteria, fungi, plants, or mammals. This process results in protein recombinants of hormones, antibodies, vaccines, or diagnostic substances, which are produced for therapeutic purposes in many fields of medicine.[14]

Continuous work is underway to improve methods of manufacturing biopharmaceuticals, such as the blood plasma protein alpha-1-antitrypsin. Many naturally occurring first-generation protein therapeutics produced by innovative technological systems are being converted into product proteins, like enzymes. The resulting products are relatively superior to their natural equivalents. Protein engineering has been very successful in, among other things, modifying their bioactivity, stability, or other characteristics.[2,16]

It is estimated that about 75% of enzyme consumption is used in the production of chemicals, food, and starch. Hydrolytic enzymes, such as proteases, amylases, lipases, and cellulases, account for the largest share. They are enjoying increasing recognition in the pharmaceutical industry, drug development, and many other analytical applications. It is assumed that among the enzymes used for commercial purposes, about 60% are precisely the products obtained from modern biotechnology.[17]

Pharmaceutical manufacturers through enzyme and biocatalysis are able to reduce the number of synthesis steps that would be used in conventional synthesis. The reduction in production steps translates into a significant increase in the efficiency of manufacturing operations. In addition, biocatalysts make it possible to produce and perform biotransformation of single molecules derived from optically active compounds.[18,19]

The pharmaceutical industry is constantly under considerable pressure, as it has the responsibility of ensuring that the end products they obtain contain only the pharmacologically active enantiomer of the active substance, and not the racemate. The ability to selectively produce a single unit of the desired enantiomer saves on expensive precursor materials, but this aspect is nevertheless often overlooked. The advantage of this method is, first of all, the possibility of negating the cost of subsequent purification of the obtained product and the absence of traces of contamination.[5,14,20]

Typically, classical medicinal substances are purified from natural products or synthesized using proven chemical processes. Their main task is to nullify the symptoms of disease by interacting with specific proteins. Biological drugs are otherwise protein molecules of quite sizable size, whose main purpose is causal treatment, that is, the elimination of any ailment from the human body. The production of such active substances is carried out with the participation of living organisms. Sometimes, bacteria are genetically modified to increase the efficiency of the production processes of the desired protein and to give it the appropriate tertiary structure, so that the entire purification process does not lead to its destruction. In the production of protein drugs, the most commonly used are *E. coli* bacteria, which are intentionally modified, yeast or Chinese hamster ovary (CHO) cells. These therapeutic substances have already made or will make breakthroughs in the treatment of many diseases or conditions.[21]

Advances in disease treatment have been noted with the start of mass production in bacteria recombinant human insulin, erythropoietin, growth hormone, and other types of protein drugs. Previously, these were purified using animal or human organs. However, it was soon noticed that direct contact with them caused serious sensitization reactions and even led to infection of the patient with various diseases. Nowadays, in addition to many types of human insulins, other protein hormones are synthesized, such as growth hormone, erythropoietin; clotting factors, interferons, growth factors, and other proteins that show therapeutic effects.[5]

Extensive progress is being made in the targeted treatment of cancer. Pathological formations diagnosed on the basis of symptoms differ significantly only at the molecular level. It should be noted that only some of their genetic features may be the same in two different patients. Moreover, commonly used cancer drugs act in a reproducible manner—they affect dividing cells without distinguishing between healthy cells and cancerous ones. Therefore, the ability to distinguish between cancer cells and transport the drug only to them was added to another premise of biotechnology.[2,22]

Another huge breakthrough is being noted in the treatment of inflammatory and autoimmune diseases. In recent years, new therapeutic substances such as Raptiwa®, or efalizumab, which is used in the treatment of psoriasis, Kineret®, an agent used in rheumatoid arthritis, Enbrel®, a drug for arthritis, and interferon alpha-1a, which has gained recognition in the treatment of multiple sclerosis, have been introduced into circulation.[23,24]

Traditional vaccine technologies are based on the multiplication of pathogenic (virulent or non-virulent) bacteria in animals (the first antiviral vaccines), in animal cell cultures (second-generation antiviral antidotes) and in bacterial cultures (antimicrobial remedies). The technological process of vaccine production involving animals has always raised ethical concerns and, above all, has been considered dangerous. The processes involved in the multiplication of pathogens in bioreactors are also called dangerous. Modern and safe vaccine production technologies, mainly antiviral, are patented on recombinant DNA methods.[22] Thus, it is possible to grow weakened strains of viruses that have lost their virulence as a result of changes in genetic material at specific locations in the genome. Microbial biosynthesis of antigenic proteins is increasingly being carried out in bacterial or yeast cells, which have no adverse effect. One of the first vaccines produced with yeast, was a recombinant hepatitis B vaccine (1986).[25] The gene for the surface antigen, HBsAg, was cloned in *E. coli* bacteria, *S. cerevisiae* yeast, and various mammalian cell lines. Eventually, a yeast biotechnological process system was developed, by means of which it was possible to obtain good expression of the gene and the correct structure of the antigen molecule.[18]

14.2 BIOTECHNOLOGY IN POLYMER INDUSTRY

Biodegradable plastics, such as polyhydroxyalkanoic acids (PHAs) in bacteria, have been known since the early 20th century. Unfortunately, the high costs associated with bioplastics production and the availability of versatile, low-cost plastics derived from petroleum processing led to bioplastics being overlooked for some time. Concern over the environmental durability of standard plastics has renewed interest in biopolymers.

Japan's Institute of Physics and Chemical Research has created a microorganism that is capable of producing up to 96% dry weight known as biodegradable plastic. Today, a wide variety of natural plastic and non-plastic polymers is available. Although they are relatively expensive, their use and production process do not show negative effects on the environment.[26,27]

Increased work is underway on improving the production of polyhydroxyalkanoates (PHAs) and many other biodegradable biopolymers. Scientists are constantly developing natural polymers with improved properties, and they are also working to develop special strains of bacteria for their production. More efficient fermentation and product recovery processes are being explored. Use of mixed cultures and inexpensive substrates can significantly reduce PHA production costs.[28,29]

The conversion of acrylonitrile to acrylic acid is observed in the production of anionic polyacrylamides. This is an example of large-scale biotransformation with commercial and environmental benefits. The Ciba Specialty Chemicals specializes in obtaining macromolecular compounds based on acrylamide and acrylic acid by means of biological processes. The traditional way of obtaining acrylic acid was a dangerous, multi-step, energy-intensive process, requiring high concentrations of toxic acrylonitrile, conducted at elevated temperatures, and generating dangerous emissions. The Ciba biotransformation route is believed to have the following advantages such as it is simple, consists of a one-step cost-effective process, and ensures that the final product will be of good quality. In addition, the production steps take place at room temperature and atmospheric pressure. The new method featured a low concentration of dangerous acrylonitrile volatilized during production. Unfortunately, it contained several by-products close to quantitative conversion. Another example is the bioprocess of the well-known Mitsubishi Rayon Company that produces acrylamide. In the next step, the acrylamide is polymerized into regular plastic polyacrylamide. In the UK, the Baxenden Company is known for producing polyesters, acrylic polymers, emulsions, and many other specialty chemicals. The British company is one of the few to have succeeded, using biocatalytic processes, in lowering the reaction temperature from 200 to 60°C.[30]

DuPont, together with Genencor International, developed a method that used a genetically modified *E. coli* bacterium to convert sugar from corn starch. The result was an organic compound, 1,3-propanediol, in high-efficiency fermentation. According to DuPont, the technological system created is the world's cheapest way to obtain a key chemical intermediate.

DuPont quickly supported this belief and built new processing plants based on this technology. Propanediol is the main starting material for polypropylene terephthalate, a new type of polyester fiber. This new polyester differs from others in properties such as resilience, strength, and the ability to be easily dyed without the need for chemical modifiers.[31]

In addition to bioplastics, a wide range of non-plastic biopolymers is available for use as thickening agents, gels, or lubricants, among others. It is interesting to note that by 1999 about 30,000 tons of xanthan polysaccharide biopolymer worth $408 million had been produced.[6]

Lactic acid is a precursor to produce polylactic acid (PLA). With its help, it was possible to produce polymers that decompose by biochemical transformations. These transformations are environmentally harmless and atypical of synthetic plastics obtained from the refining industry. Another valuable compound is L-polylactic acid (PLA), which is widely used in the chemical industry in the production processes of plastics such as films, containers, and food packaging. Moreover, it is included in adhesives, solvents, removers, or chemicals. It is a valuable material used in the production of plastics for the medical industry: dissolvable threads, implants, and much more. In the pharmaceutical industry, lactic acid is used in the production of many ointments, liquids, dermatological drugs, and even enters into the composition of medicinal anti-cancer substances.[32] Such polymers can be then mixed in compositions with other polymers, just like their fossil-fuel varieties.[33,34]

The aforementioned advantages of lactic acid extracted through biological processes involving microorganisms (fungi of the genus *Rhizopus*) prove that it plays a key role in the aspect of global environmental problems. Through its use, it contributes to the reduction of post-production plastic waste. In addition, it is widely used in the manufacture of everyday products, which are harmless to the environment unlike chemical plastics. The global market of lactic acid is approx. 270,000 tons per year, from which 90% is produced microbiologically. The development and commercialization of biopolymers of the acid are helping to increase its use. It is estimated that the annual growth of new plastics in global production will be about 10%.[35]

Commonly known polymeric compounds such as poly: -saccharides, -esters, and -phenols are of great interest due to their biodegradability and the possibility of obtaining them from renewable raw materials, which are abundant today. Commercially available lipases, from various sources, are increasingly used in catalytic polyreactions of structurally complex monomers having different types of substituents. Enzyme-mediated polyreaction is characterized by higher efficiency compared to other compounds.

The ever-increasing interest in ester synthesis, catalyzed by organic compounds such as acylglycerol hydrolases (HAGs), is also affecting the chemical industry. Hydrolases catalyze the esterification of diols with carboxylic, acrylic, and methacrylic acids, yielding monomers used in the synthetic polymer industry. Enzymatic synthesis of macromolecular compounds involving HAG due to its advantages was described in detail later. Shortly thereafter, using the polytransesterification reaction of a prodrug—fumaric acid diester with an organic compound: 1,4-butanediol, carried out in the presence of a catalyst, lipase, and alkyds (unsaturated polyesters) were obtained. The obtained compounds are used, among others, in the production of plastics.[36,37]

TABLE 14.1 The Main Methods and Producers of Polymeric Materials

	Polymers			
Name of sample	**PLA**	**Polyesters**	**PHB**	**PHBV**
Production process	Microbial process	Fermentation process	Fermentation process	Metabolic biosynthesis
Main producer	INGEO™ (NATUREWORKS)[38]	DuPont[31]	Biomer[39]	Tianan[40]

Although biotechnology helps with plastics synthesis, the problems with plastics usage and waste remain. Conventional polymers such as PE, PP, or PET have projected life span of more than 500 years.[41] The degradation of plastics can occur with different mechanisms under distinct conditions. The most important factors are summarized in Figure 14.2. The most important part in real-life degradation of plastic waste is that all the factors function and influence the process at the same time, thus it is very complicated process. Depending on the polymer, different factors can play higher role in the process. It was proven that some environmental factors significantly accelerate the degradation process—mainly UV irritation and heat. Those factors induce formation of radicals, which shorten polymer chain and produce active centers, thus increases susceptibility to enzymatic degradation. The typical biodegradation stages include biodeterioration, biofragmentation, bioassimilation, and mineralization.

Abiotic	Biotic	Polymer
Temperature	Nutrients	Crystalinity
Moisture	Enzymes	Molecular weight
UV radiation	Surfactants	Surface properties
pH	Biomass	Size
Oxygen access	Present organisms	Addtitives

FIGURE 14.2 Factors affecting biodegradation of polymers.

There are more than 100 bacteria types capable of degrading polymers, although there is no single one which is capable to degrade all types of plastics. The most gifted bacteria are *Bactillus* sp., which have the potential to degrade PE, PVC, PP, PS, and PU. The most refractory polymer for degradation by bacteria is PVC (probably because of its cytotoxicity)—there are only two bacteria types which degrades it in addition to the previously mentioned *Bactillus* sp., namely *Bacillus flexus* and *Pseudomonas citronellolis.*

Bacteria are one of the main themes of biotechnology research in topic of polymer pollution. It can be divided into two main parts: degradation and recycling. The scheme of most important topics in biotechnology for end-life of polymers is presented on Figure 14.3.

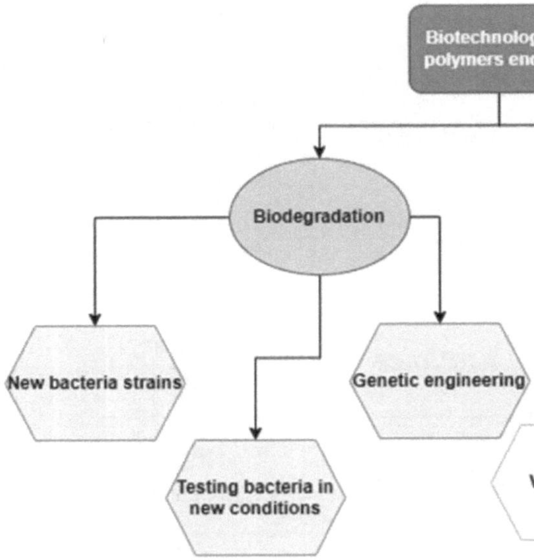

FIGURE 14.3 Most important topics of biotechnology applications in field of end-life of polymers.

The degradation part of the research can be split into three main routes of the research: discovering novel microorganisms in the environment, testing known bacteria in modified conditions of degradation and genetic engineering of the bacteria.

The first route focuses on examination of landfill and composting sites and isolating strains which are grooving there. Skariyachan *et al.* found and described three strains described as *Enterobacter* sp. bengaluru-btdsce01, *Enterobacter* sp. bengaluru-btdsce02, and *Pantoea* sp. bengaluru-btdsce03, which in its individual application exhibited 70 ± 4, 68 ± 4, and 64 ± 4% weight reduction for LDPE strips and 21 ± 2, 28 ± 2, 24 ± 2% weight reduction for LDPE pellets over a period of 120 days ($p < 0.05$).[42] Roberts *et al.* isolated five strains from petroleum-polluted soils containing *Pseudomonas* and *Bacillus* species and were capable of partially converting PET to the metabolically usable monomers terephthalic acid (TPA) and ethylene glycol in the presence of PET and the cleavage product bis(2-hydroxyethyl) terephthalic acid (BHET) as sole sources of carbon.[43] Meyer-Cifuentes *et al.* isolated marine microbial enrichment culture and studied the biodegradation of an aromatic-aliphatic copolyester blend.[44] The consortium used plastic film as the sole carbon source and reached maximum conversion to CO_2 and biomass in around 15 days.

The second route focuses on changing the conditions of degradation sites. Chen *et al.* investigated known hypethermophilic bacteria *Thermus*, *Bacillus*, and *Geobacillus* using high temperature composting technology and reached nine times improvement in degradation rate comparing to standard procedure.[45] Pathak and Navneet suggest that potentially micro-organisms from cold habitats could be employed in open area landfill, due to lower costs associated with heating up the composting sites. The most often named spices in this regard are *Pseudomonas* and *Micrococcus*.[46] The other attempt is to increase affinity of bacteria to plastic surface. As most of the polymers are hydrophobic, then the Tribedi and Sil investigated enhancing the attachment of *Pseudomonas* sp. AKS2 to a PE surface by usage of mineral oil and proved 200% increase in degradation rate.[47]

Third, the most complicated route is the genetic engineering of the bacteria. Yan *et al.* engineered *Clostridium thermocellum*, which was able to degrade PET at rate of 60% after 14 days in 70°C. Gamerith *et al.* mutated *P. pastoris* with cutinase 1 from *Thermobifida cellulosilytica* and achieved 93% degradation of poly(butylene succinate) (PBS).[48] Gamerith *et al.* also fused polyamidase from *Nocardia farcinica* to polyhydroxyal-kanoate depolymerase from *Alcaligenes faecalis*.[49] Obtained enzymes were successfully used to cleave the urethane bond in polyurethane-polyesters with different degree of crystallinity.

Biotechnology can be used not only for enhancing degradation of polymers but also to recycle them. Chemical recycling of plastics using biotechnological processes is quite expensive, although it can be one of few available options in case of vanishing fuel sources. Those processes can be divided into two paths: creation of monomers or other chemicals.

The first path of recycling polymers into their monomers, which could be then used in re-synthesis, is effective solution for the recycling of polyester waste. PET is the most investigated polymer in this manner with its depolymerization to terephthalic acid and ethylene glycol. Variety of bacteria applicable for PET could be also used for other polyesters. *Thermobifida fusca* DSM43793, *Ideonella sakaiensis, Rhodococcus fascians,* or *Ideonella sakaiensis* 201-F6[50] are just examples of many bacteria used for PET decomposition. For PET hydrolyzation, the temperature resistance of selected spices is crucial, as the glass transition temperature, thus ideal temperature for the process is around 75°C. Tournier *et al.* reported improved PET hydrolase that 90% PET depolymerization into monomers after 10 h using modified leaf-branch compost cutinase.[51] Xi *et al.* reported that the secretory expression in *Bacillus subtilis* a PETase (BhrPETase) from the bacterium HR29 has stability up to 101°C, while sustaining high conversion rates.

The other path for recycling polymers is producing value-added chemicals, which can contribute to sustainable usage of plastics. Such approach is known as valorization. Sadler and Wallace reports modified *Escherichia coli* application for conversion of terephthalic acid obtained from PET to vanillin with 79% conversion.[52] Kim *et al.* used different modified *Escherichia coli* to produce five different aromatics and aromatic-derived compounds from PET hydrolysis products. Utomo *et al.* proposed biotechnological route to utilize PU hydrolysate as a carbon source for a defined microbial mixed culture, where *Pseudomonas putida* KT2440 derivatives utilized adipic acid, 1,4-butanediol, and ethylene glycol, and *Pseudomonas* sp. TDA1 was using 2,4-toluenediamine (TDA) as carbon source and monorhamnolipids were produced at the end of process.[53] Some other researches use pyrolysis products to produce new chemicals. Pyrolysis is a method of depolymerizing plastic waste into aliphatic hydrocarbons under anaerobic conditions and high temperature. Mihreteab *et al.* reports 80% conversion of PP pyrolysis residues into lipids with usage of *Yarrowia lipolytica* strain 78-003.[54]

Overall, there is enormous interest in the field of biotechnology usage in pharmaceutical and polymer industries. In the pharmaceutical sector, biotechnology is already widely used on industrial scale—production of antibiotics, vitamins, or enzymes is already developed and used by companies. Although, substantial amount of money and effort is put on research, because biotechnology offers increased yields, flexibility, and lowered environmental impact for existing, traditional processes. On the other hand, in polymer industry, the biotechnology did not make the big difference on the market yet. There are many studies about biotechnological methods of polymer production, but there are only few companies using it on industrial scale. This is mainly due to low price of petroleum-based polymers and properties of biopolymers, which are not fully matched to all applications. The situation will probably change in near future due to legislative changes, especially in Europe (Fit for 55, Directive (EU) 2019/904 of the European Parliament and of the Council) regarding reduction of CO_2 emission and plastic pollution. Biotechnology offers less environmentally burdening approaches for polymer production and recycling, but with higher costs in comparison to conventional polymers.

KEYWORDS

- **biotechnology**
- **pharmaceutical sector**
- **polymer industry**
- **environmental burden**
- **natural resources**
- **management**

REFERENCES

1. Mallela, K. Pharmaceutical Biotechnology—Concepts and Applications. *Human Genom.* **2010,** *4* (3), 218. https://doi.org/10.1186/1479-7364-4-3-218.
2. Gavrilescu, M.; Chisti, Y. Biotechnology—a Sustainable Alternative for Chemical Industry. *Biotechnol. Adv.* **2005,** *23* (7–8), 471–499. https://doi.org/10.1016/j.biotechadv.2005.03.004.
3. Allison, L. A. *Podstawy Biologii Molekularnej*; Szweykowska-Kulińska, Z., Jarmołowski, A., Eds.; University of Warsaw Press, 2009. https://doi.org/10.31338/uw.9788323537793.
4. Fernandes, P.; Cabral, J. M. S. Biotransformations. In *Basic Biotechnology*; Ratledge, C., Kristiansen, B., Eds.; Cambridge University Press: Cambridge, 2006; pp 579–626. https://doi.org/10.1017/CBO9780511802409.026.
5. Gad, S. C. *Handbook of Pharmaceutical Biotechnology*; Gad, S. C., Ed.; John Wiley & Sons, Inc.: Hoboken, 2007. https://doi.org/10.1002/0470117117.
6. Demain, A. L. Small Bugs, Big Business. *Biotechnol. Adv.* **2000,** *18* (6), 499–514. https://doi.org/10.1016/S0734-9750(00)00049-5.
7. Vilches Ferron, M. A.; Casas Lopez, J. L.; Sanchez Perez, J. A.; Fernandez Sevilla, J. M.; Chisti, Y. Rapid Screening of Aspergillus Terreus Mutants for Overproduction of Lovastatin. *World J. Microbiol. Biotechnol.* **2005,** *21* (2), 123–125. https://doi.org/10.1007/s11274-004-3045-z.

8. Sanchez, S.; Demain, A. L. Metabolic Regulation of Fermentation Processes. *Enzyme Microb. Technol.* **2002,** *31* (7), 895–906. https://doi.org/10.1016/S0141-0229(02)00172-2.
9. Parmar, A.; Kumar, H.; Marwaha, S. S.; Kennedy, J. F. Advances in Enzymatic Transformation of Penicillins to 6-Aminopenicillanic Acid (6-APA). *Biotechnol. Adv.* **2000,** *18* (4), 289–301. https://doi.org/10.1016/S0734-9750(00)00039-2.
10. Kurzątkowski, W.; Gębska-Kuczerowska, A. Compartmentalization in Cephalosporin C Biosynthesis by Industrial Strains Acremonium Chrysogenum. *Postępy Mikrobiologii, Polskie Towarzystwo Mikrobiologów* **2015,** *54* (4), 374–379.
11. OECD. *The Application of Biotechnology to Industrial Sustainability*; OECD, 2001. https://doi.org/10.1787/9789264195639-en.
12. Vandamme, E.; Bienfait, C. G. Industrial Biotechnology and Sustainable Chemistry. *Brussels R. Belgian Acade. Council Appl. Sci.* **2004.**
13. Chu, R.; Li, R.; Wang, C.; Ban, R. Production of Vitamin <scp>B2</Scp> (Riboflavin) by <scp> *Bacillus Subtilis* </Scp>. *J. Chem. Technol. Biotechnol.* **2022,** *97* (8), 1941–1949. https://doi.org/10.1002/jctb.7017.
14. Sekhon, B. S. Biopharmaceuticals: An Overview. *Thai J. Pharm. Sci.* **2010,** *34* (1), 1–19.
15. Melmer, G. Biopharmaceuticals and the Industrial Environment. In *Production of Recombinant Proteins*; Wiley-VCH Verlag GmbH & Co. KGaA: Weinheim, FRG, 2005; pp 361–383. https://doi.org/10.1002/3527603670.ch16.
16. Truppo, M. D. Biocatalysis in the Pharmaceutical Industry: The Need for Speed. *ACS Med. Chem. Lett.* **2017,** *8* (5), 476–480. https://doi.org/10.1021/acsmedchemlett.7b00114.
17. Bhatia, S.; Sharma, K.; Dahiya, R.; Bera, T. *Modern Applications of Plant Biotechnology in Pharmaceutical Sciences*; Academic Press, 2015.
18. Blaser, H.-U. Enantioselective Catalysis in Fine Chemicals Production. *Chem. Commun.* **2003,** *3*, 293–296. https://doi.org/10.1039/b209968n.
19. Malerba, F.; Orsenigo, L. The Evolution of the Pharmaceutical Industry. *Business Hist.* **2015,** *57* (5), 664–687. https://doi.org/10.1080/00076791.2014.975119.
20. Bhatia, S. Application of Plant Biotechnology. In *Modern Applications of Plant Biotechnology in Pharmaceutical Sciences*; Elsevier, 2015; pp 157–207. https://doi.org/10.1016/B978-0-12-802221-4.00005-4.
21. Carro, L. Protein–Protein Interactions in Bacteria: A Promising and Challenging Avenue Towards the Discovery of New Antibiotics. *Beilstein J. Org. Chem.* **2018,** *14*, 2881–2896. https://doi.org/10.3762/bjoc.14.267.
22. L. von Ranke, N.; M. Fierro, I.; Antunes, A. Trends in Biotechnological Drugs for Cancer Treatment. *Recent Patents Anti-Cancer Drug Discov.* **2016,** *11* (1), 112–120. https://doi.org/10.2174/1574892810666150928160202.
23. Bonnekoh, B.; Böckelmann, R.; Pommer, A. J.; Malykh, Y.; Philipsen, L.; Gollnick, H. The CD11a Binding Site of Efalizumab in Psoriatic Skin Tissue as Analyzed by Multi-Epitope Ligand Cartography Robot Technology. *Skin Pharmacol. Physiol.* **2007,** *20* (2), 96–111. https://doi.org/10.1159/000097982.
24. Shannon, E.; Daffy, J.; Jones, H.; Paulson, A.; Vicik, S. Etanercept (Enbrel®) Alternative Storage at Ambient Temperature. *Clin. Pharmacol.: Adv. App.* **2017,** *Volume 9*, 87–99. https://doi.org/10.2147/CPAA.S131832.
25. Bai, D.-M.; Li, S.-Z.; Liu, Z. L.; Cui, Z.-F. Enhanced l -(+)-Lactic Acid Production by an Adapted Strain of Rhizopus Oryzae Using Corncob Hydrolysate. *Appl. Biochem. Biotechnol.* **2008,** *144* (1), 79–85. https://doi.org/10.1007/s12010-007-8078-y.

26. Janik, H.; Kucińska-Lipka, J.; Sienkiewicz, M.; Kosakowska, A. Biodegradable Polymer Packaging Materials in Seawater Environment. In *Advanced Materials Polymers and Composites-New Research on Properties, Techniques and Applications*; Mukbaniani, O. v., Tatrishvili, T., Abadie, M. J. M., Eds.; Apple Academic Press, 2022; pp 163–187.
27. Soldo, A.; Miletić, M.; Auad, M. L. Biopolymers as a Sustainable Solution for the Enhancement of Soil Mechanical Properties. *Sci. Rep.* **2020,** *10* (1), 267. https://doi.org/10.1038/s41598-019-57135-x.
28. Javaid, H.; Nawaz, A.; Riaz, N.; Mukhtar, H.; -Ul-Haq, I.; Shah, K. A.; Khan, H.; Naqvi, S. M.; Shakoor, S.; Rasool, A. et al. Biosynthesis of Polyhydroxyalkanoates (PHAs) by the Valorization of Biomass and Synthetic Waste. *Molecules* **2020,** *25* (23), 5539. https://doi.org/10.3390/molecules25235539.
29. Vu, D. H.; Wainaina, S.; Taherzadeh, M. J.; Åkesson, D.; Ferreira, J. A. Production of Polyhydroxyalkanoates (PHAs) by *Bacillus Megaterium* Using Food Waste Acidogenic Fermentation-Derived Volatile Fatty Acids. *Bioengineered* **2021,** *12* (1), 2480–2498. https://doi.org/10.1080/21655979.2021.1935524.
30. Varga, V.; Štefuca, V.; Mihálová, L.; Levarski, Z.; Struhárňanská, E.; Blaško, J.; Kubinec, R.; Farkaš, P.; Sitkey, V.; Turňa, J. et al. Recombinant Enzymatic Redox Systems for Preparation of Aroma Compounds by Biotransformation. *Front. Microbiol.* **2021,** *12*. https://doi.org/10.3389/fmicb.2021.684640.
31. Zhou, W.; Zhou, P.; Tang, Y.; Zhou, Z. The Properties and Application Progress of Sorona Fiber. *J. Phys.: Conf. Ser.* **2021,** *2079* (1), 012019. https://doi.org/10.1088/1742-6596/2079/1/012019.
32. Liu, C.; Zheng, J.; Ou, X.; Han, Y. Anti-Cancer Substances and Safety of Lactic Acid Bacteria in Clinical Treatment. *Front. Microbiol.* **2021,** *12*. https://doi.org/10.3389/fmicb.2021.722052.
33. Przybytek, A.; Sienkiewicz, M.; Kucińska-Lipka, J.; Janik, H. Preparation and Characterization of Biodegradable and Compostable PLA/TPS/ESO Compositions. *Ind. Crops Prod.* **2018,** *122*, 375–383. https://doi.org/10.1016/j.indcrop.2018.06.016.
34. Janik, H.; Sienkiewicz, M.; Przybytek, A.; Guzman, A.; Kucinska-Lipka, J.; Kosakowska, A. Novel Biodegradable Potato Starch-Based Compositions as Candidates in Packaging Industry, Safe for Marine Environment. *Fibers Polym.* **2018,** *19* (6), 1166–1174. https://doi.org/10.1007/s12221-018-7872-1.
35. Alexandri, M.; Schneider, R.; Mehlmann, K.; Venus, J. Recent Advances in D-Lactic Acid Production from Renewable Resources. *Food Technol. Biotechnol.* **2019,** *57* (3), 293–304. https://doi.org/10.17113/ftb.57.03.19.6023.
36. Kundys, A.; Białecka-Florjańczyk, E.; Fabiszewska, A.; Małajowicz, J. Candida Antarctica Lipase B as Catalyst for Cyclic Esters Synthesis, Their Polymerization and Degradation of Aliphatic Polyesters. *J. Polym. Environ.* **2018,** *26* (1), 396–407. https://doi.org/10.1007/s10924-017-0945-1.
37. Sánchez, D. A.; Tonetto, G. M.; Ferreira, M. L. Valorization of Glycerol Through the Enzymatic Synthesis of Acylglycerides with High Nutritional Value. *Catalysts* **2020,** *10* (1), 116. https://doi.org/10.3390/catal10010116.
38. Samsudin, H.; Iñiguez-Franco, F. Packaging and Consumer Goods. In *Poly(Lactic Acid)*; Wiley, 2022; pp 605–618. https://doi.org/10.1002/9781119767480.ch26.

39. Ashby, M. F. Material Profiles. In *Materials and the Environment*; Elsevier, 2013; pp 459–595. https://doi.org/10.1016/B978-0-12-385971-6.00015-4.
40. Guo, M.; Stuckey, D. C.; Murphy, R. J. Is It Possible to Develop Biopolymer Production Systems Independent of Fossil Fuels? Case Study in Energy Profiling of Polyhydroxybutyrate-Valerate (PHBV). *Green Chem.* **2013,** *15* (3), 706. https://doi.org/10.1039/c2gc36546d.
41. Ojeda, T. Polymers and the Environment. In *Polymer Science*; InTech, 2013. https://doi.org/10.5772/51057.
42. Skariyachan, S.; Manjunatha, V.; Sultana, S.; Jois, C.; Bai, V.; Vasist, K. S. Novel Bacterial Consortia Isolated from Plastic Garbage Processing Areas Demonstrated Enhanced Degradation for Low Density Polyethylene. *Environ. Sci. Pollut. Res.* **2016,** *23* (18), 18307–18319. https://doi.org/10.1007/s11356-016-7000-y.
43. Roberts, C.; Edwards, S.; Vague, M.; León-Zayas, R.; Scheffer, H.; Chan, G.; Swartz, N. A.; Mellies, J. L. Environmental Consortium Containing *Pseudomonas* and *Bacillus* Species Synergistically Degrades Polyethylene Terephthalate Plastic. *mSphere* **2020,** *5* (6). https://doi.org/10.1128/mSphere.01151-20.
44. Meyer-Cifuentes, I. E.; Werner, J.; Jehmlich, N.; Will, S. E.; Neumann-Schaal, M.; Öztürk, B. Synergistic Biodegradation of Aromatic-Aliphatic Copolyester Plastic by a Marine Microbial Consortium. *Nat. Commun.* **2020,** *11* (1), 5790. https://doi.org/10.1038/s41467-020-19583-2.
45. Chen, Z.; Zhao, W.; Xing, R.; Xie, S.; Yang, X.; Cui, P.; Lü, J.; Liao, H.; Yu, Z.; Wang, S. et al. Enhanced in Situ Biodegradation of Microplastics in Sewage Sludge Using Hyperthermophilic Composting Technology. *J. Hazard. Mater.* **2020,** *384*, 121271. https://doi.org/10.1016/j.jhazmat.2019.121271.
46. Pathak, V. M.; Navneet. Review on the Current Status of Polymer Degradation: A Microbial Approach. *Bioresour. Bioprocess.* **2017,** *4* (1), 15. https://doi.org/10.1186/s40643-017-0145-9.
47. Tribedi, P.; Sil, A. K. Low-Density Polyethylene Degradation by Pseudomonas Sp. AKS2 Biofilm. *Environ. Sci. Pollut. Res.* **2013,** *20* (6), 4146–4153. https://doi.org/10.1007/s11356-012-1378-y.
48. Gamerith, C.; Vastano, M.; Ghorbanpour, S. M.; Zitzenbacher, S.; Ribitsch, D.; Zumstein, M. T.; Sander, M.; Herrero Acero, E.; Pellis, A.; Guebitz, G. M. Enzymatic Degradation of Aromatic and Aliphatic Polyesters by P. Pastoris Expressed Cutinase 1 from Thermobifida Cellulosilytica. *Front. Microbiol.* **2017,** *8*. https://doi.org/10.3389/fmicb.2017.00938.
49. Gamerith, C.; Herrero Acero, E.; Pellis, A.; Ortner, A.; Vielnascher, R.; Luschnig, D.; Zartl, B.; Haernvall, K.; Zitzenbacher, S.; Strohmeier, G. et al. Improving Enzymatic Polyurethane Hydrolysis by Tuning Enzyme Sorption. *Polym. Degrad. Stab.* **2016,** *132*, 69–77. https://doi.org/10.1016/j.polymdegradstab.2016.02.025.
50. Austin, H. P.; Allen, M. D.; Donohoe, B. S.; Rorrer, N. A.; Kearns, F. L.; Silveira, R. L.; Pollard, B. C.; Dominick, G.; Duman, R.; el Omari, K. et al. Characterization and Engineering of a Plastic-Degrading Aromatic Polyesterase. *Proc. Natl. Acad. Sci.* **2018,** *115* (19). https://doi.org/10.1073/pnas.1718804115.
51. Tournier, V.; Topham, C. M.; Gilles, A.; David, B.; Folgoas, C.; Moya-Leclair, E.; Kamionka, E.; Desrousseaux, M.-L.; Texier, H.; Gavalda, S. et al. An Engineered

PET Depolymerase to Break down and Recycle Plastic Bottles. *Nature* **2020,** *580* (7802), 216–219. https://doi.org/10.1038/s41586-020-2149-4.
52. Sadler, J. C.; Wallace, S. Microbial Synthesis of Vanillin from Waste Poly(Ethylene Terephthalate). *Green Chem.* **2021,** *23* (13), 4665–4672. https://doi.org/10.1039/D1GC00931A.
53. Utomo, R. N. C.; Li, W.-J.; Tiso, T.; Eberlein, C.; Doeker, M.; Heipieper, H. J.; Jupke, A.; Wierckx, N.; Blank, L. M. Defined Microbial Mixed Culture for Utilization of Polyurethane Monomers. *ACS Sustain. Chem. Eng.* **2020,** *8* (47), 17466–17474. https://doi.org/10.1021/acssuschemeng.0c06019.
54. Mihreteab, M.; Stubblefield, B. A.; Gilbert, E. S. Microbial Bioconversion of Thermally Depolymerized Polypropylene by Yarrowia Lipolytica for Fatty Acid Production. *Appl. Microbiol. Biotechnol.* **2019,** *103* (18), 7729–7740. https://doi.org/10.1007/s00253-019-09999-2.

Index

D

E

F

G

H